McGRAW-HILL TELECOMMUNICATIONS

Build Your Ow

Trulove	jects)

Crash Course

Louis	
Vacca	*Course*
Louis	*M-Commerce Crash Course*
Shepard	*Telecom Convergence, 2/e*
Shepard	*Telecom Crash Course*
Bedell	*Wireless Crash Course*
Kikta/Fisher/Courtney	*Wireless Internet Crash Course*

Demystified

Harte/Levine/Kikta	*3G Wireless Demystified*
LaRocca	*802.11 Demystified*
Muller	*Bluetooth Demystified*
Evans	*CEBus Demystified*
Bayer	*Computer Telephony Demystified*
Hershey	*Cryptography Demystified*
Taylor	*DVD Demystified*
Bates	*GPRS Demystified*
Symes	*MPEG-4 Demystified*
Camarillo	*SIP Demystified*
Shepard	*SONET / SDH Demystified*
Topic	*Streaming Media Demystified*
Symes	*Video Compression Demystified*
Shepard	*Videoconferencing Demystified*
Bhola	*Wireless LANs Demystified*

Developer Guides

Vacca	*I-Mode Crash Course*
Guthery	*Mobile Application Development with SMS*
Richard	*Service and Device Discovery: Protocols and Programming*

Professional Telecom

Bates	*Broadband Telecom Handbook, 2/e*
Collins	*Carrier Grade Voice over IP*
Chernock	*Data Broadcasting*
Harte	*Delivering xDSL*
Held	*Deploying Optical Networking Components*
Minoli/Johnson/Minoli	*Ethernet-Based Metro Area Networks*
Benner	*Fibre Channel for SANs*
Bates	*GPRS*
Sulkin	*Implementing the IP-PBX*
Lee	*Lee's Essentials of Wireless*
Bates	*Optical Switching and Networking Handbook*
Wetteroth	*OSI Reference Model for Telecommunications*

Russell	*Signaling System #7*, 4/e
Minoli/Johnson/Minoli	*SONET-Based Metro Area Networks*
Nagar	*Telecom Service Rollouts*
Louis	*Telecommunications Internetworking*
Russell	*Telecommunications Protocols,* 2/e
Minoli	*Voice over MPLS*
Karim/Sarraf	*W-CDMA and cdma2000 for 3G Mobile Networks*
Bates	*Wireless Broadband Handbook*
Faigen	*Wireless Data for the Enterprise*

Reference

Muller	*Desktop Encyclopedia of Telecommunications,* 3/e
Botto	*Encyclopedia of Wireless Telecommunications*
Clayton	*McGraw-Hill Illustrated Telecom Dictionary*, 3/e
Radcom	*Telecom Protocol Finder*
Pecar	*Telecommunications Factbook*, 2/e
Russell	*Telecommunications Pocket Reference*
Kobb	*Wireless Spectrum Finder*
Smith	*Wireless Telecom FAQs*

Security

Nichols	*Wireless Security*

Telecom Engineering

Smith/Gervelis	*Cellular System Design and Optimization*
Rohde/Whitaker	*Communications Receivers*, 3/e
Sayre	*Complete Wireless Design*
OSA	*Fiber Optics Handbook*
Lee	*Mobile Cellular Telecommunications*, 2/e
Bates	*Optimizing Voice in ATM / IP Mobile Networks*
Roddy	*Satellite Communications*, 3/e
Simon	*Spread Spectrum Communications Handbook*
Snyder	*Wireless Telecommunications Networking with ANSI-41*, 2/e

BICSI

Network Design Basics for Cabling Professionals
Networking Technologies for Cabling Professionals
Residential Network Cabling
Telecommunications Cabling Installation

Hotspot Networks

Wi-Fi for Public Access Locations

Daniel Minoli

McGraw-Hill
New York Chicago San Francisco Lisbon
London Madrid Mexico City Milan New Delhi
San Juan Seoul Singapore Sydney Toronto

The McGraw-Hill Companies

Library of Congress Cataloging-in-Publication Data

Minoli, Daniel, date
Hotspot networks : Wi-Fi for public access locations / Daniel Minoli.
p. cm.
Includes bibliographical references and index.
ISBN 0-07-140978-5 (alk. paper)
1. Personal communication service systems. I. Title.

TK5103.485.M56 2002
384.5'3—dc21 2002026530

1 2 3 4 5 6 7 8 9 0 DOC/DOC 0 8 7 6 5 4 3 2

ISBN 0-07-140978-5

The sponsoring editor for this book was Marjorie Spencer and the production supervisor was Pamela Pelton. It was set in Century Schoolbook by MacAllister Publishing Services, LLC.

Printed and bound by RR Donnelley.

This book is printed on recycled, acid-free paper containing a minimum of 50 percent recycled, de-inked fiber.

CONTENTS

PREFACE

From airport lounges and hotel meeting rooms to coffee shops and restaurants across the globe, a wireless world is being built for mobile professionals to stay connected.

IEEE Institute Magazine, July 2001

One of the most exciting developments amidst many exciting occurrences in the wireless world is the actual deployment of wireless personal area networks (WPANs), public access locations (PALs), and hotspot services. This book will focus on the pragmatic issues related to the deployment of inexpensive hotspot services, not the infinitude of Post Telephone and Telegraph wireless architectural alternatives and "chart-ware designs" offered by pre-telecom crash administrations and carriers. According to a Wall Street Journal article of July 26, 2002, "European mobile-phone companies are retreating from their investments in third-generation (3G) wireless technology . . . After spending more than $150 billion on 3G licenses and infrastructure, European operators have grown much less optimistic about potential returns of offering 3G services . . . Some of those . . . are likely to pull out or drastically reduce rollout costs to minimize their losses . . .". Frankly, the alphabet soup for the near-infinitude of 2G, 2.5G, 2.75G, 2.82G, . . . 3G, 4G, 00 G variants and the 100 or so books on this topic, are, in our opinion, a deafening cacophany. We look at standards, technology, performance, security, the Internet and other forms of service delivery, end-to-end networking, and, of course, engineering and deployment. In addition, we will necessarily touch on some integration efforts such as broadband services, 2.5G/3G concepts, and Voice over IP (VoIP) over wireless. This book has been written with carriers, service providers, and enterprise environments in mind, where generating a positive bottom-line is of interest.

Chapter 1, "Introduction to Wireless Personal Area Networks (WPANs), Public Access Locations (PALs), and Hotspot Services," maps WPAN, wireless local area network (WLAN), and wireless wide area network (WWAN) technologies to the market potential for these services. Chapter 2, "Standards for Hotspots," discusses key standards applicable to hotspot services. Chapter 3, "Technologies for Hotspots," covers each contributing technology in greater technical depth. Chapter 4, "Security Considerations for Hotspot Services," addresses the critical issue of security, particularly given that there has been early negative press about the first-generation security systems in place. Chapter 5, "IEEE 802.11," discusses in some detail the Institute of Electrical and Electronics Engineers (IEEE) 802.11 standard, while Chapter 6, "IEEE 802.11b and IEEE 802.11a," covers IEEE 802.11b and IEEE 802.11a.

In Chapter 7, "Wireless Application Protocol (WAP)," we'll consider the WAP protocol, a de facto standard that has been around for several years. Chapter 8, "Designing Nomadic and Hotspot Networks," covers major considerations for designing networks, and Chapter 9, "Migrating to 3G WWANs," concludes with a look at future directions in the short term.

Although hotspot technologies are neither unfamiliar nor exotic, I hope readers will find inspiration in the promise of these networks and new ways to think "outside the box."

Acknowledgments

The author wishes to thank Mr. Roy Daniel Rosner, CEO and founder of Global Nautical Networks, for his seminal help and assistance.

The author wishes to thank the following individuals for making direct and/or indirect contributions to this book:

D. Molta provided materials on security. His assistance is appreciated.

T. B. Zahariadis, K. G. Vaxevanakis, C. P. Tsantilas, and N. A. Zervos provided materials on 3G. Their assistance is also appreciated.

Ms. K. Getgen of RSA Security provided an insightful section of the issue of security in Chapter 4. Her assistance is much appreciated.

The author also wishes to thank Mr. Raymond Taluy, Cisco Systems, for his insightful help on a number of WLAN technology-related matters. The author also thanks Cisco Systems for inputs regarding the technology, particularly for Chapter 8.

Dedication

I would like to dedicate this book to Anna and her help with Global Nautical Networks.

Daniel Minoli

CHAPTER 1

Introduction to Wireless Personal Area Networks (WPANs), Public Access Locations (PALs), and Hotspot Services

There is an increasing trend toward being always on, always active, and always connected. Already over 600 million people worldwide have wireless telephones, a number that is approximately twice the number of people who have Internet access. It is predicted that by 2006, 1.3 billion people will have wireless service. Although the initial entry of wireless was to support voice services, there is now a major thrust (with hundreds of billion dollars of investments already behind it) to deliver high-speed data and Internet applications to these subscribers.

The trend toward wireless data services has already resulted in major deployments of wireless local area networks (WLANs), wireless personal area networks (WPANs), early-generation wireless wide area networks (WWANs), which we also call *nomadic networks* with service available anywhere in a metro area, and public access locations (PALs) that support hotspot services[1] in specific local environments (which are sometimes called *smart spaces*).[2] Approximately 5 percent of all business personal computers (PCs) (about 10 million) are already on WLANs.

The desire of being always on, always active, and always connected will drive the deployment of location-specific hotspot services. This book focuses on two areas: designing and deploying nomadic Internet Protocol (IP) networks with data (telematic) service available continuously everywhere in the country and designing and deploying hotspot networks that support hotspots with data service available in various locations that may or may not be contiguous. Both nomadic and hotspot networks are expected to see major deployment in the next one to three years. Mobility is viewed by proponents as the killer application. Information is only valuable to the potential consumer when it can be readily accessible. Information can have location-based value and/or time-based value. Hotspots support both kinds of values, but focus on the former. Providers of hotspot services are known as Wireless Fidelity (Wi-Fi) operators.

After researching the Wi-Fi market, Gartner/Dataquest estimates that there were over 4,000 public wireless access points (APs) in the United States by the end of 2001 and that in the next three years, 30 percent of professional notebook PCs will have WLAN cards. Wi-Fi, along with other emerging license-exempt communications standards, will gain significant market share against licensed frequency systems, radically reforming several industries.[3]

At the time of this writing, it was reported that Korea Telecom planned to roll out about 10,000 802.11 hotspots in Korea in Phase 1 and more later. Hanaro Telecom, which is also in Korea, plans to launch 15,000 hotspots. In Japan, thousands of 802.11 APs also are being installed; one service, called *WIS-net*, had 9,000 subscribers in the first month. Other companies in

Japan, including J-Phone, are actively exploring WLAN-based hotspot service. Boingo Wireless, which was recently launched in the United States, has about 400 hotspots.[4] Other carriers are reportedly looking into these services.

Clearly, there is a major opportunity for service providers in the United States and elsewhere because there is a major market opportunity and a long way to go.

Figure 1-1 depicts the taxonomy of the various technologies that are discussed in this book. The WPAN space is a subset of the WLAN space. Nomadic services rely on the continuum of WLAN-MAN-WAN connectivity, whereas PALs tend to rely more on WLAN connectivity (see Table 1-1). The term *mobile* refers to an entity that is in motion during a (data) transmission or session; movement could be at a low speed (for example, a pedestrian) or high speed (for example, a car or train). The term *portable* refers to the ability to access information while at a remote location; typically, there is no motion during the session. The term *nomadic* (wanderer) has the connotation of not being part of a fixed community, such as a specific company (enterprise); however, the user will eventually have to pass some sort of authentication test, implying that the user is ultimately a member of the

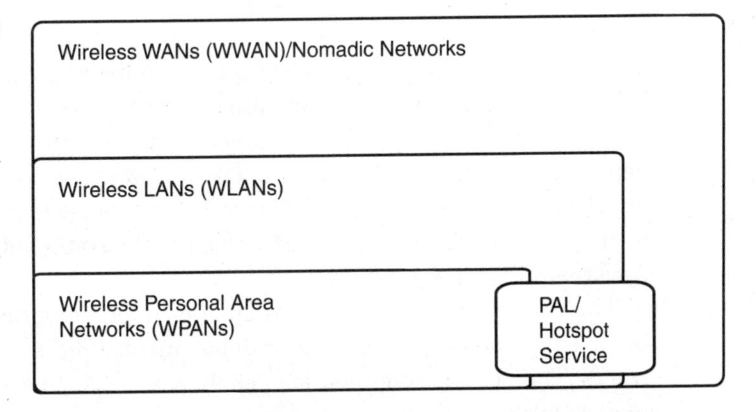

Figure 1-1 Technologies addressed in this book

Table 1-1 Basic technologies and services

Services	Technologies
Hotspot services	WPANs WLANs
Mobile/nomadic services	WWANs Enhanced second generation/third generation (2.5G/3G)

community of registered subscribers. The term *fixed wireless* refers to point-to-point or point-to-a-few-points transmission over a (directional) radio link. The term *wireless* can apply to all these scenarios, but it most often implies mobility. Wireless communication can take place within a building (the range is called a *picocell*), within a city (the range is called a *microcell*), in greater metropolitan areas including suburban locations (the range is called a *macrocell*), or globally (the range is called a *worldcell*).[5]

A recent article[6] opens with the following quote:

> If you want an easy way to get high-speed Internet access, forget your local phone company or cable company and head to a coffee shop or an airport. Seriously. The reason: wireless Net access, which is showing up more and more in public places.

The article continues:

> Somehow, the people developing wireless technology have figured out how to make getting online at broadband speeds painless, and they're getting better all the time . . . the system known as Wi-Fi is generating the kind of excitement marked by the early years of the World Wide Web.

This book focuses on the pragmatic aspects of designing, deploying, and maintaining hotspot networks and less on the infinitude of architectural alternatives that have been advanced, particularly for WWANs. The emphasis is on hotspot services that are delivered via WLAN technology. Lower-speed WWAN and lower-range WPANs are also addressed. We look at standards, technology, performance, security, Internet and other service delivery, and end-to-end networking. Broadband services are addressed and 2.5G/3G concepts are discussed at an intermediary level. References 23–28 at the back of this book provide a short bibliography of the wireless topic. Also see references 1–22.

This book is targeted to planners, designers, engineers, and managers involved in the current or future deployment of hotspot and nomadic networks. The book is intended for people who want to know more about the underlying technology. It is also aimed at venture capitalists and financiers who may look for a value proposition in this space. Educators and students can also benefit from this book.

This chapter provides an overview of the WPAN, WLAN, and WWAN technologies and the market potential for these services. Chapter 2, "Standards for Hotspots," discusses the key standards that are applicable to hotspot services. Chapter 3, "Technologies for Hotspots," covers each of these technologies in greater technical depth. Chapter 4, "Security Considerations for Hotspot Services," addresses the critical issue of security, particularly given that there has been early negative press about the first-generation security systems in place. Chapter 5, "IEEE 802.11," discusses in some detail the Institute of Electrical and Electronics Engineers (IEEE) 802.11 standard, whereas Chapter 6, "IEEE 802.11b and IEEE 802.11a," covers the newer IEEE 802.11b and IEEE 802.11a. Chapter 7, "Wireless Application Protocol (WAP)," discusses the WAP, a de facto standard that has been around for several years. Chapter 8, "Designing Nomadic and Hotspot Networks," covers design aspects of nomadic network. Finally, Chapter 9, "Migrating to 3G WWANs," addresses the migration to 3G WWANs and the future direction of the technology.

The following is an interesting quote from an article written by Rick Perera. The purpose of this text is to address the issue raised:

> . . . He [an executive at top wireless vendor] said that what he called Bellhead-based mobile networks need an infusion of Nethead thinking. "We have got to stop thinking about this network-centric, very defined world," he (the executive) said, pointing out that IP systems are highly decentralized. Even 3G is a very traditional, network-centric concept, he said, one that "assumes the terminal is not much more intelligent than (an ordinary telephone). This is not how Netscape (Communications Corp.'s Internet browser) got on everyone's PC in 1994 and 1995 . . ."[7]

The Potential Opportunity

Figure 1-2 depicts the expected growth in subscribers in the United States for major wireless technologies based on information from 2002. Approximately 105 million people will use mobile data services in 2005 (with 80 million on 2G/2.5G systems and 25 million on 3G systems). As can be seen in this figure, there are major opportunities for all types of data-enabled wireless services, spanning the WPAN, WLAN, WWAN, and hotspot arena. 3G services are expected to be widely available in the United States from 2005 on. It is expected that there will be 25 million subscribers in 2005 in the United States out of a base of 201 million mobile users who will be on 3G systems in 2005, and the number is expected to grow to 171 million out of a base of 229 million by 2010. 3G will support location-based services for business and residential customers.

Figure 1-3 provides global numbers. The demand for wireless services has experienced major growth in the past 20 years. Since the introduction

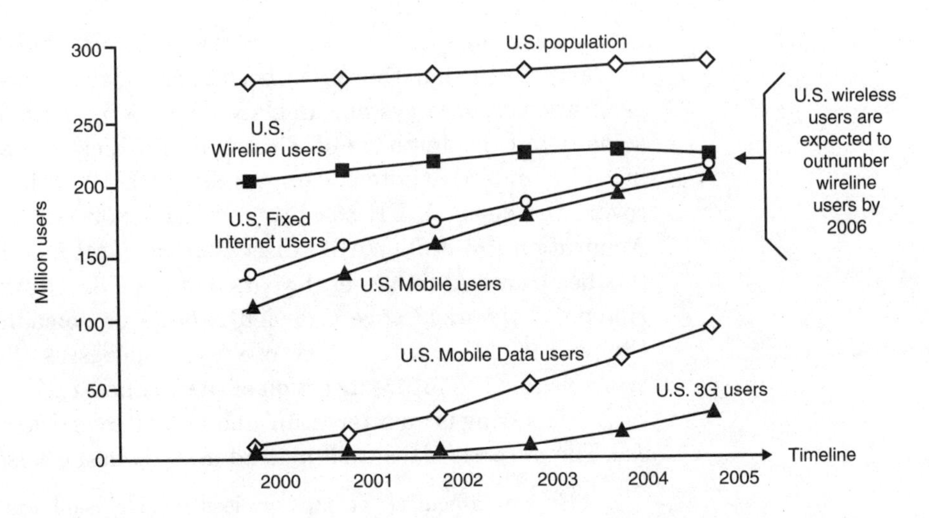

Figure 1-2 Anticipated growth of mobile data/3G users in the United States (Source: Dresdner Kleinwort & Siemens Mobile Network marketing data)

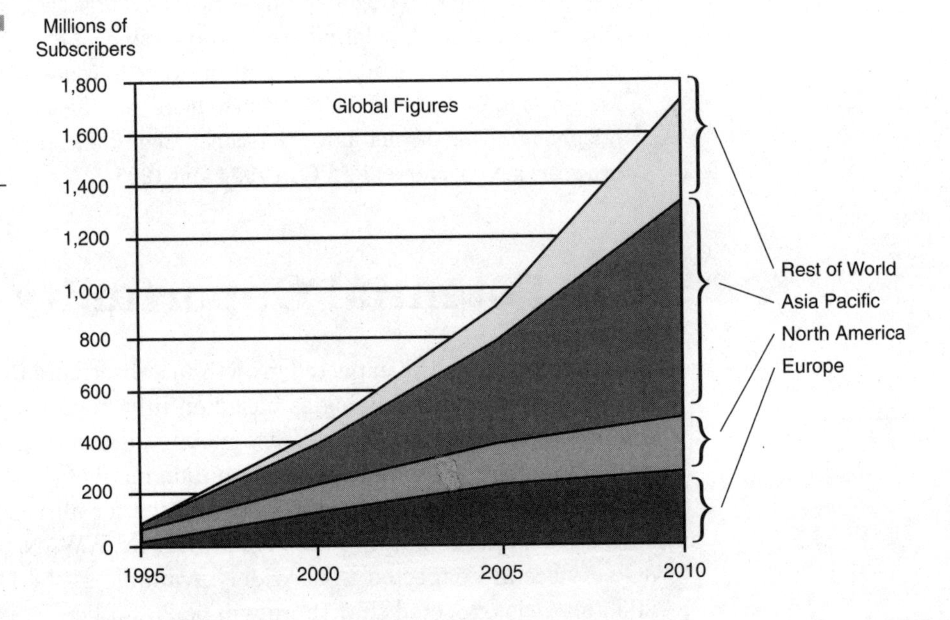

Figure 1-3 Anticipated growth of wireless users worldwide (Source: UMTS Forum)

of generally available cellular service in 1981, it has experienced a compounded annual growth around 40 percent; by comparison, regular telephone service has only experienced a 5 percent annual growth. The majority (almost totality) of traffic on WWAN networks today is voice oriented, but the demand for data and Internet services is becoming pronounced (see Figure 1-4).

The WLANs and hotspot service now being deployed will eventually drive the evolution to 3G. According to research conducted by Siemens Mobile, a plethora of wireless-oriented services are of interest, as shown in Table 1-2. Revenue services that are being considered include advertisement (personalized, location specific, and push/pull), mobile commerce (including the commissions of products sold), and event-driven services (also including transmission charges). Both business and residential customers have expressed a "high interest in location-based services,"[8] as illustrated in Figure 1-5, which is also based on market research data from Siemens. Industry leaders hope that the decreasing mobile voice average revenue per user (ARPU) will be compensated by increasing mobile data ARPU (see Figure 1-6).[9]

Major opportunities continue to exist in the data wireless space, as implied by the following quote:

> . . . Cisco is betting that tornado markets will spur new business this decade, much the way the Internet opened a torrent of new opportunities during the past one. The tornado markets include Internet-based phone systems for businesses, wireless networks for home and offices, Internet-based systems to

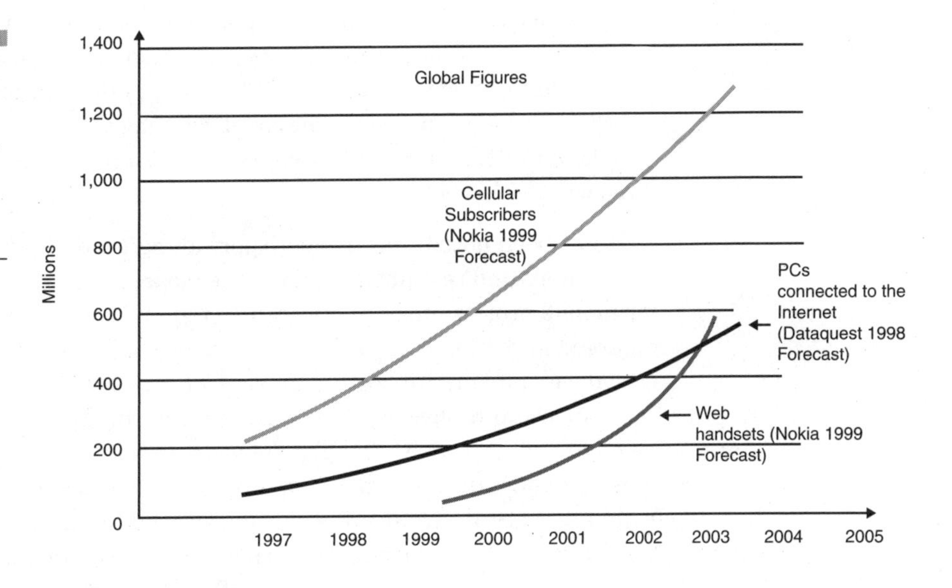

Figure 1-4 Data-enabled wireless units will equal the number of PCs in the near future. (Source: Nokia)

Table 1-2

Wireless services of interest

Residential User Segment	Business User Segment
Map-based local information	Mobile corporate Intranet access (laptop/personal digital assistant [PDA])
Map-based traffic information (via vehicular display)	Video conferencing (laptop/PDA)
Reservation service	Map-based local information
Multimedia messaging	Enterprise Resource Planning (ERP) system online access
Mobile house monitoring	Mobile Internet access
Mobile Internet access	Multimedia booking and reservation
Videophone	Traffic information service
Entertainment option: gaming	Mobile personal organizer
Shopping via mobile phone	
Mobile auction service	

Source: Siemens Mobile Networks Marketing, end-user survey 2001

streamline the way corporations store data, and gear that reduces Internet traffic congestion by putting copies of popular digital content on computers in multiple geographic locations. A fifth twister is the gear for building metropolitan fiber-optic networks, the speedy on-ramps that connect the long haul to local traffic centers. Cisco conservatively estimates these markets at $20 billion, and $40 billion by 2004.[10]

It has been noted that "from airport lounges and hotel meeting rooms to coffee shops and restaurants across the globe, a wireless world is being built for mobile professionals to stay connected."[11] Now that key standards have emerged and other ancillary standards are being developed, enabling low-cost development, supportive devices are poised to rapidly penetrate the consumer and professional networking market. Many end users are interested in mobile devices that can deliver web content to smaller form factors, whether they are in their home base or on the move. According to recent Cahners In-Stat Group's market data,[12] public area network services in hotels, convention centers, and airports are poised to flourish in the coming years. For example, they projected $180 million of service revenues in 2002.

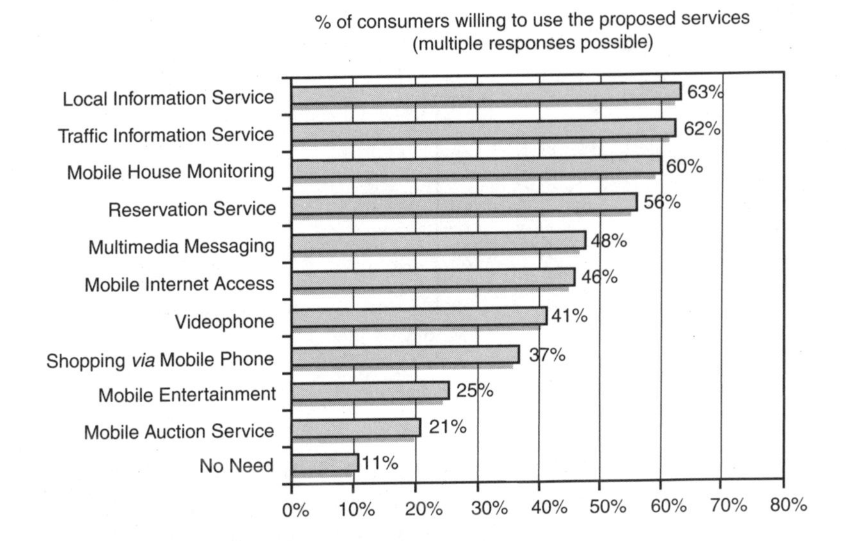

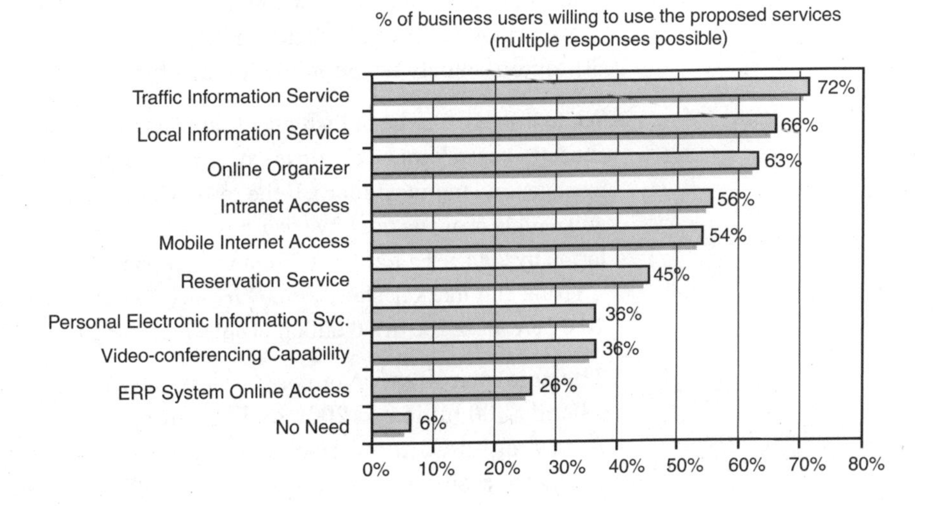

Figure 1-5 Interest in location-based and other wireless/nomadic services (Source: Siemens Mobile Networks [2001 data])

There will be significant investment in the future wireless data space. Gartner put the WLAN market at $1.4 billion in 2001, and estimates that it will grow to $1.7 billion in 2002 and $3.8 billion in 2006. Wireless network interface cards (NICs) were counted at 5 million in 2001 and are forecast to

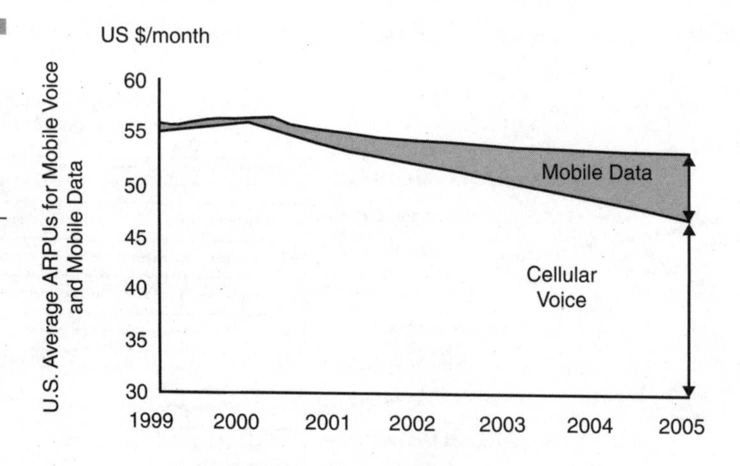

Figure 1-6 Revenues for location-based and other wireless/nomadic data services

be 9 million in 2002 and 40 million in 2006.[13] The per-unit cost of WLAN connection hardware has been dropping rapidly. Although WLANs continue to see penetration in vertical applications, interest is also growing in horizontal settings; some see the eventual emergence of a totally wireless campus. The overall 802.11b market is expected to keep growing at a healthy rate despite the economic slowdown of the early 2000s, according to an April 2001 report by Cahners In-Stat Group.

Statements such as the following are typical:

> While it is hard to be bullish about anything these days, the Boston, Massachusetts-based Yankee Group is extremely gung-ho about wireless field (in its most general manifestation). If the research firm is right, wireless penetration will double over the next five years—reaching 21 percent of the world's population by 2006, or a total of 1.3 billion subscribers. That is a bold forecast considering the market slowdown. Currently, wireless penetration amounts to 10.6 percent of Earth's 6 billion people, according to the Yankee Group.[14]

The total wireless revenues worldwide were placed at $200 billion in 1998 and $400 billion in 2002 by ITU sources.

According to Cahners In-Stat Group, the business WLAN market was remarkably resilient over the course of 2001; it experienced a growth rate that exceeded expectations, even in the midst of the technological economic downturn. The year was buoyed by strong sales to verticals, including education, retail, financial, and healthcare. From a horizontal standpoint, Fortune 500 companies built out their existing WLAN networks, and small and medium businesses moved to embrace WLANs. Asia Pacific and Europe remained strong geographic segments, but North America grabbed the lion's percentage of the market.[15]

Although the first half of 2001 was marked by indications of an economic downturn, Cahners In-Stat Group expected the WLAN market to show healthy growth throughout the forecast period. In the LAN space, wireless networking is the next step, and Cahners In-Stat Group apparently believes that shipments will continue to grow as prices fall and new technologies bring increased speeds. They state that by 2005, the total enterprise WLAN end-use revenues will be $4.6 billion. Throughout the forecast period, revenues will grow significantly less than shipments as prices continue to fall steeply. The driving force in the WLAN market will be Wi-Fi products, including 802.11b and 802.11g products. These products have been especially successful in penetrating the education, hospitality, healthcare, and financial environments and, as prices fell in 2001 and the beginning of 2002, other industries have adopted them at a high rate. Also, the push of these products into areas outside of North America will continue. In 2002, the growth rate in end-use revenues was expected to increase as comparatively high-priced 5 GHz products are introduced into the market.

As noted, there is a longer-term trend for always-on, always-active, and always-connected environments created by wearable computers. In the not too distant future, wearable computers will be met with the same interest that handheld computers are met with today, according to proponents. Loosely predicted, by 2006, progress made in the fields of wireless short-range communication and battery capacity will make it possible for anyone to carry an always-on link to the Internet or any other network.[16] Gartner predicts that by 2010, 40 percent of adults and 75 percent of teenagers will be utilizing wearable devices, and 70 percent of the population will spend 10 times longer a day interacting with people in the cyberworld rather than in the physical world.[2]

Some companies like Levi-Strauss are already advancing the wearable concept by offering a jacket that incorporates a hidden MP3 player and mobile telephone connected to a microphone and remote control in the collar. Some wearable devices are already available for less than $300.[17] As another example, Mitsubishi Electric's R&D lab recently showed a prototype of a new wearable display that, unlike most other wearable displays, positions the liquid crystal display (LCD) piece under the user's eyes. The idea behind this concept is to enable users to switch more easily between looking at the display and to have a full visual range than what is possible with most current solutions, where the user's view is most often occupied entirely by a display piece placed in front of the user's eye. The near-term goal is to connect these wearable computers' IP appliances to the Internet.[18]

Approaches to WLANs, WWANs, WPANs, and PALs

The major types of data-oriented wireless systems that are currently getting the most attention include the following (see also Figure 1-7):[19]

- WLANs
  - IEEE 802.11b, which is sometimes called *wireless Ethernet* or Wi-Fi
  - IEEE 802.11a, high-performance radio local area network type 2 (HIPERLAN/2),[20] IEEE 802.11g, and Home Radio Frequency (HomeRF) (alternatives to 802.11b)[21]
- WPANs (for example, IEEE 802.15 and Bluetooth)
- WWANs, also known as *mobile wireless* (voice/telephony with increasing integration of data and video)

Figure 1-8 depicts wireless technologies by locus. Table 1-3 captures some of the design goals of wireless solutions (particularly in a 3G context).

Forward-looking carriers and companies plan to use Bluetooth, IEEE 802.1a, 802.11b, and IEEE 802.15 to enable top-of-the-line locals and public spaces to achieve global reach over an intranet, extranet, or the Internet. This matches a hotspot service concept. Top-of-the-line buildings, hotels, airports, shopping malls, universities, hospitals, marinas, and other locales can avail themselves of wireless data technology and achieve global reach. Some of these companies support the planning, engineering, deployment, and operations of in-building networks (building PALs) as well as the metro-access, metro-core, long-haul, and Internet connectivity required for end-to-end services. Unlike the Building Local Exchange Carrier (BLEC) concept of the late 1990s, forward-looking companies intend to spend only a relatively small amount of money in each building (not to exceed $5,000 at most and often as little as $1,000) and in the network (for example, by using virtual LAN [VLAN] technology), thereby making the undertaking financially cost effective with a relatively small number of subscribers.

Unlike the technologies used in 3G, where European telecom firms have recently had to invest more than $100 billion to secure spectrum, leaving such companies cash strapped, the kinds of technologies used by hotspot

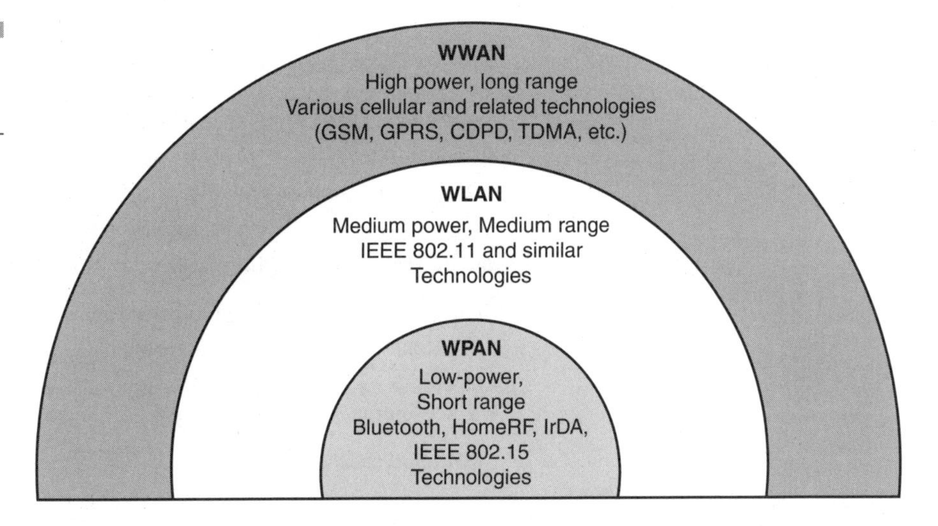

Figure 1-7 Continuum of wireless technologies

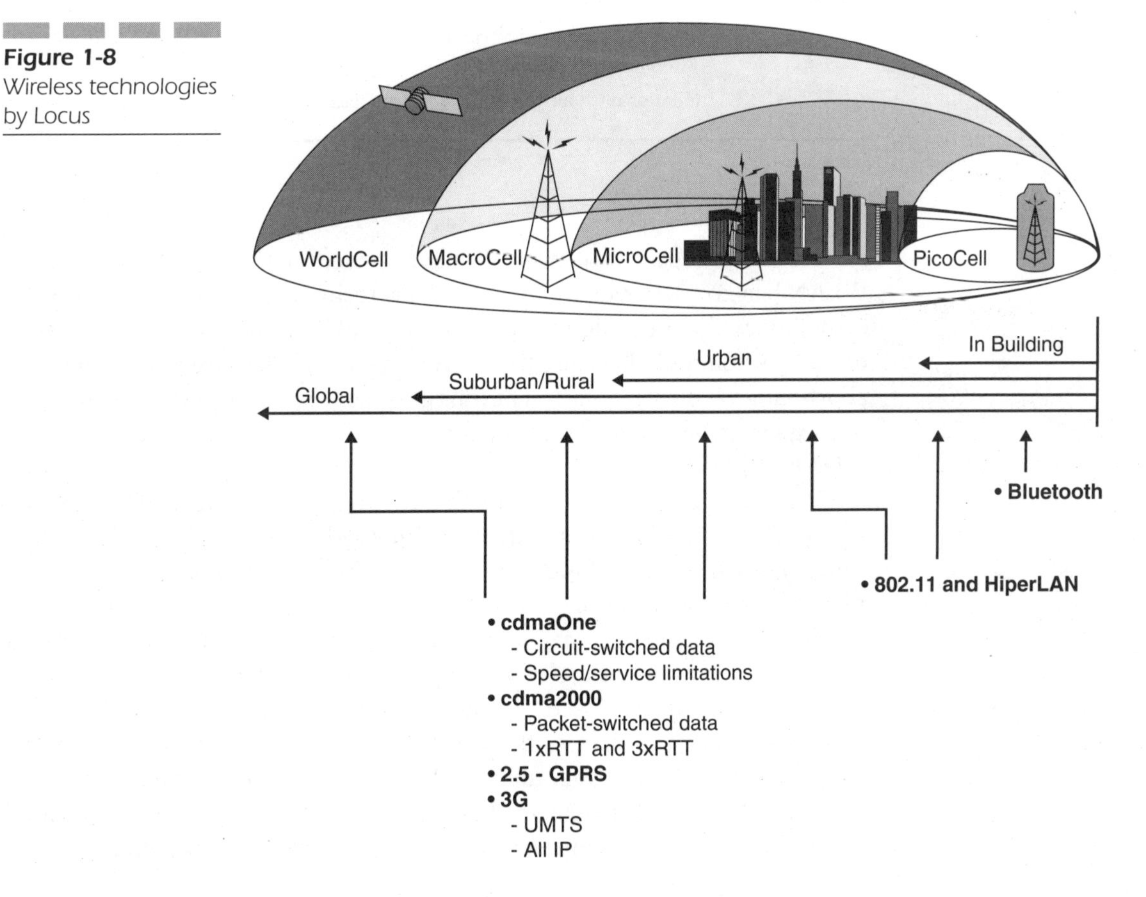

Figure 1-8 Wireless technologies by Locus

Table 1-3
Design goals of 3G

Feature	Goals
Services	Stationary users or pedestrian applications with speeds of 0 to 10 km/h
	Vehicular applications with speeds up to 100 km/h
	High-speed rail applications with speeds up to 500 km/h
	Aeronautical applications with speeds up to 1,500 km/h
	Satellites applications with speeds up to 27,000km/h
Data rates	Up to 144 Kbps (more in the future) in vehicular applications
	At least 384 Kbps for pedestrian (nonhotspot) applications
	2.048 Mbps for indoor or low-range outdoor hotspot applications for WWAN solutions and 55 Mbps for WLAN solutions
Connection mode	Circuit and packet mode
Cell sizes	Regional megacells/worldcell greater than 35 km in radius
	Macrocells of radius 1 to 35 km
	Indoor or outdoor microcells of radius up to 1 km for WWAN solutions and 5 km for WLAN solutions
	Indoor or outdoor picocells of radius less than 50 m

service companies do not need licensing (whether the service is WPAN or WLAN based). However, in fairness, these technologies are short range and do not allow the multiple-mile links of the 2.5G/3G alternatives (at least for omnidirectional applications); however, the hotspot service is achieved with short-range radios in selected buildings or spaces connected to the network at large over terrestrial links such as a T1, Synchronous Optical Network (SONET), and/or fiber.

Nearly all hotspot applications to date are based on IEEE 802.11b technology. Some, however, also advance Bluetooth technology in this context. The Bluetooth Local Infotainment Point (BLIP) is a free-standing communications platform that uses Bluetooth technology to receive and transmit information. It is locally targeted, giving users access to situation- and location-specific services, thereby acting as a filter in the media clutter, according to proponents. Accessed by users' mobile phones or PDAs, it lets users download images, music, and video files. The first version of BLIP had a range of general functions and is now commercially available, and more specialized versions are expected to be developed in the future. In recent trials in Japan, visitors equipped with Bluetooth-ready handheld computers or PCs and passengers in trains were able to gain location-based useful information with the push technology. Users gained access to contents such as streaming video and the Internet via the Bluetooth network.

Bluetooth proponents see the vast implications for the hotspot and believe that it represents a potential solution for service providers and location partners who want to communicate with the end users more cost effectively and attractively. IEEE-802.11b- and IEEE-802.15-enabled services are also contemplated for hotspots with 802.11b currently having a lead. However, compared to Bluetooth wireless technology, WLANs based on the IEEE 802.11 are more expensive and power consuming, and the hardware requires more physical space; the counterpoint is that the former supports distances of 1,500 feet (or even several miles in narrow beams), while the latter only supports 10 meters. Bluetooth is therefore more suitable for small mobile devices. Bluetooth technology is discussed in more detail in Chapter 3. Figure 1-9 shows connectivity ranges for various technologies.

The convenience of wireless access, however, comes with security concerns that must be properly addressed.[22] When data is broadcast over radio waves, it can be intercepted and compromised. Thus, it is imperative to utilize robust security mechanisms to protect wireless communications. Security for IEEE 802.11 networks can be boiled down into three components: the authentication mechanism (or framework), the authentication algorithm, and data frame encryption. *Authentication* is the act of verifying a claimed identity in the form of a preexisting label from a mutually known name space as the originator of a message (message authentication) or the

Figure 1-9 Ranges of various technologies

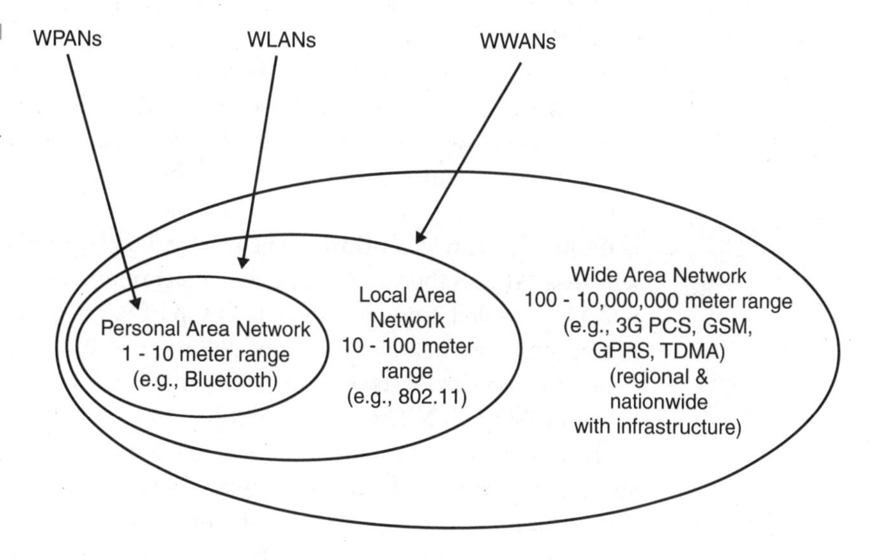

endpoint of a channel (entity authentication).[23] *Encryption* is the process of rendering the data impervious to scrutiny, except to those that possess the decryption key. In addition to security and privacy, other issues regarding wireless include input/output mechanisms, invisibility and social acceptance, battery power, and usability.

Wireless Local Area Networks (WLANs)

WLANs provide a cost-effective way to interconnect computers and mobile devices in defined environments such as enterprises, small offices/home offices, airports, hotels, coffee shops, hospitals, libraries, schools, and residences. The approval of the IEEE 802.11 standard for WLANs in the late 1990s along with the ongoing progress of increasing transmission speeds have made wireless systems a useful technology for businesses and consumers, offering mobile computing and untethered network access. Office applications are already shifting from niche use to more general-purpose use, as computer vendors such as IBM, Dell, Apple, and others build the wireless capability directly into the laptop.

Wireless networks have recently gained popularity. By the end of 2002, an estimated 10 million 802.11 radio-based Network Interface Cards (NICs) were deployed.[24] Adapter card vendors are also bringing out 802.11b cards for handheld computers, such as those using Microsoft PocketPC software. Vendors of WLAN technology include 3Com, Agere, Avaya, Buffalo Technologies, Cisco, Compaq, Dell, D-Link, Enterasys, Intel, Intersil, Linksys, MobileStar, Proxim, Resonext, and Symbol Technologies.

WLANs based on IEEE 802.11b are becoming more widespread in enterprise applications where mobility in the office and campus is needed. This technology is the leading hotspot technology by actual deployment; it is likely that it will continue to be the leader. Equipping office spaces with wireless APs is becoming relatively common; as a result, users have 802.11b-enabled laptops or even desktop PCs. The price of the wireless APs is coming down (they cost around $400 for a 20 to 40 person workgroup, depending on the features supported) and NICs are also reasonably affordable. An 802.11b NIC costs about $80 to $150. If you compare that figure to the cost of wiring up an office or cubicle (usually around $200 per seat), you can see why network planners are looking into this technology, particularly because this wireless technology also brings more flexibility compared to a tethered network. Data rate ranges from standard Ethernet performance at 11 Mbps down to perhaps 2 Mbps if/when there is significant interference or if/when the user moves too far away from an AP.

The technology is typically based on shared access methods. Therefore, an 802.11b network achieves less performance than a wired Ethernet connection when, for example, considering the switched arrangements typical of today's wired networks; however, it offers mobile capabilities for users. If the NIC and AP support roaming (which most implementations do), a user can wander nomadically around a building or campus; the NIC will automatically switch between APs based on the strength of the beacon signal it receives from them. APs are usually connected to the wired LAN, enabling users to have complete connectivity. Advances in IEEE 802.11b technology allow for more reach (approximately 30 percent more reach) and higher performance (about 70 percent more in area) than original systems. This should further facilitate penetration. IEEE 802.11a systems provide increased performance to 54 Mbps. The following are some factors that affect range:

- Modulation technique (such as, complimentary code keying [CCK] at 2.4 GHz and OFDM at 5 GHz)
- Cell design (for example, 802.11b has three nonoverlapping channels, whereas 802.11a has eight nonoverlapping channels in U.S. indoor applications. This means that 802.11a has less co-channel interference (CCI) and, consequently, higher cell throughput)[25]
- Environment (for example, path loss/absorption, multipath/echoes, interference, and collisions indoors and outdoors)
- Hardware/software system (for example, radio quality, antenna type and gain, computer speed, protocol efficiency, and so on)

An AP is the size of a portable CD player and has an antenna connected to it. A variety of high-gain and sector antennas can be used along with amplifiers up to 1 Watt (30 dB gain) in the United States and higher abroad. Travelers carrying equipment with an Internet browser and wireless Personal Computer Memory Card International Association (PCMCIA) card can pass within 30 to 500 feet of a wireless AP and log onto their corporate intranet or check their e-mail or portfolios. Behind this device is a network/Internet infrastructure.

As an example of hotspot/WLAN application (based on Cisco System's sources), the University of Akron has begun installing approximately 1,200 Cisco Aironet 350 Series APs to create a wireless infrastructure that encompassed all instructional space, residence halls, houses, the library, campus centers, the alumni association, arenas, and other places where students congregate. Upon completion, the WLAN system will cover 75 buildings and serve 35,000 people. All 24,000 students who attend the university and all

6,000 full- and part-time employees will have the ability to connect to the university's information system infrastructure and the Web through the Cisco 802.11b wireless standard. The wireless coverage will spread into campus green areas such as tree-lined courtyards, commons, and park-like open spaces; ultimately, only areas such as parking decks and tennis courts that are too far from the campus will remain outside the wireless network. Virtually every other part of the 5.4 million square feet of building space will be included—even the football stadium.

As another other example from Cisco sources, the University of Missouri is testing a WLAN system that will eventually be available to 25,000 users at its main campus in Columbia by using Cisco Aironet products. Approximately 30 Cisco Aironet 340 Series APs were installed on the Columbia campus in August 2000 as part of a pilot program involving a cross-section of 20 to 30 students, administrators, and faculty members. The APs bridge a wireless and wired network to help create a standalone wireless network. Most participants connect to the wireless system via laptops, which gives them access to the campus enterprise network in the designated areas of eight buildings. Some, however, use desktops set up in computer labs or open study areas. The university is in the process of expanding the pilot program by more than doubling the number of APs to increase the roaming space. As previously implied, amplifiers can extend the range; however, the problem is constrained by the return signal (from the client), so that amplifiers have a practical limit in the design of hotspot systems. The author has built several dozen hotspot locations with sizes of 2,000 by 1,000 feet; to accomplish this, the author has developed the concept of Redundant Array of Inexpensive Radios (or Repeaters) (RAIR). This entails using 2, 4, or 6 distributed radios operating in repeater mode placed clos(er) to the user compared with a centralized omnidirectional or sectorized antenna. Using this design approach, the distance from the client to the radio can be kept in the 300 foot range, which provides the best and most reliable hotspot service with the IEEE 802.11b technology (again, because the problem is constrained by the return path).

Key questions regarding WLANs relate to security, the potential interference in the unlicensed 2.4 GHz Industrial, Scientific, and Medical (ISM) band, and obsolescence when higher-speed services arrive.

As already noted, security has been a major concern of late. IEEE 802.11 defined the Wired Equivalent Privacy (WEP) protocol to address some of the security issues. The goals of WEP are to maintain the confidentiality of data from eavesdroppers, guard against the modification of data (integrity), and provide access control to the WLAN infrastructure (this function is not an explicit goal in the 802.11 standard, but it is frequently considered to be a feature of WEP). WEP utilizes a static (secret) key to authenticate, associate, and transmit encrypted frames. It employs the well-known Ron's Code 4 Pseudorandom Number Generator (RC4 PRNG) algorithm, which is a symmetric key encryption algorithm from RSA Security, Inc.[26] However, flaws in the overall scheme permit several passive and active attacks that allow eavesdropping and the modification of wireless transmissions.[27]

WEP is easy to administer, but is problematic, as noted. WEP relies on a secret key that is shared between a mobile station (for example, a laptop with a wireless Ethernet card) and an AP; the secret key is used to encrypt packets before they are transmitted, and an integrity check is used to ensure that packets are not modified in transit. The IEEE standard does not discuss how the shared key is established. Consequently, most installations use a single key that is shared between all mobile stations and APs.[28] The device using the IEEE 802.11 card is configured with a key, which in practice usually consists of a password or key derived from a password. The same key is deployed on all devices, including the APs. The idea is to protect the wireless communication from devices that do not know the key.[29] More sophisticated key management techniques can be utilized to help defend from the attacks; however, no commercial systems currently support such techniques.

Improvements to WEP will be applied in three phases over time, as follows:

- **Short term** Filter out bad initialization vectors.[30]
- **Near term** Use dynamic key management—keys are rotated every few seconds or minutes.
- **Long term** Use new methods such as IEEE 802.1x (EAP-MD5) (higher-layer authentication mechanisms and key management) and IEEE 802.11i (improved security in multiple areas).

As implied by the last observation, improvements to WEP are under way. The near-term idea, therefore, is to use dynamic key management by using a static key only for authentication and association; a dynamic key is used for routine traffic. You should automatically update (without user intervention) the dynamic key on a fairly regular basis. This prevents the potential hacker from collecting a large number of frames that have been encrypted with the same key. In addition, organizations can use layer 3 virtual private networks (VPNs) to address the issue. The issue of security is revisited in Chapter 3.

The issue of other users/applications on the same band should be noted. HomeRF, Bluetooth, and 802.11g all share the 2.4 GHz band with 802.11b. There is the potential for interference at 2.4 GHz. Several networks share this bandwidth such as other 802.11b networks, Bluetooth, HomeRF, microwave ovens, cordless phones, fixed wireless, amateur radio, and so on. A number of experts worry about eventual overcongestion. For indoor applications, cordless phones and microwave ovens can be an issue. However, HomeRF systems do not have a large base and do not impact corporate environments. Bluetooth is a low-power technology that does not appear to represent a major interference problem; rollout up to the early 2000s had been slow. Adaptive frequency hopping will address the issue.

Regarding obsolescence, many people believe that the current solutions are already too slow. 802.11b only supports 11 Mbps and operates in an 83.5 MHz band at the 2.4 GHz range (in the Ultra High Frequency [UHF] band). Bluetooth only supports 1 Mbps and has a much shorter range. These limitations are proving impetus for IEEE 802.11a work at 5 GHz (5 GHz is in the Super High Frequency [SHF] range). This portion of the spectrum is available throughout much of the world. The allocated spectral bandwidth is 300 MHz in the United States, 100 MHz in Japan, and 455 MHz in Europe; this bandwidth supports much higher throughput. There is no interference from microwave transmissions that are at 4 and 6 GHz or higher. IEEE 802.11a uses OFDM techniques that further improve performance. IEEE-802.11a-based technology increases the range and results in fewer APs. 802.11a was already shipping at the time of this writing. Its logic fits onto a single-sided PC card. It is similar pricewise to 802.11b, and the power drain is also similar. Some claim that it will replace 802.11b as the dominant WLAN standard in the near future; however, the embedded base of 802.11b is significant.[13]

Figure 1-10 compares line rate and throughput between 802.11a and 802.11b. Figure 1-11 shows the topological cell layout topology possible with these two technologies. Finally, Figure 1-12 illustrates the advantages of 802.11a in terms of reducing costs (requiring less APs) for about the same throughput or increasing the throughput by keeping the same number of APs.[31] At the lower end of the data rate, observers expected an increased penetration of Bluetooth technology from 2003 to 2004 for cell phones and PDAs. Having made these observations regarding newer technology, however, you should note that any well-designed business plan has a payback of 24 to 26 months. This implies that carriers rolling out a given generation of technology can recover their costs if they upgrade their systems every 2 to 3 years. Particularly in the case of hotspot systems, the carrier's equipment is fairly inexpensive and can be depreciated relatively quickly.

The field of WLANs is being tracked by the Wireless Ethernet Compatibility Alliance (WECA). The mission of WECA is to provide certification of compliance with the IEEE 802.11 standard, thereby advancing the oppor-

Figure 1-10 Comparison of line rate and throughput

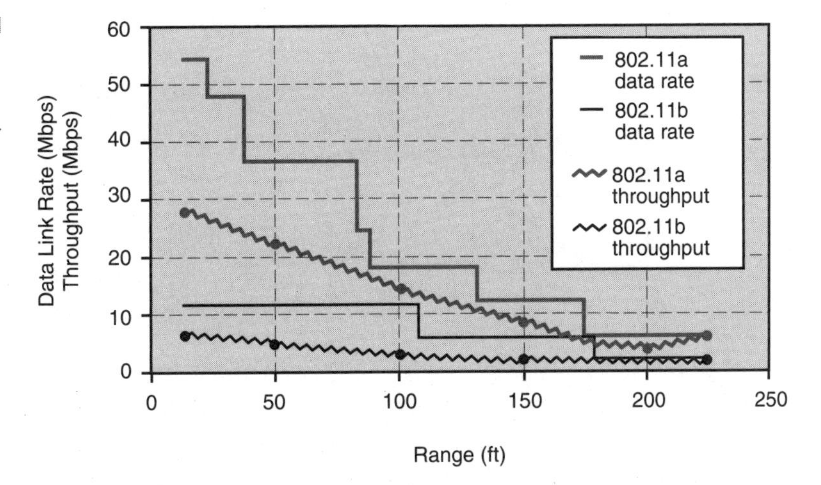

Notes: 802.11a link rates are 2 to 5 times higher cut (at the same distance) than 802.11b

802.11a throughputs are 2.5 to 4.5 times that of 802.11b

Figure 1-11 Typical cell layout

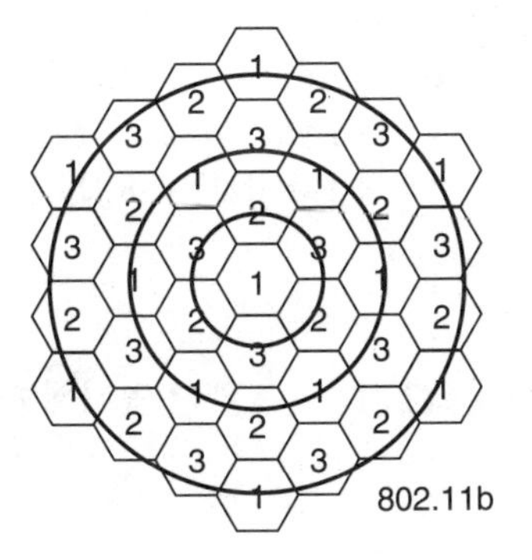

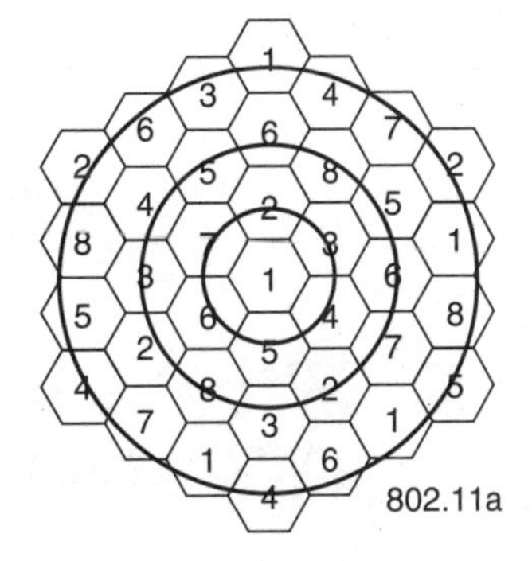

802.11a suffers less CCI than 802.11b

tunity for interoperability. The founding members include Cisco, IBM, Intel, 3Com, and Microsoft. WECA is looking to forge relationships and network standards among wireless Internet service providers (WISPs) and eventually carriers that will enable roaming for IEEE 802.11b WLAN users. According to WECA members, the public access WLANs that will be deployed in airports, marinas, convention centers, and restaurants "will cre-

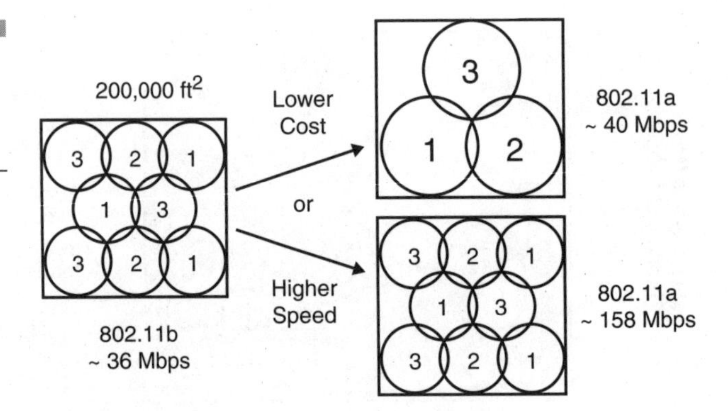

Figure 1-12 Advantages of 802.11a based on cell topology

ate a burgeoning web of WLAN hotspots. These hotspots will let mobile workers with 802-11b-equipped computers connect over a shared 11 Mbps link to Internet-based services and corporate networks."[12] The carriers are watching the project closely according to WISP roaming (WISPr) members —a contingent within WECA that is preparing roaming proposals/specifications. "There is a tremendous amount of work going on by all carriers; they are quiet about it, but they are all doing it."[12] For public access, installations at places such as Starbucks, Admiral's Clubs, hotels, conference centers, and airports (Dallas-Fort Worth, Austin, San Jose, and Sea-Tac) illustrate the technical and market feasibility of these services.

Wireless Personal Area Networks (WPANs)

PAN refers to very small-area and low-power radio-transmission-based networking systems. Typically, these systems operate at 1 Mbps or below and have low power requirements; the distance is measured in feet (for example, 33 feet). Bluetooth technology is a prime example of this kind of technology; there are IEEE groups addressing the further standardization of the technology (the IEEE 802.15 committee is now working on an IEEE-level standard).

WPAN's Bluetooth wireless technology has received considerable media attention. Proponents position it as a de facto standard as well as a specification for small-form-factor, low-cost, short-range (LAN-level) radio links between mobile PCs, mobile phones, and other portable devices. The technology is just now becoming available. In support of WPANs, inexpensive

PCMCIA cards are being developed that can fit in PCs and PDAs. Bluetooth chips that cost under $5 are also being targeted. Other equipment has the radio chip built in (especially in the case of telephone equipment). The ultimate goal is to incorporate these chips in virtually every kind of device: PDAs, mobile phones, computers, and desktop phones.[32] Figure 1-13 provides a comparison between WLANs and WPANs.[33]

Home Radio Frequency (HomeRF)

The other localized 2.4 GHz radio technology is HomeRF, which has many similarities to the Bluetooth wireless technology alluded to in the previous section. HomeRF can operate ad hoc networks (data only) or be under the control of a connection point coordinating the system. The system also provides a gateway to the telephone network (data and voice). The technology uses frequency hopping. The hop cycle is 8 Hz, whereas a Bluetooth link hops at 1,600 Hz.

In the simplest terms, HomeRF is wireless home networking—a way to connect PCs, peripherals, cordless phones, and many other consumer electronic devices so they can communicate with each other, sharing resources and access to the Internet. Wireless home networking makes this possible

Figure 1-13
WLAN versus WPAN

Feature	WLAN	WPAN
Interaction	Interaction is with a wired LAN infrastructure	Personal objects interact with each other over a wireless system
Frame	Header size is 240 bit	Header size is 54 bit
Type of devices	Portable devices	Mobile devices
Network timeframe build - out	Hours to days (the LAN portion that needs to be built-out, generally takes longer)	Seconds to hours
Relative cost	More expensive	Less expensive
Network topology	Tree topology 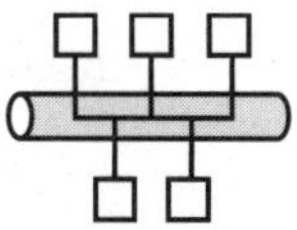	Star or Ring topology 

without the expense and complexity associated with running wires. In more technical terms, HomeRF is an open industry specification that defines how devices share and communicate voice, data, and streaming media in and around the home. HomeRF-compliant products operate in the license-free 2.4 GHz frequency band and utilize frequency-hopping spread spectrum (FHSS) RF technology for secure and robust wireless communications. HomeRF blends technologies from several worldwide standards because none of them alone could meet the market requirements. Data-networking technologies based on carrier sense multiple access with collision avoidance (CSMA/CA) protocols (essentially wireless Ethernet) are derived from the OpenAir and IEEE 802.11 standards, and cordless phone technologies based on Time Division Multiple Access (TDMA) are adapted from Digital Enhanced Cordless Telecommunications (DECT).[34] HomeRF extends beyond WLAN home support—with 10 Mbps performance, a cordless telephone can support up to eight lines and QoS can support media streaming including music and TV and standardized roaming. Table 1-4 provides a comparison between a number of PAN technologies.[35] It appears, however, that IEEE 802.11b will be the technology of choice for home networking.

HomeRF was developed by the HomeRF Working Group (WG), which initially included five leading computer companies, but has since expanded to over 50 companies made up of leaders across the PC, consumer electronics, networking, peripherals, communications, software, retail channel, home control, and semiconductor industries worldwide. This group was launched in 1998 to promote the mass deployment of interoperable consumer devices that share and communicate voice, data, and streaming media in and around the home without the complication and expense associated with running new wires. The HomeRF WG is dedicated to developing wireless home networking products that are simple, secure, reliable, and affordable to the consumer. At the beginning of 2001, it announced its interest in increased transmission speeds and support for additional device types, applications, and services. In fall 2001, it announced the formation of a HomeRF European WG to focus on the European marketplace.

Table 1-4
GHz PAN radios

Parameter	Bluetooth	HomeRF	IEEE 802.11 WPAN
Distance	10 meters	50 meters	Not determined
Hop rate	1,600 hertz	50 hertz	2.5 hertz
Transmit power	1 milliwatt	100 milliwatts (North America)	< 1 watt (United States) 100 milliwatts (Europe and Japan)

HomeRF is a blend of wireless Ethernet, cordless telephony, and streaming media. Using only one Internet connection, whether it is dial up, Digital Subscriber Line (DSL), or cable, each member of the family can wirelessly access the Internet and PC resources simultaneously, using his or her different laptops, PCs, or Web-based devices. Because HomeRF is derived from a blend of DECT[36] (for TDMA voice networking) and IEEE 802.11 (for CSMA/CA data networking) and uses the globally available 2.4 GHz frequency band, it is the logical evolution of DECT for high-speed home networking and will benefit from DECT's large market and mature technology. Over 50 million DECT devices were sold in 2000, and by 2002, the installed base was expected to be over 200 million. That is 10 times the size of the WLAN market. The DECT technology is very mature; semiconductors are in their fifth generation and chip sets are sold for well under $10.

HomeRF products first entered the market in 2000 and had 1.6 Mbps performance, which is a logical match for DSL and cable modems. Nearly all of the early products were PC related and supported data applications. HomeRF 2.0 (introduced in 2001) has 10 Mbps performance and supports more phone lines and new features for digital music and Internet-based TV. Products that could enter the market include the first HomeRF cordless phones, web tablets, music devices, and the increased use of HomeRF in home gateways. Because HomeRF 2.0 is compatible with first-generation HomeRF products, the roadmap is evolutionary. HomeRF 3.0, which has an even greater bandwidth of more than 25 Mbps, will support streaming video with DVD quality. By then, we could see broader use of HomeRF in TV set-top boxes, video tablets, and multimedia servers.

Wireless Wide Area Networks (WWANs)

WWANs have received a lot of attention in recent years. WWANs/mobility services give the user the ability to move around town or around the country while still remaining connected to the Internet or the company intranet. As such, they require extensive infrastructure, like cellular telephony with its numerous tall towers[37] all over town, switches, central offices, and gateways. These networks and systems tend to be the purview of major carriers such as AT&T Wireless, Sprint, Verizon, and so on. WWANs have been around for a number of years (see Figure 1-14),[38] but data support was rather limited and expensive until the late 1990s. Mobile telephones phones are also becoming portable Internet terminals.

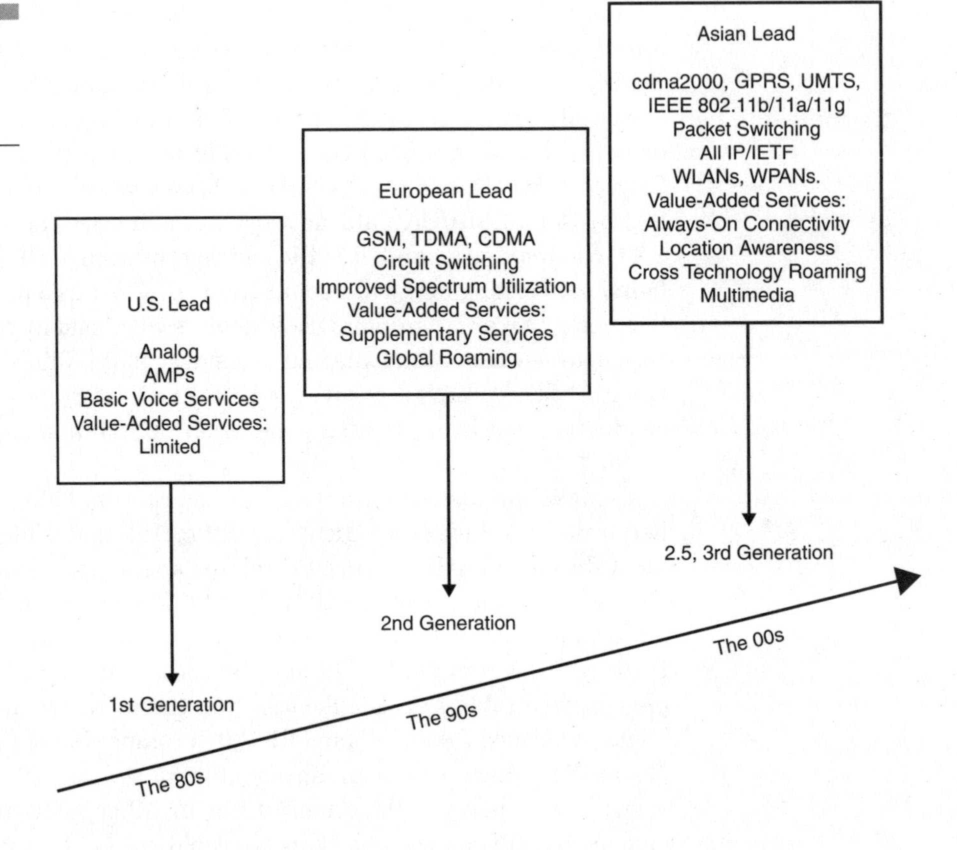

Figure 1-14 The evolution of WWLAN over three decades

PDAs, laptops, and data-ready phones benefit from mobile Internet access, although the speeds started out at the low end (compared to LAN connections).

As of yet, no single killer application has emerged beyond the basic convenience of mobility because individual interests are varied. For example, e-mail has limits (not everyone wants to read e-mails while shopping at the mall and so on); stock quote/trading via wireless has been available since the mid-1980s. Enterprise applications (secure Web-enabled intranet access to ERP systems via wireless) may be of increasing importance in the near future. Figure 1-15 provides nominal and actual data rates for some WWAN technologies.

The range of a WAN is typically measured in miles. Communication over such distances requires relatively high-power transmissions and, because of that, a license for a specific frequency band. In most instances, carriers pay a fee for a license to transmit at certain power levels in a particular frequency spectrum. High-power transmission also leads to trade-offs between

Figure 1-15 Nominal/actual data rates for various technologies

System → / Aspect ↓	UMTS (Universal Mobile Telecom-munication System)	EDGE (Enhanced Data Rates for Global Evolution)	CDMA (Code Division Multiple Access) 3G-1X RTT (Radio Trans-mission Technology)	CDMA 3G-1X EV (DO) Evolution (Data Only)	CDMA 3G-3X
Access	Code Division	Time Division	Code Division	Code Division	Code Division
Nominal Data Rate (kbps)	2,000 384/144	167/200/ 349/470	144 - 300	2,400 (downstream)	2,000
Actual Rate (Kbps)	500/100	90-140	60- 120	500	200 - 500

power consumption and data rates in WWANs. Typical data rates for today's cellular networks are relatively slow due largely to the transmission power needed to reach the cellular tower from a handset. Higher data rates at these same levels of power transmission are impractical using today's battery technologies. 3G cellular systems will have significantly faster data rates; however, to maintain power consumption at reasonable levels, 3G systems will require cellular towers to be much closer together—on the order of hundreds of meters—approaching the coverage of WLANs. Some of these WWAN technologies are those in common use for cellular communications, such as Global System for Mobile Communications (GSM), TDMA, Code Division Multiple Access (CDMA), and others (see Figure 1-16).[39] Unfortunately, there is a real cornucopia of standards, making interworking a challenge, particularly for data applications. Table 1-5 depicts the set of such standards[40] and Figure 1-17 identifies some of the key standards bodies working on 3G.

Environments with a highly developed wired network lower the demand for wireless services, just as the availability of cable TV in major urban locations lowers the demand for Direct TV for city dwellers (however, in more rural environments, Direct Broadcast Satellite services are more in demand). This is why demand for web access via cell phones is more pronounced in Asia and Europe. Also, the presence of highly developed wired networks has a downward price pressure on cell phone and wireless charges. Wireless has taken off in countries where telecom charges are high because it is more available and less expensive than the wireline alternatives.

WWANs piggyback on wireless telephone systems. Hence, any discussion of WWANs will entail references to these systems, so a quick review is in order.

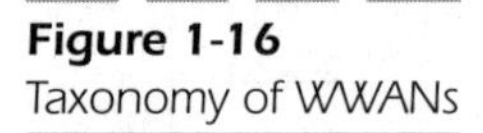

Figure 1-16
Taxonomy of WWANs

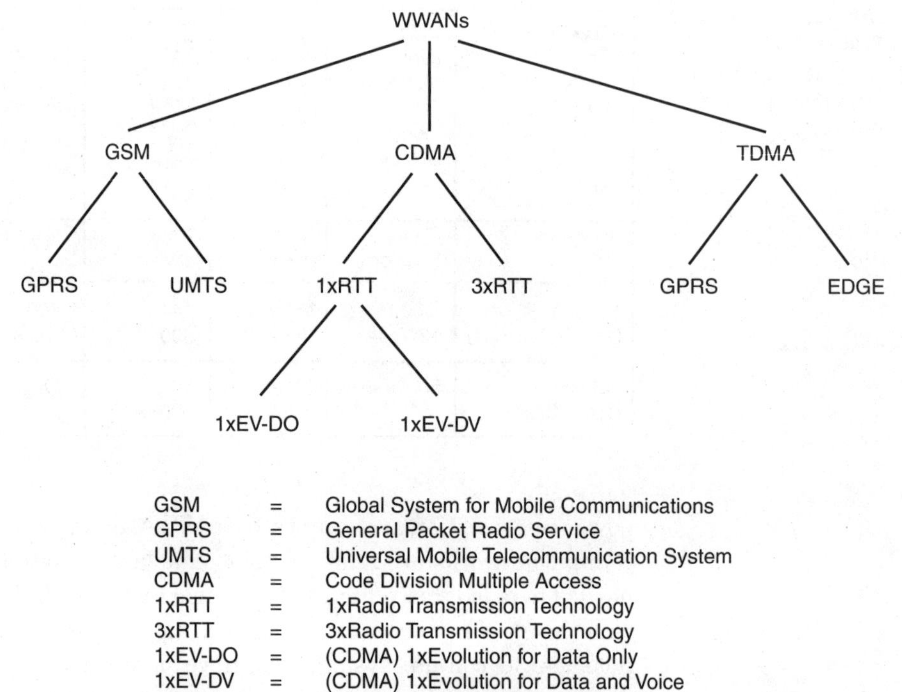

European providers have deployed systems based on the GSM standard.[41] In the United States, however, carriers have utilized a variety of standards including CDMA, TDMA, and GSM. TDMA is the most widely used wireless technology in the Americas, but other systems are also used. The plethora of standards forces web developers to duplicate their work to support each protocol and the specific types of devices that use each protocol.[42]

This paragraph provides a snapshot of the GSM history.[43] In the early 1980s, there were several different cellular solutions in Europe, such as NMT 450, Total Access Communications Service (TACS), C-Netz, Radiocom, and RTMI/RTMS. In 1982, the Conférence des Administrations Europèenes des Postes et Télécommunications (CEPT), a group of 26 European telecommunication carriers, established the Groupe Spéciale Mobile (GSM), which is also known as Global System for Mobile Communications, for a pan-European collaboration. In 1984, the GSM project was endorsed by the European Commission, which paved the way for a digital wireless solution for the 900 MHz frequency band. With the rapid increase of mobile communications, pressure was put on the European Communities to adopt

Table 1-5

Cornucopia of standards for WWANs

	2G		2.5G		3G			MWIF
Radio technology	GSM	CDMA IS-95	General Packet Radio Services (GPRS)	CDMA2000 (1xRTT)	UMTS (Wideband CDMA [W-CDMA])	Universal Mobile Tele-communications Service (UMTS) (W-CDMA)	CDMA2000 (HDR, . . .)	ALL/Any wireless + wireline
Core network architecture	TDM circuit switched (CS)	TDM CS	TDM CS + Packet Switched (PS) overlay	TDM CS + PS overlay	Third-Generation Partnership Project (3GPP) R'99 (CS + PS)	3GPP R'00 All-IP	3GPP2 All-IP	All-IP
Voice equipment	Mobile Switching Center (MSC)	MSC	MSC	MSC	MSC	SGSN/GGSN MGW	WAG (PDSN) MGW	Access gateway
Data Equipment	Inter-working function (IWF)	IWF	Serving GPRS Service Node/ Gateway GPRS Support Node (SGSN/GGSN)	PDSN	SGSN/ GGSN	SGSN/GGSN	WAG (PDSN)	Access gateway
Signaling equipment	MSC	MSC	MSC + SGSN/GGSN	MSC + PDSN	MSC + SGSN/GGSN	CSCF SGSN/GGSN	Session manager	CSM
Signaling protocols	SS#7 based	SS#7 based	SS#7 + IP based	SS#7 + IP based	SS#7 + IP based	Session Initialization Protocol (SIP) and others	SIP and others	SIP, Real-Time Protocol (RTP), Resource Reser-vation Protocol (RSVP), Multi-proto-col Label Switching (MPLS), megaco, *diffserv*
Mobility Management for data	Through MSC	Through MSC	GPRS mobility	Mobile IP	GPRS mobility	GPRS mobility	Mobile IP	Mobile IP

Source: WaterCove Networks

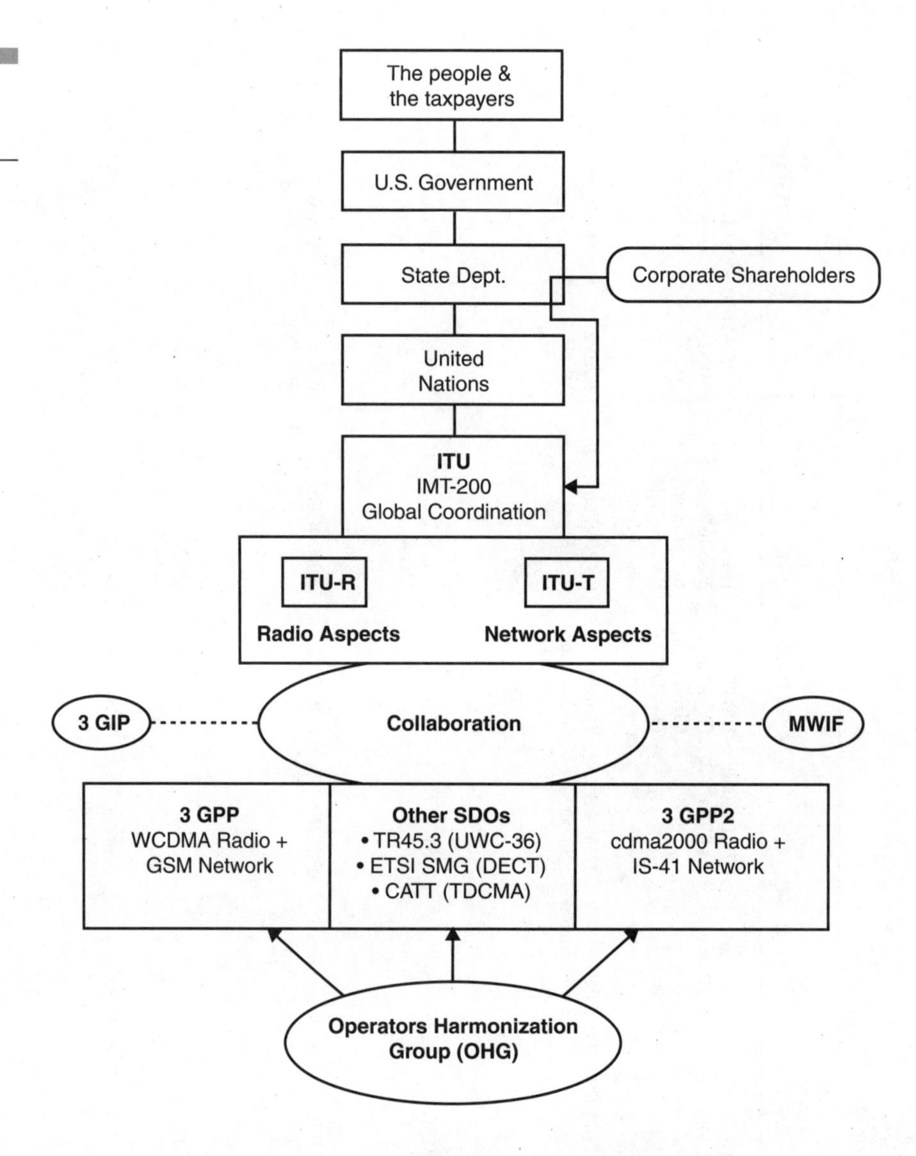

Figure 1-17 Standards bodies involved in 3G

the GSM solution quickly. An agreement was reached to reserve the 900 MHz frequency block in member states for a rollout in major European cities in 1993 and a linking in 1995.

On September 7th, 1987, operators from 13 countries signed a Memorandum of Understanding (MoU) in Copenhagen. The new standard was supposed to employ TDMA, a technology supported by Nokia, Ericsson, and Siemens. After validation tests, the MoU operators signed an invitation-to-tender in 1988. In 1989, the European Telecommunications Standards

Institute (ETSI) was formed and accorded equal status to administrators, operators, and manufacturers, which resulted in the publication of the GSM 900 specifications in 1990.

The United Kingdom's personal communications network (PCN) decided to adopt the GSM specification for their Digital Cellular System 1900 (DCS 1900) development, which was later renamed GSM 1800. A major problem with the network launches in 1991 was that no GSM terminals were available. Each terminal had to undergo a rigorous approval regime that was not available at that time, but that was needed for international roaming. This resulted in the introduction of interim type approval (ITA) to ensure that terminals would not create problems for the networks, which resulted in the wide availability of terminals in 1992.

In 1992, Australia's Telstra signed on with the MoU and just four years later they had 1 million subscribers, which was 5.6 percent of the population. One year later, GSM had also expanded to India, Africa, Asia, and the Arab world. In 1994, the U.S. Federal Communications Commission (FCC) made the 1,900 MHz band available in the United States. In 1995, the MoU had 156 members serving 12 million customers in 86 countries. At the same time, Phase 2 of GSM and demonstrations of fax, video, and data communications via GSM appeared. One of the most attractive features of GSM is network security, which has encryption for both speech and data. GSM subscribers are identified by their Subscriber Identity Module (SIM) card, which means that a user can put his or her SIM card in any phone and be recognized in the network. At the end of 2001, there were 600 million GSM subscribers worldwide, which is 65 percent of the world's wireless market.

The GSM-ANSI-136 Interoperability Team (GAIT) is working on developing mobile phones that work on both GSM and TDMA networks while providing overseas roaming. This is of interest to companies such as Cingular and AT&T Wireless, which are operators who still have TDMA networks and are currently deploying GSM networks. GAIT Phase 1 is the integration of GSM and TDMA voice and data technology into one handset with the option to select which system to use. The supporting organization is the Universal Wireless Communications Consortium (UWCC). The Consortium was founded in 1996 and works closely with other global organizations like the ITU, 3GPP, and the UMTS Forum.

GPRS is a packet-switched wireless data network that enables data to be sent and received using GPRS devices in a more cost-efficient and quicker way than was possible over the GSM cellular system. Users can enjoy data download rates up to 53.6 Kbps over GPRS compared to 14.4 Kbps via circuit-switched data over GSM. GPRS is a new (2.5G) wireless technology standard that is expected to alter and improve the data services that can be added to GSM-based PCNs. ETSI defined GPRS in 1997 with the goal of providing packet-mode data services in GSM. GPRS is an over-the-air system for transmitting data on GSM networks that converts data into standard Internet packets, enabling interoperability between the Internet and GSM network.

In GPRS, a single time slot may be shared by multiple users to transfer packet data. GSM service providers in North America (which does not have a large market share at this time) and in the rest of the world (ROW) are aggressively enhancing their service capabilities through the support of GPRS for packet data. With GPRS, high-speed wireless access to the Internet, e-mail, and corporate databases can be supported at speeds up to 53.6 Kbps. GPRS wireless technology employs authentication and encryption via standard GSM algorithms. The SIM card (a component of the GSM network) stores the authentication and encryption algorithms, preventing unauthorized users from capturing and deciphering messages.[44]

CDMA is a high-capacity cellular technology that employs spread spectrum technology and a unique digital coding system rather than separate channels to differentiate subscribers. CDMA was developed by Qualcomm and introduced commercially in 1995. In 1999, CDMA was adopted as the basis for 3G wireless systems by the ITU. The CDMA Development Group (CDG)[45] has been established as an industry consortium of companies to develop products and services to assist the adoption of CDMA. CDMA is used as the basis of UMTS and W-CDMA. Future 3G phones will be based on CDMA technology. Enhancing this data capability is the 1xRTT CDMA standard, which forms the basis for one of the 3G wireless standards. 1xRadio Transmission Technology (1xRTT) is designed to support always-on data transmission speeds up to 10 times faster than typically available today up to a maximum of 153.6 Kbps.[46]

With the new GSM, GPRS, and CDMA 1xRTT packet data networks in the United States on track in 2002, a fair number of interesting and useful products should become available to end users in 2003.[47]

Other systems are available and/or evolving. High-speed circuit-switched data (HSCSD) is an enhancement to GSM networks that enables users to send and receive data at higher speeds. This improvement is achieved by increasing the data rate of one channel and providing the option of utilizing several channels at the same time. These enhancements result in speeds of about 60 Kbps. The improved performance makes HSCSD useful to people who want to access the Internet at higher speeds than the current GSM phones offer. The higher data speed per channel is achieved by utilizing a new coding scheme that has lower bandwidth requirements for error correction. A GSM channel provides a bit rate of 22.8 Kbps, but about 13.3 Kbps of this bandwidth is used for error correction. With HSCSD, only 8.6

Kbps is used for error correction. This results in higher speeds in cases where users are close to GSM cells where there is high-quality reception (the need for error correction increases with distance/noise). Several manufacturers already have HSCSD-enabled mobile phones on the market.

Personal digital cellular (PDC) is the dominant cellular network technology used in Japan. PDC is a variation of TDMA. The technology was introduced in Japan by NTT DoCoMo in 1991 and operates in the 800 and 1,500 MHz bands. The upgrade path for PDC to 3G is W-CDMA.

Above the bearer radio channel, carriers and device manufactures have for the most part agreed to support WAP, but WAP-complaint phones and pagers had not made major market inroads by early 2002.

As a specific example of WWAN-based services, recent work has been done to add wireless services to (Palm-based) PDAs. For example, a Wells Fargo's wireless system was contemplated to give customers the ability to check balances, transfer funds, and get alerts and stock quotes. Many people want *one* converged PDA rather than *several* devices. Palm OS took the initial lead in the endeavor of adding connectivity to PDAs and published an open operating system (OS). Next Microsoft brought a credible alternative to the market with its PocketPC platform, which offers seamless integration with other Microsoft products and has greater performance and better networking than Palm OS. However, at the time of this writing, Palm still has significant market share. Potential providers have developed the business case that shows acceptable return on investment (ROI) numbers for investing in a wireless infrastructure, predicated on finding some killer app. Real-time stock quotes and transactions on wireless devices or cell phones have been possible for a number of years (over 17 years in the former case). All of these applications are low speed. Figure 1-18 depicts examples of 2G/2.5G data systems.

3G mobile communications is a concept outlined in a set of proposals called the International Mobile Telecommunications-2000 (IMT-2000) to define an anywhere, anytime standard for the future of universal personal communications. The ITU has given support to two 3G technologies: W-CDMA and CDMA2000. W-CDMA appears to be poised to take the lead in high-speed wireless services. North American providers have leaned in favor of CDMA2000 and Enhanced Data Rates for GSM Evolution (EDGE). The ITU has recommended that 3G wireless devices work on W-CDMA and CDMA2000 networks to make worldwide data roaming possible.

By implementing 3G wireless services, carriers plan to upgrade their infrastructures to high-speed data. 3G supports bit rates as fast as 2 Mbps. Although this is less than the 11 Mbps possible with an IEEE 802.11b WLAN or the 54 Mbps possible with the IEEE 802.11a WLAN, the reach (distance) is much greater with 3G. 3G wireless data services could bring

Figure 1-18 Examples of portable terminals supporting hotspot services (Source: GSM World, http://www.gsmworld.com)

Today's GSM Platform

EDGE Platform

many of the benefits of wireline bandwidth broadband to mobile devices.[48] 3G, however, is not a trivial service to deploy. Particularly in Europe, carriers have made significant investments to acquire the spectrum needed for the services, which has left them cash-strapped for a period of time. Carriers were planning commercial rollouts in 2003 or 2004. The global deployment of wireless has been hampered recently by a downturn in the telecom industry. Market fragmentation is also an issue impacting ubiquitous deployment, particularly in North America. Another retarding factor is lack and/or cost of spectrum.

UMTS is commonly used alongside the term 3G in Europe. This can be confusing, however, as UMTS refers to a specific implementation of W-CDMA within the 2.1 GHz band, which is the frequency allocated for 3G in Europe and other parts of the world. It is an implementation of 3G. UMTS is being developed within a framework defined by ITU, IMT-2000, and the UMTS Forum.[49]

As noted, existing 2G systems offer low data rates, such as 14.4 Kbps circuit-switched services. Besides throughput, other 2G limitations include the fact that the link is not always on and that applications do not have access to a packetized stream, which is intrinsically more desirable for data applications than a circuit-mode link. Some of the early packet-mode services such as Cellular Digital Packet Data (CDPD) and Ericsson's Mobitex have proven to be well suited for application support; these systems will be subsumed by more widely adopted standards, such as GPRS and CDMA 1X. For example, 1x Evolution for Data Only (1xEV-DO) is a CDMA data protocol for faster speeds over dedicated channels. The maximum theoretical data rate is 2.4 Mbps. Although it is unlikely you can achieve the maximum rate, 1xEV-DO should indeed offer significantly faster data rates than its precursors. Currently, 1xRTT (voice plus data) provides speeds of about 40 to 70 Kbps, depending upon the implementation and the operator that is implementing it. Korean subscribers are able to receive the higher data rate. In the United States, Verizon Wireless and Sprint PCS have indicated that they will offer 40 to 60 Kbps at the outset. Korea is pioneering CDMA development and is also manufacturing CDMA hardware.[47] It is possible for Sprint and/or Verizon to offer 1xEV-DO service late in 2003; these carriers are still, however, in the early stages of 1xRTT. This will be the main priority for some time (Sprint PCS will offer 1xRTT service in the second half of 2002—there are various challenges with cellular data, but the industry is progressing).

CDMA2000 is the name used by the Telecommunications Industry Association (TIA) to refer to 3G CDMA. CDMA2000 now is an IMT-2000 standard and is included in the evolution part from cdmaOne. CDMA2000 1x will double the voice capacity and allow packet-based data transfer speeds of up to 307 Kbps. CDMA2000 1xEV, the next step, will allow data transfer rates up to 2.4 Mbps (1xEV-DO—data only) and 4.8 Mbps in Phase 2 (1xEV-DV—data and voice). The CDG is the advocacy group.[49] To support roaming, carriers need to work out billing and access agreements. Users with EDGE devices may have a harder time achieving roaming.

The evolution process will be complex. The CDMA path to 3G is through CDMA2000 1X and then to 1xEV-DO and 1xEV-DV. The first phase of CDMA2000, also known as 1X, enables operators with existing IS-95 systems to double the overall system capacity yielding data rates up to 153.6 Kbps.[50] Figures 1-19 and 1-20 depict possible migration scenarios and basic system arrangements.[51]

As carriers enhance their networks to prepare for 3G, they can offer early adopters upgraded data services. Labeled 2.5G because they fall between current 2G and 3G technology, these intermediary technologies (namely, GPRS and CDMA2000 1X) deliver data at rates between 115 and 307 Kbps. (2.5G WWAN systems supporting 50 to 150 Kbps will be available through 2003 and systems supporting 300 to 1,000 Kbps are expected to be available between 2003 and 2005.) When 3G is finally deployed, 2.5G systems can be

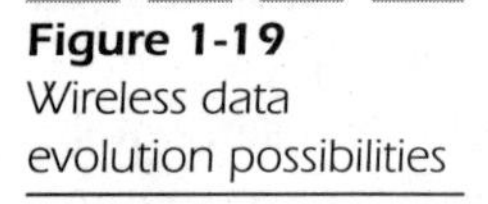

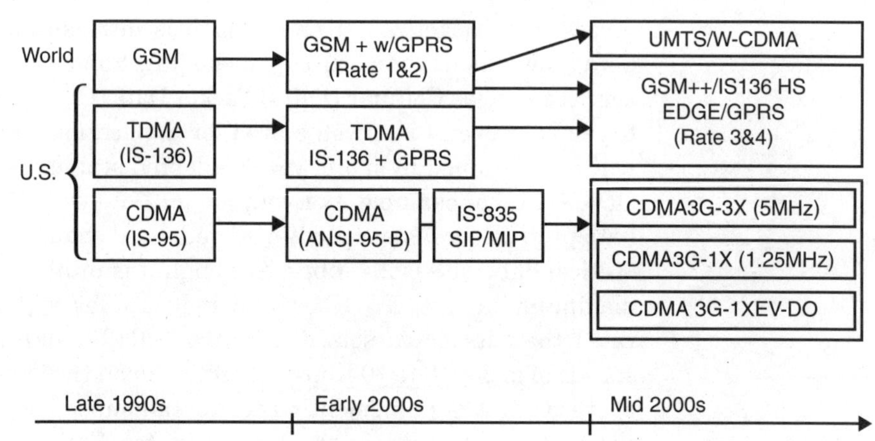

Figure 1-19 Wireless data evolution possibilities

replaced, but both carriers and end users must budget to replace the 2.5G devices and infrastructure.

As noted, GPRS is a packet-based technology for data transfer over an IP-based network and is used as an enhancement for GSM networks. With rapid session setup (0.5 to 1 second) and a data rate between 9 and 171.2 Kbps, GPRS is often said to be a very important step toward 3G. The speed of a GPRS-capable device depends on how many time slots (one to eight, shared by active users) are used within a TDMA frame. Uplink and downlink are treated separately and various radio channel coding schemes allow bit rates from 9 to over 171.2 Kbps per user (if a device were to use all eight channels without error correction). To make the most efficient use of network resources, GPRS provides dynamic sharing between speech and data services as well as for several profiles to make QoS decisions on how to treat packages (for example, video packages might be more time critical than packages resulting from normal WAP browsing).[49]

GPRS also supports the option for bandwidth-based billing. GPRS utilizes the same method for authentication and encryption used by GSM, but its techniques are optimized for packet data transmission. It also enables Short Message Service (SMS) transfer over GPRS radio channels. The technical specifications for GPRS are generated by the 3GPP. GPRS supports three modes of operating:

- **Class A** GPRS and GSM services at the same time (one person can be on the phone with another individual while browsing WAP sites).
- **Class B** Control channels for GPRS and GSM are monitored at the same time, but you can only use one service at one time.
- **Class C** Exclusive GPRS operation.

System	CDMA2000	GSM, TDMA, UMTS, EDGE
Data Approach	Mobile IP-based	GPRS (General Packet Radio Service) GTP Mobility-Based
Notes	• Network layer IETF standard for IP mobility (RFC 2002) • Based on Foreign Agent (FA) and Home Agent (HA) • Adopted by TIA for CDMA2000 (PDSN in CDMA2000 performs FA function) PDSN (Packet Data Serving Node) • Mobility management • PPP termination point • Mobile IP tunnels termination • IP address allocation	• ETSI-defined GSM packet data overlay • Later adopted to TDMA and UMTS • Mainly based on SGSN and GGSN SGSN (Serving GPRS Support Node) GGSN (Gateway GPRS Support Node) • Mobility management, SS7 • PPP termination point • GTP tunnel termination point • Allocation of dynamic addresses
Diagram	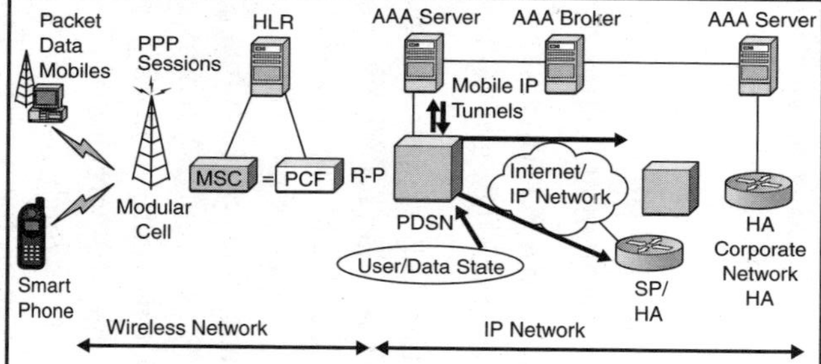 Packet Data Mobiles PPP Sessions HLR AAA Server AAA Broker AAA Server Mobile IP Tunnels Smart Phone Modular Cell MSC = PCF R-P PDSN Internet/ IP Network HA Corporate Network HA User/Data State SP/ HA Wireless Network IP Network	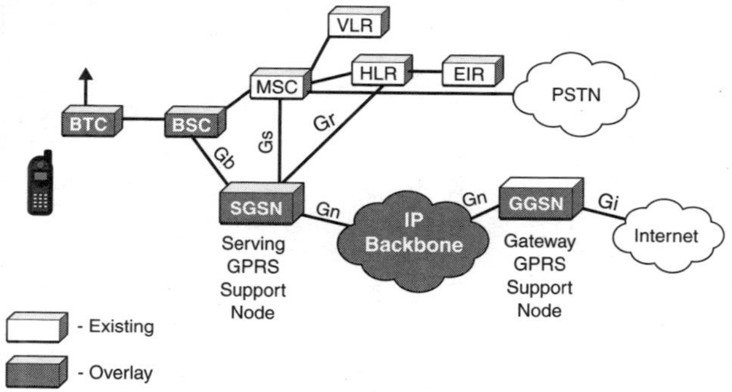 VLR MSC HLR EIR PSTN BTC BSC Gb Gs Gr SGSN Gn IP Backbone Gn GGSN Gi Internet Serving GPRS Support Node Gateway GPRS Support Node - Existing - Overlay

MSC = Mobile Switching Center

AAA: Authentication, Authorization and Accounting
HLR: Home Location Register

PCF: Packet Control Function
R-P: Radio Protocol
SP/HA: Service Provider Home Agent

BTS = Base Transceiver Station
VLR = Visitor Location Register

MSC = Mobile Switching Center
EIR = Equipment Identity Register

Adopted from Steve Akers, Lucent Technologies

Figure 1-20 3G examples

Applications of GPRS include laptops or handhelds connected through a GPRS-capable cell phone or dedicated GPRS mode, mobile phones with a WAP browser, and dedicated equipment such as mobile credit-card swipers.

ETSI has also defined yet another standard called EDGE to support higher-speed data in GSM environments. EDGE is meant to be the basis for a 3G wireless standard that can be used by TDMA and GSM operators.[49] EDGE is an evolution of GPRS that will allow up to three times higher throughput compared to GSM using the same bandwidth. EDGE, in combination with GPRS, will deliver data services up to 384 Kbps in the near future in specific areas.[52] It works with a new modulation technique that enables better usage of existing frequencies. EDGE can be deployed on existing GSM networks and is part of the UWC-136 standard that TDMA carriers have proposed as their 3G standard of choice. EDGE enables data transfer speeds of up to 384 Kbps. This results in 48 Kbps per time slot enabling three times the data speeds of GPRS.

With GPRS implemented, the network parts that facilitate EDGE are located in the base station system (BSS). All you need is EDGE-capable transceiver unit (TRU) plug-ins and some software updates.[53] The upgrades that will be done on existing hardware are relatively small, implying the possibility of fast and cost-effective deployment of EDGE over GSM networks. EDGE-ready terminals are expected by the second half of 2002.

In summary,

- GSM, TDMA, and cdmaOne IS-95A are 2G technologies.
- GPRS and cdmaOne IS-95B are 2.5G technologies.
- EDGE, W-CDMA, CDMA2000 1x, CDMA2000 1x EV-DO, and CDMA2000 1xEV-DV are 3G technologies.

The commercially reported migration paths to 3G in North America are as follows.[54] Cingular, AT&T Wireless, and Voice Stream have made a commitment to GSM/UMTS; this represents 48 percent of the U.S. wireless market at the time of this writing (Cingular has 20 percent, AT&T Wireless has 16 percent, Voice Stream has 5 percent, and other regional TDMA systems have 7 percent). Verizon and Sprint PCS have committed to CDMA2000; this represents 45 percent of the market (Verizon has 28 percent, Sprint has 10 percent, and other regional CDMA companies have 7 percent). Nextel, which uses iDEN, has not yet announced a 3G (it has, however, conducted tests with CDMA2000). The 3G era has already begun in the United States, but it's likely that the initial technology and services will fall short of expectations. In particular, very high-speed applications (see Figure 1-21) won't be well served until wideband 3G is deployed on a larger scale, which probably won't happen until after 2003.[55]

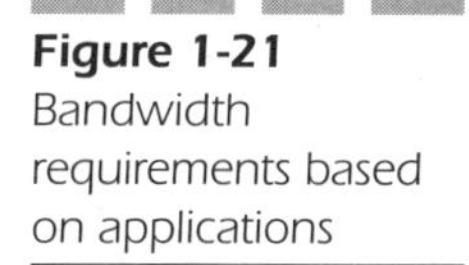

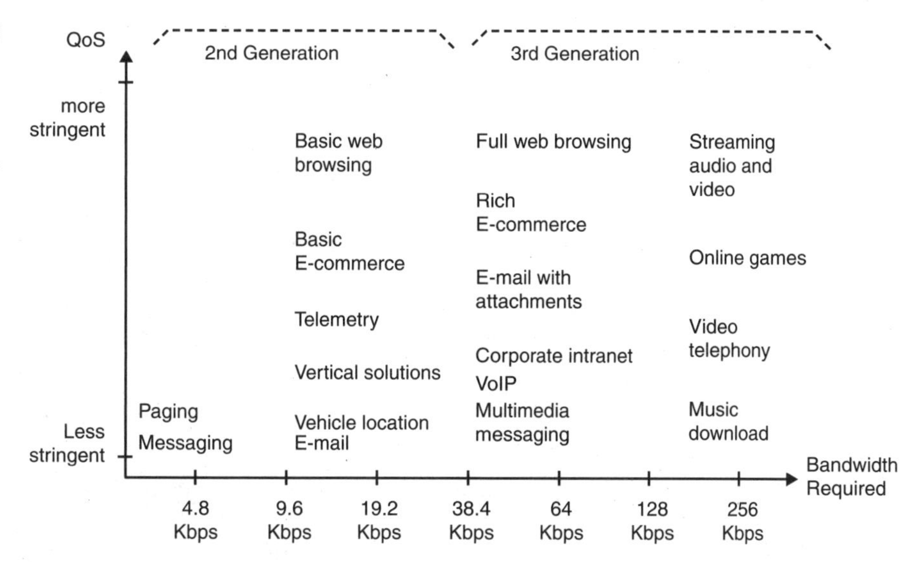

Figure 1-21 Bandwidth requirements based on applications

While Universal Mobile Telecommunications System (UMTS) has been championed by companies such as NTT DoCoMo, AT&T Wireless, VoiceStream, and European Operators (bound to this technology by terms of their government licenses), CDMA2000 (backed by KDDI, Qualcomm, Sprint, and Verizon Wireless) has achieved an early lead. As of June 2002, 10 million subscribers were using CDMA2000 worldwide, compared with 112,000 for UMTS. There were 100 handset models for CDMA2000, but only 7 handset models for UMTS; the former cost $65 to 120, while the latter cost $240 to 400. Nonetheless, the world's largest operators and equipment vendors have invested billions of dollars in UMTS, so that the industry still favors UMTS as the long-term leader.

As we move to 3G, there is a strong desire to move to an all-IP environment. Figure 1-22[56] depicts one example of such a network from a topology standpoint. The transition from the current embedded base, however, will not be quick, inexpensive, or easy. With millions and even billions of mobile users, IP address space could become an issue. IP version 6 (IPv6) provides more space, but the adoption of IPv6 by service providers has been slow. Support in servers and in the core of the network (via IPv6/IPv4 encapsulation methods) is critical. Some people think that IPv6 has a heavier processing burden (for example, a wider address lookup, memory bandwidth issue, authentication/encryption processing, and so on).

Transmission standards are only a part of the 3G wireless data solution; delivery and data presentation technologies, including messaging,

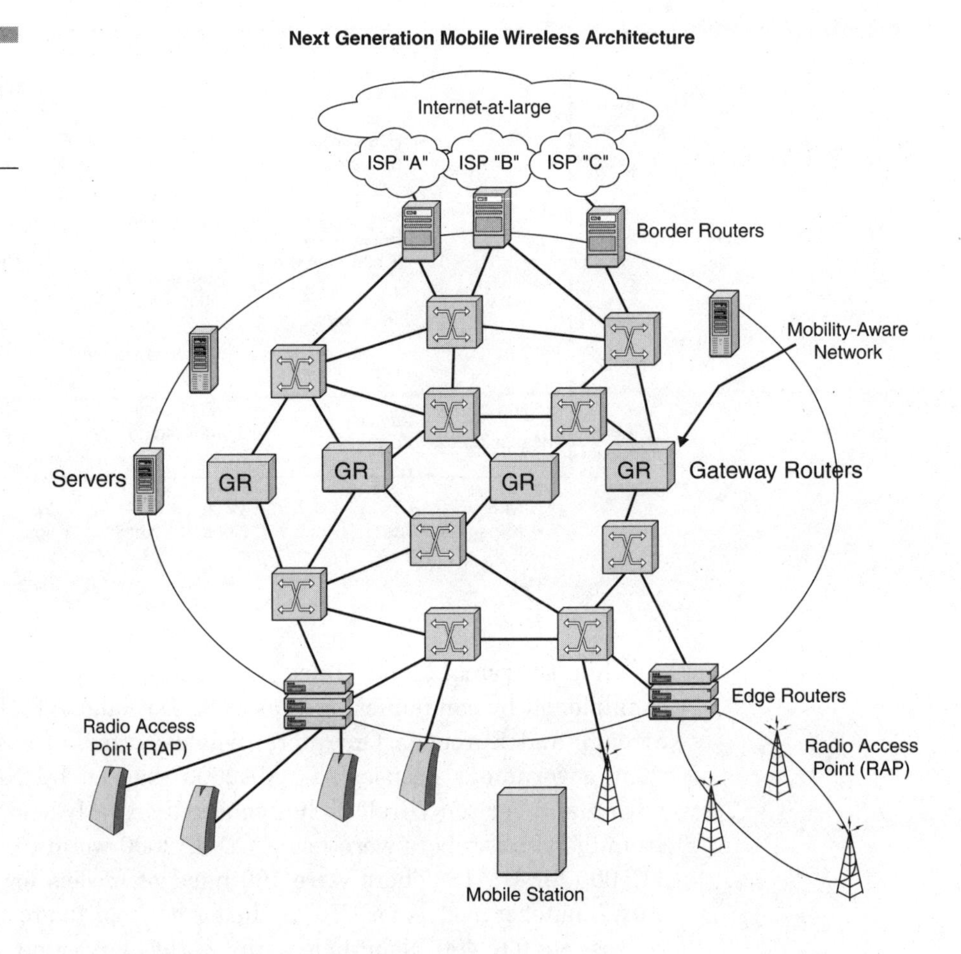

Figure 1-22 Future all-IP wireless WWANs (Source: Nortel Networks)

multimedia, and document representation standards, are still being developed. Successful new revenue opportunities for carriers depend on the carrier's ability to stimulate end-user interest (data services have to be perceived as being necessary) and the ability to make it simple for users to acquire and use the technology. (Integrated Services Digital Network [ISDN] and DSL have turned out to be the antithesis of user-friendliness in terms of being a plug-and-play technology.) Make the system work everywhere it is needed and articulate the value proposition. In an age of the positive bottom line, it is worth remembering that customer demand is enhanced by usefulness, not by technical sophistication of the core technology or the underlying protocols.

The good news is that hotspot operators do not have to wait for all of the complex and costly transition to take place (as summarized in Figure 1-23,

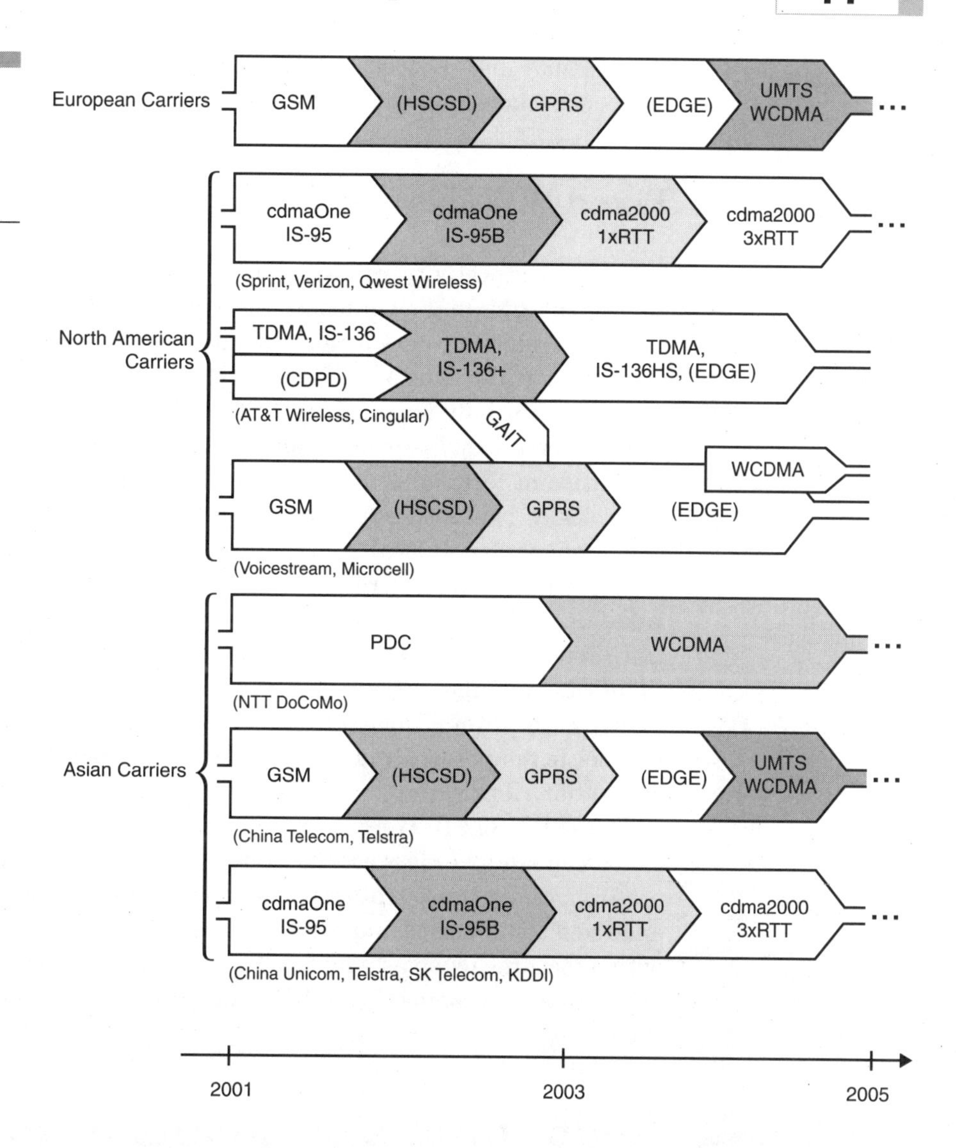

Figure 1-23 Hotspot services need not wait for complex evolution scenarios.

which encapsulates the discussion of this entire subsection) to start collecting revenues from hotspot and nomadic networks. 3G will not replace the need for WLAN/WPAN-based hotspot technologies. 3G systems are wide area technologies, and although they have the advantage of seamless roaming, they offer much less bandwidth (2 Mbps at most and, more likely, at the practical deployment level, 384 Kbps for a long time to come). Hotspot services can also be designed to eventually support roaming with other wide area services when these become available. In summary, IEEE 802.11b and

802.11a may prove to be reasonable, presently available, and relatively inexpensive hotspot solutions for the short and intermediate term.

Fixed WWANs

Last-mile connectivity continues to be of interest. Only approximately 20,000 buildings in the United States have fiber facilities, and DSL services have seen relatively slow penetration in the past five years. Proponents are positioning wireless as a viable alternative to (wired) DSL, cable, and fiber-optic lines. Initially, line of sight (LOS) technology was more often than not point to point; today's advances allow for point-to-multipoint services. Point-to-multipoint techniques lower the cost of providing service. Some of these technologies can even support obstructed transmission paths, which are more common in typical communities.[32] Some vendors have coined the phrase "wireless DSL" to refer to their service offering because they can provide fairly high-speed connections that are similar to various flavors of DSL. Some vendors (such as Sprint, StarTouch, and Air2LAN) also offer the significantly higher-speed connections via wireless, but at DSL pricing. Many of these offerings provide downlink speeds of 1 Mbps, whereas uplinks, in many cases, may be provided via a separate telephone line operating at 56 Kbps speeds.

The IEEE 802.16 Work Group is in the process of standardizing high-rate broadband-wireless access services to buildings through rooftop antennas from central base stations. The hope is that by integrating both mobile and fixed technologies, individuals will eventually be able to move around with a machine (such as a laptop, PDA, phone, and so on) and always have access to a data connection or VPN connection to the company LAN.

Economics and Realities of Hotspot Services—A Provider's Perspective

In 2001 and 2002, the telecom space saw carnage with dozens of companies filing for Chapter 7/11 and billions of dollars in investments and loans lost. Does this imply that telecom has no future? Far from it! One personal fact may illustrate why. I spent $33 a month in telecom-related services in the mid-1980s, but now spend more than $333 a month in all sorts of new services. The telecom industry is nearly a trillion-dollars-a-year industry (see Figure 1-24).

The problem is that the Competitive/DSL/Radio/Fiber Local Exchange Carriers (CLEC/DLEC/RLEC/FLEC) proponents and executives made erroneous statements about the business proposition, promising 80 percent gross margins with 40 percent net positive bottom lines in four years. Yet venture capitalists and financiers failed to practice due diligence and engage competent telecom professionals as project consultants who would have immediately challenged these erroneous statements. The reality is much more as follows. Telecom is a utility business, and investors who have an eight-year outlook will do well (if/when they choose companies/management/business propositions appropriately). The realities are more like 35 to 40 percent gross margins with 15 to 20 percent net positive return lines in seven to eight years.

Hotspot networks fall in the same category. Investors will not have 80 percent gross margins with 40 percent net positive bottom lines in four years; they will have 35 to 40 percent gross margins with 15 to 20 percent net positive return lines in seven to eight years.

Figure 1-24 Worldwide revenues and percentage allocation

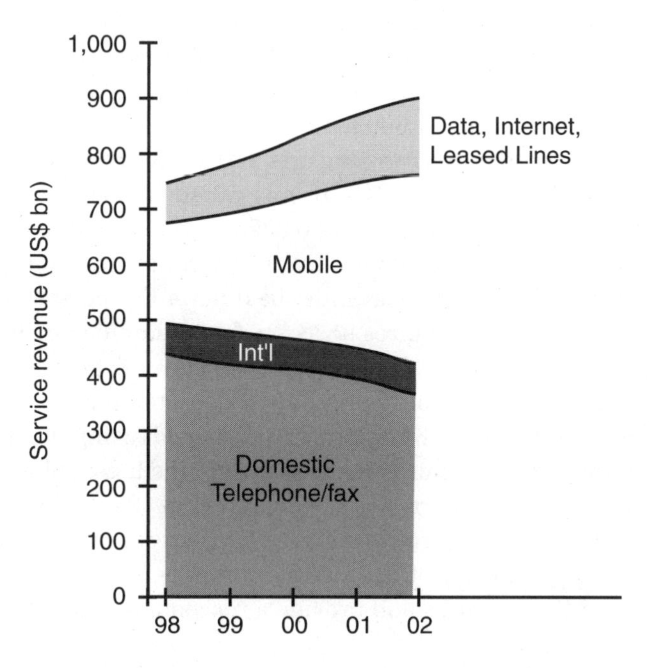

ITU Projections

Examples of Usage

Hotspot services can be designed to support indoor or outdoor applications, or a combination of both. Figures 1-25 and 1-26 give two examples.

Basic Economics

When utilizing the RAIR concept development by the author, the equipment to support an outdoor hotspot network costs about $1,500 per location (for a simple indoor application, this figure is cut in half). Without using the RAIR concept (but using large centralized sectorized antennas), a typical 2,000 by 2,000 feet hotspot would cost $15,000 to $20,000 in equipment. Assuming a 30-month amortization, this is $50 a month. The T1/FT1 Internet line is approximately $1,000 a month. Hence, the network (transmission plus amortized equipment) cost per month is around $1,050. The operator could decide to use DSL services instead of the T1 line; this would result in a transmission cost figure of $50 to $100 a month, making the overall economics much more attractive. A 20 percent contribution (of the total) must also be added to sales, general, and administrative (SGA), a 20 percent contribution must be added to operations, and at least a 10 percent contribution must be added to net income. The carrier-level cost of a hotspot location is about $1,600 a month.

If a hotspot location supports 100 subscribers, the monthly fee for these subscribers must be $16. If only 50 subscribers can be supported/found, then the monthly fee must be $32.

If the hotspot users are transient in nature and only have weekly subscriptions, then the fees must be different. A hotspot supporting 400 transients a month requires a weekly fee (assuming one-week usage out of the month) of $4.00, and so on.

It is probably not likely that an operator can secure 100 customers in a hotspot. However, when properly designed (using the RAIR concept coupled with a DSL transmission mechanism), the hotspot has a breakeven with as few as 10 customers: $1,500 amortized in 30 months plus $50 for transmission, plus a 50 percent markup for perations, SGA, and profit, leads to a monthly expense of $150—10 customers each paying $30 per month can cover the expenses and provide a reasonable return.

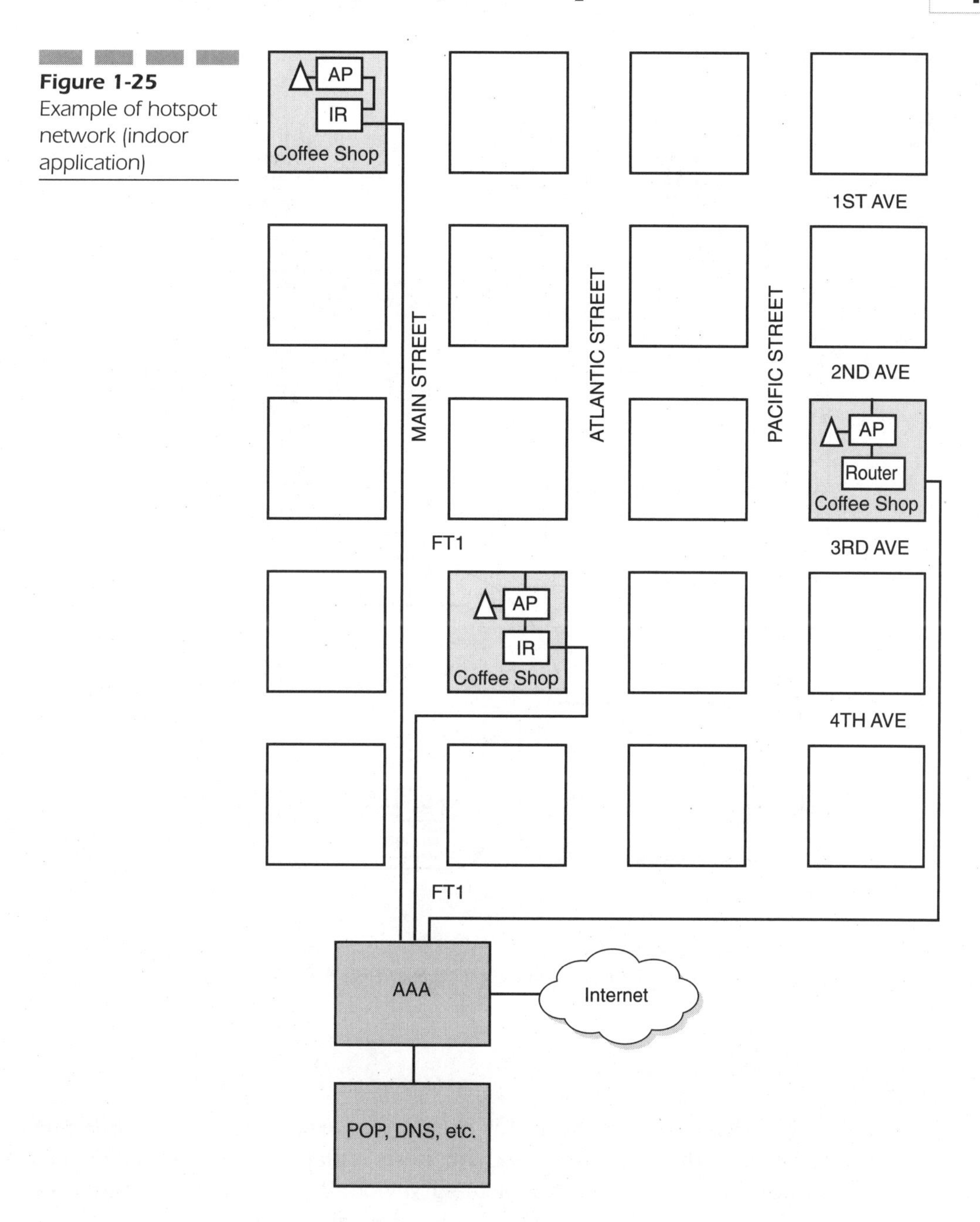

Figure 1-25 Example of hotspot network (indoor application)

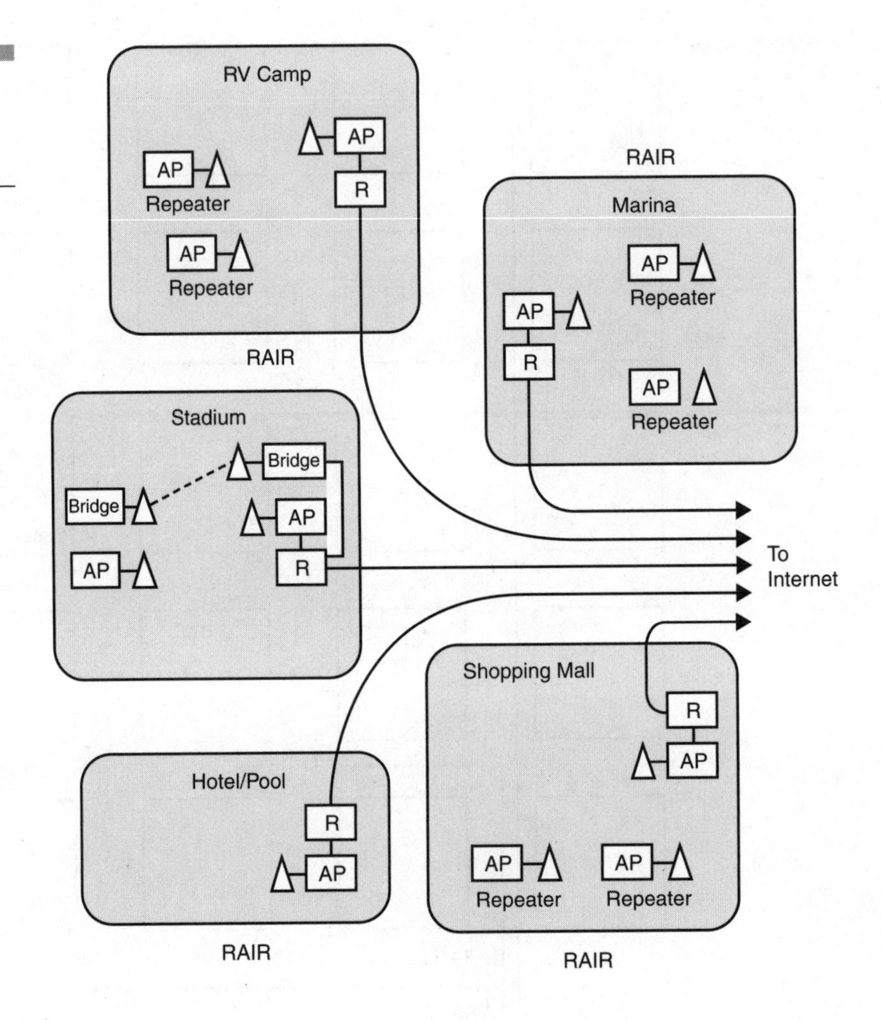

Figure 1-26 Example of hotspot network (outdoor application)

The themes discussed in this chapter are expanded and elaborated on throughout the rest of this book. Emphasis is placed on practical guidelines rather than providing a complete survey of all of the available platforms and alternatives spanning the past 20 years and the next five.

End Notes

1. Some people just use the term *hotspots*. The terms *location specific* and *location based* are also used interchangeably to discuss hotspot services.
2. S. Stemberger, "New Body Art," IBM DeveloperWorks, www-106.ibm/developerworks/library/wi-wear.html?dwzone=wireless, January 2002.
3. S. Mathison, conversation with author, March 14, 2002.
4. http://reiter.weblogger.com, February 8, 2002.
5. These terms are used by the International Telecommunication Union (ITU) in describing the IMT-2000 vision, namely its concept for future nomadic/wireless networks.
6. "High-Speed Wireless Offers Internet Access without the Hassle." *Wall Street Journal*, February 4, 2002.
7. Rick Perera, "Researchers Outline Vision of 4G Wireless World," *IDG News Service* (March 7, 2001).
8. K. Florschutz, "3G Wireless, Meeting High Expectations," Next-Generation Networks Conference (NGN), Boston, Mass., November 2001.
9. Data from Siemens, Salomon Smith Barney, and JP Morgan, 2002.
10. *Wall Street Journal*, July 19, 2001.
11. *IEEE Institute Magazine* (July 2001).
12. *Networkworld* (May 28, 2001).
13. D. Molta, "Mobile and Wireless Technology," *Network Computing* (December 17, 2001): 37 ff.
14. Dan McDonough, Jr., Wireless News Factor, July 16, 2001.
15. www.instat.com/abstracts/ln/2001/ln0110wl_abs.htm.
16. Jørgen Sundgot, "A Wearable Handheld?" www.infosync.no/show.php?id=469, September 2, 2002.
17. Matias Corporation, www.halfkeyboard.com.
18. Jørgen Sundgot, "New Wearable Display from Mitsubishi," www.infosync.no/show.php?id=1262, September 2, 2002.

19. Fixed broadband wireless multiservice WANs are also receiving some attention. Multichannel Multipoint Distribution Service (MMDS) and Local Multipoint Distribution Service (LMDS) and small dish satellite technologies can be used in this context.
20. HIPERLAN/1 is an ETSI standard offering 20 Mbps LANs over 50 meters that operates at 5.12 to 5.30 GHz and 17.1 to 17.3 GHz. HIPERLAN/2 operates at 5.2 GHz with a 100 MHz spectrum. It supports 54 Mbps using orthogonal frequency division multiplexing (OFDM). It has quality of service (QoS) capabilities.
21. New areas to track include 802.11a, 802.11e, and 802.11i.
23. National security government installations (such as Lawrence Livermore National Laboratory, Aeronautical Radio [which provides communication services to airlines], and the U.S. Department of Transportation) were reassessing their use of Wi-Fi (*USA Today*, January 29, 2002, B1) at the time of this writing. Fortunately, there are military-grade encryption technologies that can be put in place where needed. IEEE 802.11b provides a data link layer and physical layer service; many people see encryption services as presentation layer services, which can be placed onto the network as needed.
23. S. Glass, T. Hiller, S. Jacobs, and C. Perkins, "Mobile IP Authentication, Authorization, and Accounting Requirements," RFC 2977, October 2000.
24. Jared Sandberg, "Hackers Poised to Land at Wireless AirPort," www.zdnet.com/enterprise/stories/main/0,10228,2681947,00.html, ZDNet, February 5, 2001.
25. Cells that utilize the same channel are called *co-channel cells*. The interference encountered by a mobile user from its neighboring co-channel cells is called CCI. The signal-to-CCI interference at a location depends on the path loss characteristics of the radio frequency (RF) channel and on the co-channel reuse ratio.
26. Candance Grogans, Jackie Bethea, and Issam Hamdan, "RC4 Encryption Algorithm," www.ncat.edu/~grogans/main.htm, North Carolina Agricultural and Technical State University, March 5, 2000.
27. Princy C. Mehta, "Wired Equivalent Privacy Vulnerability," http://rr.sans.org/wireless/equiv.php, April 4, 2001.
28. Nikita Borisov, Ian Goldberg, and David Wagner, "Wired Equivalent Privacy (WEP)," www.isaac.cs.berkeley.edu/isaac/wep-faq.html, wep@isaac.cs.berkeley.edu, Summer 2001.

29. Adam Stubblefield, John Ioannidis, and Aviel D. Rubin, "Using the Fluhrer, Mantin, and Shamir Attack to Break WEP," August 6, 2001; Adam Stubblefield (astubble@cs.rice.edu) at Rice University, and John Ioannidis (ji@research.att.com) and Aviel D. Rubin (rubin@research.att.com) at AT&T Labs — Research, Florham Park, N.J.

30. Initialization vectors (IVs) are random numbers used as starting points for the encoding of data. WEP defines an IV as a 24-bit code generated by a 40-bit WEP seed that is transmitted with the WEP key in plaintext. WEP does not offer a recommendation as to who generates the IV. When a system starts with an IV of zero and generates additional IVs in a manner that is easy to predict, obvious problems arise. Better WEP implementations use improved random number generation techniques.

31. R. Redelfs, "IEEE 802.11a Advanced Wireless LANs," NGN Proceedings, Boston, Mass., 2001.

32. www.wkmn.com/newsite/wireless.html.

33. This figure is partially based on IBM materials by V. Malhotra, "Checking on IEEE 802.15," www-106.ibm/developerworks/library/wi-checking/? dwzone=wireless, September 2001.

34. These notes on HomeRF are based directly on material from http://homerf.org/learning_center/faq.html.

35. T. G. Zimmerman, "Wireless Networked Digital Devices: A New Paradigm for Computing and Communication," *IBM Systems Journal* 38, no. 4 (1999).

36. DECT is a digital wireless telephone technology intended for cordless phones. Formerly, it was called the Digital European Cordless Telecommunications standard; it was developed by European companies. DECT uses TDMA to transmit radio signals to phones. DECT is designed especially for a smaller area with a large number of users, such as in cities and corporate complexes. As an example of an application mode, users can have telephones equipped for both GSM and DECT (known as a *dual-mode* phone) and have them operate seamlessly.

37. This also includes infrastructures that use cellular telephony with antennas located in elevated mounts such as on tall buildings.

38. IBM field technicians used mobile data terminals as early as the mid-1980s.

39. B. Miller, "The Phony Conflict: IEEE 802.11 and Bluetooth Wireless Technology," IBM Promotional Materials, www-106.ibm.com/developerworks/wireless/library/wi-phone/, October 2001.
40. www.watercove.com.
41. In less than 10 years after the first GSM network was commercially launched, it became the world's leading and fastest growing mobile standard, spanning over 174 countries. Today, GSM technology issued by more than 1 in 10 of the world's population and growth continues to soar with the number of subscribers worldwide expected to surpass 1 billion by the end of 2003.
42. C. Moore, "The Race for Wireless Space," *InfoWorld* (September 18, 2000): 43 ff.
43. www.gsmworld.com/about/history_page1.html.
44. www.novatelwireless.com/pressreleases/2001/story108.htm.
45. www.cdg.org.
46. www.novatelwireless.com/pressreleases/2001/story103.htm.
47. reiter.weblogger.com/.
48. T. Yager, "Wireless Turbulence Ahead," *InfoWorld* (September 18, 2000): 50 ff.
49. Oliver Thylmann, www.infosync.no/show.php?id=1246, September 2, 2002.
50. www.novatelwireless.com/pressreleases/2001/story105.htm.
51. These two figures are adapted from S. Akers, CTO, Lucent Technologies Inc., Integrated Network Solutions.
52. http://www.nortelnetworks.com/solutions/providers/wireless/gsm/edge.html
53. www.ericsson.com/edge/how_it_works.shtml.
54. This is based on information from Siemens, Merrill Lynch, Dresdner Kleinwort, JP Morgan, and Legg Mason.
55. Observations by J. Grams, AT&T Wireless.
56. Observations by Dr. Al Javed, CTO, Wireless Internet, Nortel Networks.

CHAPTER 2

Standards for Hotspots

While we are just starting to see the use of 802.11b wireless local area networks (WLANs) as a widespread broadband connectivity method in the United States, wireless carriers in Korea and Japan are reportedly ramping up massive deployments of hotspot networks. These deployments punctuate the importance of the topic at hand. This chapter starts with an assessment of key standards that have applicability to hotspot services and it then looks at quality of service (QoS). QoS itself has an extensive body of literature, but given its discussion in the wireless standards bodies and its need in applications such as voice over Internet Protocol (VoIP) and video streaming, a short synopsis of this topic is provided at the end of the chapter. Some standards that have a special significance to hotspots are discussed in more detail in following chapters.

Standards

A major issue facing wireless-system designers is the fact that quite a number of wireless protocols exist (as shown in Table 2-1). There are standards for WLANs, wireless personal area networks (WPANs), and wireless wide area networks (WWANs). For hotspot networks built from WLANs and WPANs, the major standards of interest are the Institute of Electrical and Electronics Engineers (IEEE) standards. IEEE 802.11-based technology assists wireless device roaming through buildings and IEEE 802.15-based technology supports short-range links among computers, mobile telephones, peripherals, and other consumer electronics that are worn or carried. IEEE 802.16 supports high-rate broadband-wireless access services to buildings through rooftop antennas from central base stations.

IEEE 802 and Related Activities

WLANs appeared in the early to mid-1990s, but the technology was proprietary to the various vendors in that space. In the mid-1990s, the IEEE sought to produce an industry-wide standard (the committee, however, had been in existence since 1990, handling preliminary activities related to wireless).

IEEE 802.11 Table 2-2 (from IEEE sources) provides a recent view of the standardization activities underway or recently completed in the 802.11 space. IEEE 802.11b is an extension to IEEE 802.11 to support higher data rates. Both standards (initially) used the Wired Equivalent Privacy (WEP) algorithm to address security.

Table 2-1

*Wireless standards**

Standard	Significance
Bluetooth/ IEEE 802.15	Derivative of Bluetooth 1.x spec and more meaningful standards developments relate to Bluetooth application profiles.
Code Division Multiple Access (CDMA) 2000 1x	2.5 G standard for wireless WANs, this provides more efficient voice and packet-switched data services with peak data rates of 153 Kbps.
CDMA 2000 1xEV	Qualcomm is pushing 1xEV as an evolution of 1x technology. It uses a 1.25 MHz CDMA radio channel dedicated to and optimized for packet data, and has throughputs of more than 2 Mbps.
CDMA 2000 3x	Third-generation (3G) standard for WWANs, this uses the same architecture as 1X. It offers 384 Kbps outdoors and 2 Mbps indoors, but operators will likely need to wait for new spectrum.
Enhanced Data rates for Global Evolution (EDGE)	Pushes the General Packet Radio Service (GPRS) data rate to 384 Kbps, but upgrades may be costly for carriers.
General Packet Radio Service (GPRS)	The 2.5G standard for WWANs based on Global System for Mobile Communications (GSM) systems deployed throughout Europe and in other parts of the world. GPRS is an IP-based, packet-data system providing theoretical peak data rates of up to 160 Kbps.
IEEE 802.1x	Security framework for all IEEE 802 networks, this is one of the key components of future multivendor interoperable wireless security systems, but implementation will not be simple.
IEEE 802.11	Basic standard for WLANs which was developed in the late 1990s supporting speeds up to 2 Mbps.
IEEE 802.11b	Basic standard for WLANs. An extension of the IEEE 802.11 specifications, supporting speeds of 1, 2, 5.5, aand 11 Mbps. Operates at 2.4 GHz.
IEEE 802.11a	High-speed WLAN (6 Mbps through 54 Mbps ranges), operating at 5 GHz.
IEEE 802.11e	Revision of 802.11 Media Access Control (MAC) standards, this provides QoS capabilities needed for real-time applications like IP telephony and video.
IEEE 802.11g	A new standard for 2.4 GHz WLANs, this provides a bump in the data rate to 20+ Mbps, but backward-compatible products will not arrive soon.
IEEE 802.11i	Mired in technical debate and politics, this is critical to WLAN market expansion, but delays and indecisiveness may make it meaningless if de facto standards emerge.
IEEE 802.16	Its goal is to define physical and MAC standards for fixed point-to-multipoint broadband wireless access (BWA) systems.
Java 2 Platform, Micro Edition (J2ME)	Provides application-development platform for mobile devices, including cell phones and personal digital assistant (PDAs).

Table 2-1 cont.

Wireless standards*

Standard	Significance
Mobile Management Forum (MMF)	The Open Group's initiative aimed at defining standards for mobile-device management, including session management, synchronization, device-independent content, security, and accounting.
Wideband-CDMA (W-CDMA)	3G standard similar to CDMA 2000 but uses wider 5 MHz radio channels. It provides data rates up to 2 Mbps, but more spectrum needs to be allocated in some areas.
Wireless Internet Service Provider Roaming (WISPR)	Driven by the Wireless Ethernet Compatibility Association (WECA), this represents the industry's first effort to provide transparent roaming and billing across public WLANs.

*www.networkcomputing.com

Table 2-2

IEEE wireless standardization activities

Group	Label	Description	
IEEE 802.11 Working Group	WG	The Working Group is comprised of all of the Task Groups together.	
Task Group	TG	The committee(s) that is tasked by the WG as the author(s) of the standard or subsequent amendments.	
MAC Task Group	MAC	Scope of project	The scope of the project is to develop one common MAC for WLANs applications in conjunction with the Physical Layer (PHY) Task Group.
		Status	Work has been completed and is now part of the original standard published as IEEE Std. 802.11-1997.
		Update Status	Work has been completed on the International Organization for Standardization/International Electrotechnical Commission (ISO/IEC) version of the original standard published as 8802-11: 1999 (ISO/IEC) (IEEE Std. 802.11, 1999 Edition).
PHY Task Group	PHY	Scope of Project	The scope of the project is to develop three PHYs for WLAN applications using Infrared (IR), 2.4 GHz Frequency Hopping Spread Spectrum (FHSS), and 2.4 GHz Direct Sequence Spread Spectrum (DSSS) in conjunction with the common MAC Task Group.
		Status	Work has been completed and is now part of the original standard published as IEEE Std. 802.11-1997.

Table 2-2 cont.

IEEE wireless standardization activities

Group	Label		Description
		Update Status	Work has been completed on the ISO/IEC version of the original standard published as 8802-11: 1999 (ISO/IEC) (IEEE Std. 802.11, 1999 Edition).
Task Group a	TGa	Scope of Project	The scope of the project is to develop a PHY to operate in the newly allocated Unlicensed National Information Infrastructure (UNII) band.
		Status	Work has been completed and is now part of the standard as an amendment published as IEEE Std. 802.11a-1999.
		Update Status	Work has been completed on the ISO/IEC version of the original standard as an amendment published as 8802-11: 1999 (E)/Amd 1: 2000 (ISO/IEC) (IEEE Std. 802.11a-1999 Edition).
Task Group b	TGb	Scope of Project	The scope of the project is to develop a standard for a higher-rate PHY in the 2.4 GHz band.
		Status	Work has been completed and is now part of the standard as an amendment published as IEEE Std. 802.11b-1999.
Task Group b-cor1	TGb-Cor1	Scope of Project	The scope of this project is to correct deficiencies in the Management Information Base (MIB) definition of 802.11b.
		Purpose of Project	As the MIB is currently defined in 802.11b, it is not possible to compile an interoperable MIB. This project will correct the deficiencies in the MIB.
		Status	Ongoing.
Task Group c	TGc	Scope of Project	This adds a subclause under 2.5 support of the Internal Sublayer Service by specific MAC procedures to cover bridge operations with IEEE 802.11 MACs. This supplement to ISO/IEC 10038 (IEEE 802.1D) will be developed by the 802.11 Working Group in cooperation with the IEEE 802.1 Working Group.
		Purpose of Project	To provide the required 802.11-specific information to the ISO/IEC 10038 (IEEE 802.1D) standard.
		Status	Work has been completed and is now part of the ISO/IEC 10038 (IEEE 802.1D) standard.

Table 2-2 cont.

IEEE wireless standardization activities

Group	Label	Description	
Task Group d	TGd	Scope of Project	This supplement will define the physical-layer requirements (channelization, hopping patterns, and new values for current MIB attributes) and other requirements to extend the operation of 802.11 WLANs to new regulatory domains (countries).
		Purpose of Project	The current 802.11 standard defines operation in only a few regulatory domains (countries). This supplement will add the requirements and definitions necessary to enable 802.11 WLAN equipment to operate in markets not served by the current standard.
		Status	Ongoing.
Task Group e	TGe	Scope of Project	Enhance the 802.11 MAC to improve and manage QoS, provide classes of service, and enhanced security and authentication mechanisms. Consider efficiency enhancements in the areas of the distributed coordination function (DCF) and point coordination function (PCF).
		Purpose of Project	To enhance the current 802.11 MAC to expand support for LAN applications with QoS requirements and to provide improvements in security as well as in the capabilities and efficiency of the protocol. These enhancements, in combination with recent improvements in PHY capabilities from 802.11a and 802.11b, will increase overall system performance and expand the application space for 802.11. Example applications include the transport of voice, audio, and video over 802.11 wireless networks, video conferencing, media stream distribution, enhanced security applications, and mobile and nomadic access applications.
		Status	Ongoing. Note that the Security portion of the TGe project authorization request (PAR) was moved to the TGi PAR as of May 2001.
Task Group f	TGf	Scope of Project	To develop recommended practices for an Interaccess Point Protocol (IAPP) that provides the necessary capabilities to achieve multivendor access point interoperability across a distribution system supporting IEEE P802.11 WLAN links. This IAPP will be developed for the following environment(s): 1. A distribution system consisting of IEEE 802 LAN components supporting an IETF IP environment. 2. Others as deemed appropriate.

Table 2-2 cont.

IEEE wireless standardization activities

Group	Label		Description
			This Recommended Practices Document shall support the IEEE P802.11 standard revision(s).
		Purpose of Project	IEEE P802.11 specifies the MAC and PHY layers of a WLAN system and includes the basic architecture of such systems, including the concepts of access points and distribution systems. Implementation of these concepts was purposely not defined by P802.11 because there are many ways to create a WLAN system. Additionally, many of the possible implementation approaches involve concepts from higher network layers. Although this leaves great flexibility in distribution systems and access point functional design, the associated cost is that physical access point devices from different vendors are unlikely to interoperate across a distribution system due to the different approaches taken to distribution system design. As P802.11-based systems have grown in popularity, this limitation has become an impediment to WLAN market growth. At the same time, it has become clear that a small number of distribution system environments comprise the bulk of the commercial WLAN system installations. This project proposes to specify the necessary information that needs to be exchanged between access points to support the P802.11 distribution system functions. The information exchanges required will be specified for one or more distribution systems in a manner sufficient to enable the implementation of distribution systems containing access points from different vendors that adhere to the recommended practices.
		Status	Ongoing.
Task Group g	TGg	Scope of Project	The scope of this project is to develop a higher-speed PHY extension to the 802.11b standard. The new standard shall be compatible with the IEEE 802.11 MAC. The maximum PHY data rate targeted by this project shall be at least 20 Mbps. The new extension shall implement all mandatory portions of the IEEE 802.11b PHY standard. The project will take advantage of the provisions for rate expansion that are in place on the current standard PHY. The 802.11 MAC defines a mechanism for the operation of stations supporting different data rates in the same area. The

Table 2-2 cont.

IEEE wireless standardization activities

Group	Label		Description
			current 802.11b standard already defines the basic rates of 1, 2, 5.5, and 11 Mbps. The proposed project targets further developing the provisions for enhanced data rate capabilities of 802.11b networks. The 802.11 MAC currently incorporates the interpretation of data rate information and the computation of expected packet duration even if the specific station does not support the rate at which the packet was sent.
		Purpose of Project	To develop a new PHY extension to enhance the performance and the possible applications of the 802.11b-compatible networks by increasing the data rate achievable by such devices. This technology will be beneficial for improved access to fixed network LANs and internetwork infrastructures (including access to other WLANs) via a network of access points as well as the creation of higher-performance ad hoc networks.
		Status	Ongoing.
Task Group h	TGh	Scope of Project	To enhance the 802.11 MAC standard and 802.11a high-speed PHY in the 5 GHz band supplement to the standard, to add indoor and outdoor channel selection for 5 GHz license-exempt bands in Europe, and to enhance channel energy measurement and reporting mechanisms to improve spectrum and transmit power management (per the Conference of European Postal and Telecommunications Administrations [CEPT] and the subsequent European Union [EU] committee or the body ruling the incorporating CEPT Recommendation ERC 99/23).
		Purpose of Project	To enhance the current 802.11 MAC and 802.11a PHY with network management and control extensions for spectrum and transmit power management in 5 GHz license-exempt bands, enabling regulatory acceptance of 802.11 5 GHz products. Its purpose is also to provide improvements in channel energy measurement and reporting, channel coverage in many regulatory domains, and to provide dynamic channel selection and transmit power control mechanisms.
		Status	Ongoing.
Task Group i	TGi	Scope of Project	To enhance the 802.11 MAC to enhance security and authentication mechanisms.

Table 2-2 cont.

IEEE wireless standardization activities

Group	Label	Description	
		Purpose of Project	To enhance the current 802.11 MAC to provide improvements in security.
		Status	Ongoing. Note that the Security portion of the TGe PAR was moved to the TGi PAR as of May 2001.
Study Group	SG	Investigates the interest of placing something in the standard.	
Study Group 5GSG	5GSG	Presently investigating the globalization and harmonization of the 5 GHz band jointly with European Telecommunications Standards Institute-Broadband Radio Access Networks (ETSI-BRAN), and Multimedia Mobile Access Communication Systems Promotion Council (MMAC). The objective of the Council is to propose a high-performance wireless system to be used after IMT-2000. MMAC wishes to realize communication systems that allow any person to communicate "at anytime and any place".	
Ad-Hoc Publicity	PC	Looks at how IEEE 802.11 can better "publicize" the standard by collecting data related to its use and operation.	
Ad-Hoc Regulatory	R-REG	Tracks the regulatory bodies and administrations of various worldwide countries and makes sure the standard is in compliance with their rules, or lobbies for future implementations or extensions.	

IEEE 802.11 is focused on the Physical layer (PHY) and Medium Access Control (MAC) sublayer. The MAC is consistent with the IEEE 802.3 Ethernet standard. The IEEE WLAN standard developed by Working Group 802.11 was accepted by the IEEE board during the summer of 1997 and became IEEE Standard 802.11-1997 (see Table 2-3). The standard defines three different physical implementations (signaling techniques and modulations), a MAC function, and a management function. All of the implementations support data rates of 1 Mbps and, optionally, 2 Mbps. Security, roaming, and QoS are also considered. The three physical implementations are as follows:

- Direct sequence spread spectrum radio (DSSS) in the 2.4 GHz band
- Frequency hopping spread spectrum radio (FHSS) in the 2.4 GHz band
- Infrared light (IR)

Table 2-3

IEEE 802.11 standards: the "workhorses" of hotspot networks today

Standard	Description
IEEE 802.11, 1999 Edition (ISO/IEC 8802-11: 1999)	IEEE Standard for Information Technology—Telecommunications and Information Exchange between Systems—Local and Metropolitan Area Network—Specific Requirements—Part 11: Wireless LAN Medium Access Control (MAC) and Physical (PHY) Layer Specifications
IEEE 802.11a-1999 (8802-11:1999/ Amd 1:2000(E))	IEEE Standard for Information Technology—Telecommunications and information exchange between systems—Local and metropolitan area networks—Specific requirements—Part 11: Wireless LAN Medium Access Control (MAC) and Physical Layer (PHY) specifications—Amendment 1: High-speed Physical Layer in the 5 GHz band
IEEE 802.11b-1999	Supplement to 802.11—1999,Wireless LAN MAC and PHY specifications: Higher-speed Physical Layer (PHY) extension in the 2.4 GHz band

After the initial promulgation of the standard, the 802.11 Working Group then considered additions to the standard to provide higher data rates (5.5 and 11 Mbps) in the 2.4 GHz band and to allow WLANs to operate in a 5 GHz band at 54 Mbps. 802.11a uses the 5 GHz band called the Unlicensed National Information Infrastructure (UNII) in the United States; it supports 54 Mbps thanks to the higher frequency and greater bandwidth allocation. The IEEE 802.11a specification progressed toward standardization more rapidly than expected (it uses the orthogonal frequency division multiplexing [OFDM] modulation), and chipmakers quickly brought out chipsets (in early 2001). However, the higher density of hubs and the high price on early equipment still make 802.11b, and soon 820.11g, the more affordable choice for the majority of enterprise and hotspot applications at this time. In summary:

- 802.11a supports 6, 12, and 24 Mbps (mandatory), 9, 18, 36, 48, and 54 Mbps (optional) using 5 GHz OFDM.
- 802.11b supports 1, 2, 5.5, and 11 Mbps using 2.4 GHz complimentary code keying (CCK).

The differentiation with Bluetooth technology is that the latter is a low-cost, low-power, short-range radio link, while IEEE technology has higher throughput and range. The DSSS and FHSS PHY options were designed specifically to conform to FCC regulations for operation in the 2.4 GHz Industrial, Scientific, and Medical (ISM) band, which has worldwide allocation for unlicensed operation. Both FHSS and DSSS PHYs support 1 and 2 Mbps; all 11 Mbps radios are DSSS.

Table 2-4 shows global spectrum allocation around 2.4 GHz. Table 2-5 identifies some of the spectrum regulatory bodies. WLANs implemented in accordance with the IEEE standards are subject to equipment certification and operating requirements established by regional and national regulatory administrations. The Physical Medium Dependent (PMD) specs establish technical requirements for interoperability, based upon established regulations at the time this standard was issued.

The IEEE does not get involved with the marketing and promotion of the technology, which as noted in Chapter 1, "Introduction to Wireless Personal Area Networks, Public Access Locations, and Hotspot Services," is addressed by WECA's Wireless Fidelity (WiFi) efforts. IEEE 802.11b has been a successful technology with a large base, good user experience, and tested interoperability. IEEE 802.11a offers higher throughput, but there are issues related to range, battery life, cost, and spectrum. In any event, backward compatibility is a requirement (see Figure 2-1).

802.11g extends 802.11b to speeds greater than 20 Mbps for WLAN. 802.11g was under development at press time. IEEE 802.11 Task Group G reached an important milestone late in 2001 by approving its first draft. When complete, this spec will extend the IEEE 802.11 family of standards,

Table 2-4

Global spectrum allocation at 2.4 GHz

Region	Allocated Spectrum
U.S.	2.4–2.4835 GHz
Europe	2.4–2.4835 GHz
Japan	2.471–2.497 GHz
France	2.4465–2.4835 GHz
Spain	2.445–2.475 GHz

Table 2-5

Regulatory entities

Geographic area	Approval standards	Documents
United States	Federal Communications Commission (FCC)	CFR47, Part 15, sections 15.205 and 15.209; and Subpart E, sections 15.401–15.407
Japan	Ministry of Post and Telecommunication (MPT)	MPT Ordinance for Regulating Radio Equipment, Article 49.20

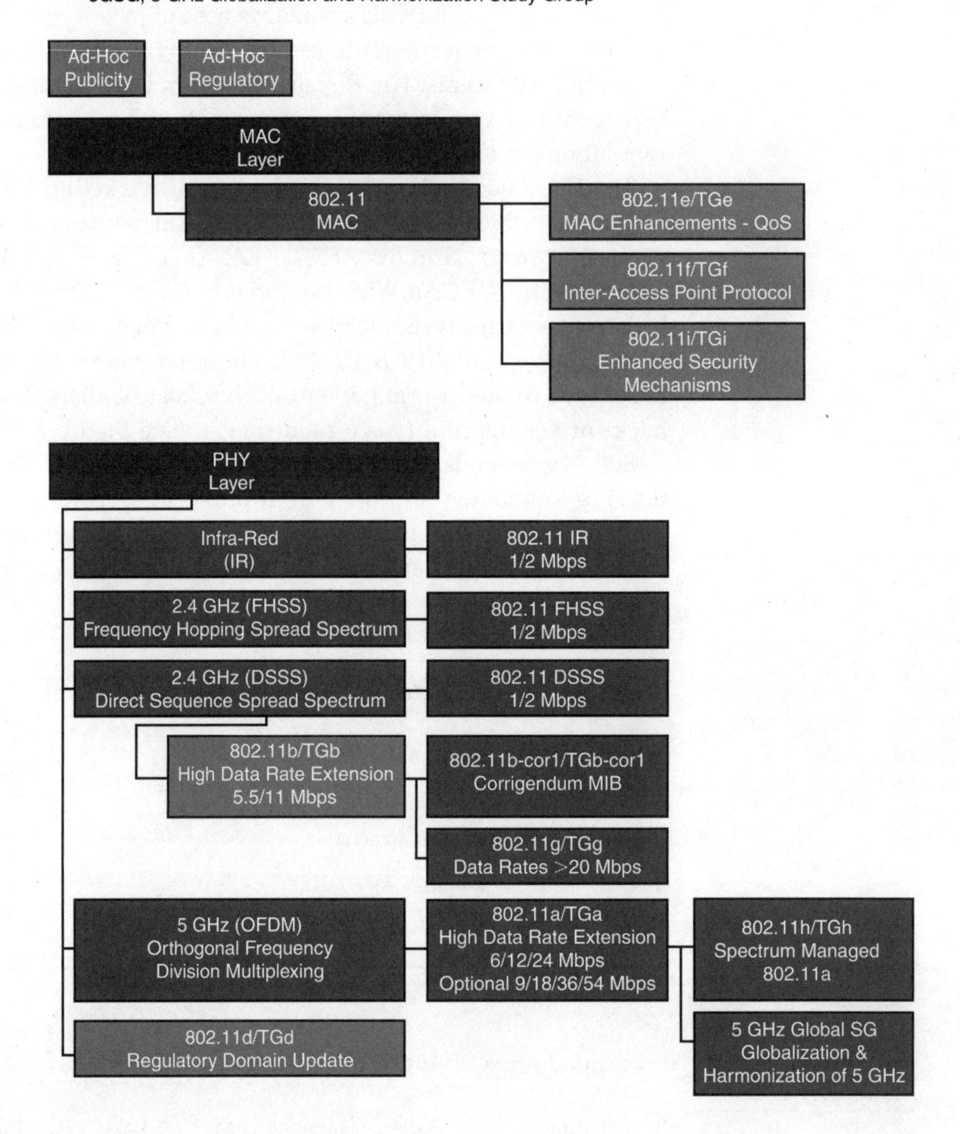

Figure 2-1 Standardization efforts

with data rates up to 54 Mbps in the 2.4 GHz band (not the 5.0 GHz band). This draft is based on CCK, OFDM, and packet binary convolutional code (PBCC) technologies. The 802.11 Task Group G initially considered a number of spread-spectrum modulations schemes to achieve higher speeds. Texas Instruments has promoted PBCC, which could offer better backward compatibility with 802.11b, while Intersil has been advocating OFDM. Both of these technologies are actually preexisting modulation schemes reinvigorated by the arrival of faster, smaller, and cheaper chips that make them practical for wireless networking.[1] The 802.11 Task Group G eventually discarded Texas Instruments' proposal of PBCC from consideration for the 802.11g protocol.[2] The Working Group expects publication by the second half of 2002. IEEE 802.11 standards are covered in more detail in Chapters 5, "IEEE 802.11," and 6, "IEEE 802.11b and IEEE 802.11a."

IEEE 802.15 The 802.15 WPAN effort aims at developing consensus standards for WPANs or short-distance wireless networks. These WPANs address the wireless networking of portable and mobile computing devices such as PCs, PDAs, peripherals, cell phones, pagers, and consumer electronics, enabling these devices to communicate and interoperate with one another. The goal of the Working Group is to publish standards, recommended practices, or guides that have broad market applicability and deal effectively with the issues of coexistence and interoperability with other wired and wireless networking solutions. IEEE 802.15 (building on Bluetooth) is a 10-meter-radius, low-power technology. The IEEE 802.15 Working Group is part of the 802 Local and Metropolitan Area Network Standards Committee of the IEEE Computer Society.

IEEE 802.15 Task Group 1 (TG1) is deriving a WPAN standard based on the Bluetooth v1.x Foundation Specification's. Approval by the Standards Board was expected by press time (2002). The scope and purpose are as follows:

1. To define PHY and MAC specifications for wireless connectivity with fixed, portable, and moving devices within or entering a Personal Operating Space (POS). A goal of the WPAN Group is to achieve a level of interoperability that could allow the transfer of data between a WPAN device and an 802.11 device. A POS is the space around a person or object that typically extends up to 10 meters in all directions and envelops the person whether stationary or in motion. The proposed WPAN standard will be developed to ensure coexistence with all 802.11 networks.

2. To provide a standard for low-complexity, low-power-consumption wireless connectivity to support interoperability among devices within or entering the POS. This includes devices that are carried, worn, or located near the body. The proposed project will address QoS to support a variety of traffic classes. Examples of devices, that can be networked, include computers, PDAs/handheld personal computers (HPCs), printers, microphones, speakers, headsets, bar code readers, sensors, displays, pagers, and cellular and personal communications service (PCS) phones.

See Figure 2-2 for a view of the protocol model.[3]

A number of technical challenges exist in regards to wireless services under discussion:

- Both IEEE 802.11 and Bluetooth operate in the same 2.4 GHz ISM band.
- Bluetooth-enabled devices will likely be portable and will need to operate in an IEEE 802.11 WLAN environment.
- There will be some level of mutual interference.

The IEEE 802.15 Coexistence Task Group 2 (TG2) for WPANs is developing recommended practices to facilitate the coexistence of WPANs (802.15) and WLANs (802.11). The Task Group is developing a coexistence model to quantify the mutual interference of a WLAN and a WPAN. The Task Group is also developing a set of coexistence mechanisms to facilitate the coexistence of WLAN and WPAN devices.

The IEEE P802.15.3 High Rate (HR) Task Group 3 (TG3) for WPANs is chartered to draft and publish a new standard for high-rate (20 Mbps or greater) WPANs. Besides a high data rate, the new standard will provide for low-power, low-cost solutions addressing the needs of portable consumer digital imaging and multimedia applications. In addition, when approved, the new WPAN-HR standard may provide compatibility with the TG1 draft standard based upon the Bluetooth specification.

The IEEE 802.15 Task Group 4 (TG4) is chartered to investigate a low data rate solution with multimonth to multiyear battery life and very low complexity. It is intended to operate in an unlicensed, international frequency band. Potential applications are sensors, interactive toys, smart badges, remote controls, and home automation.

IEEE 802.16 The activities of this Working Group are summarized later, directly from IEEE sources. These standards are more applicable to fixed wireless applications and may support the interconnection of

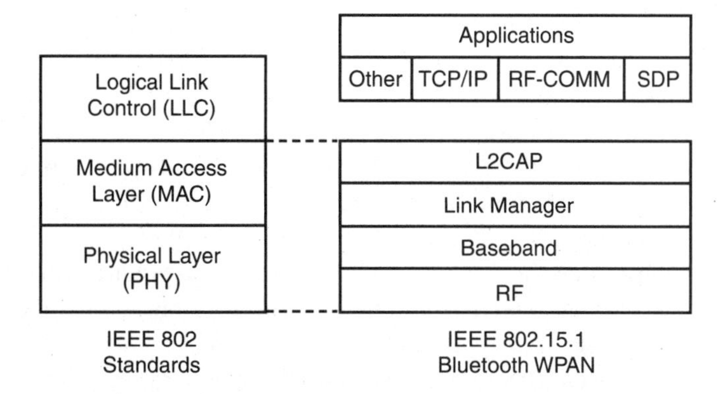

Figure 2-2
802.15 protocol view

Logical Link Control and Adaptation Protocol (L2CAP)
This layer provides the upper layer protocols with connectionless and connection-oriented services. The services provided by this layer include protocol multiplexing capability, segmentation and reassembly of packets, and group abstractions.

Link Manager
The link manager sets up the link between Bluetooth devices. Other functions of the link manager include security, negotiation of Baseband packet sizes, power mode and duty cycle control of the Bluetooth device, and the connection states of a Bluetooth device in a piconet.

Baseband layer
The Baseband layer establishes the Bluetooth physical link between devices forming a *piconet* -- a network of devices connect in an ad hoc fashion using Bluetooth technology. A piconet is formed when two Bluetooth devices connect and can support up to eight devices. In a piconet, one device acts as the master and the other devices act as slaves.

RF layer
The air interface is based on antenna power range starting from 0 dBm up to 20 dBm. Bluetooth operates in the 2.4 GHz band and the link range is anywhere from 10 centimeters to 10 meters.

hotspot cells (such as picocells and microcells). They also have more general applicability.

The 802.16-2001 *standard*[4] specifies the PHY and MAC layer of the air interface of interoperable, fixed, point-to-multipoint broadband wireless access (BWA) systems. The specification enables the transport of data, video, and voice services. It applies to systems operating in the vicinity of 30 GHz but is broadly applicable to systems operating between 10 and 66 GHz. The project purpose is to enable the rapid worldwide deployment of innovative, cost-effective, and interoperable multivendor BWA products. The goal is to facilitate competition in broadband access by providing alternatives to wireline broadband access. Another goal is to facilitate coexistence studies, encourage consistent worldwide allocation, and accelerate the commercialization of BWA spectrum.

The 802.16.1b[5] *standard* specifies the MAC layer and PHYs of the air interface of interoperable, fixed point-to-multipoint BWA systems. The specification enables the transport of data, video, and voice services. Physical layers are specified for both licensed and license-exempt bands. The amendment expands the scope of the original project by extending it to license-exempt bands (thereby defining the Wireless High-Speed Unlicensed Metropolitan Area Network [WirelessHUMAN™] Standard). It specifies the PHY and MAC layer of the air interface of interoperable fixed broadband wireless metropolitan area network (MAN) systems, including point-to-multipoint. The standard enables access to data, video, and voice services with QoS in unlicensed (license-exempt) bands designated for public network access. It focuses on the 5 to 6 GHz range and may be applied to unlicensed bands between 2 and 11 GHz. It will address strategies for coexistence with other unlicensed applications. The project utilizes or modifies applicable elements from the following: MAC: IEEE 802.16; PHY: IEEE 802.11a; and ETSI Broadband Radio Access Networks (BRAN) High-Performance Radio Local Area Network type 2 (HIPERLAN/2).[6]

The project purpose is to enable the rapid worldwide deployment of innovative, cost-effective, and interoperable multivendor BWA products. The goal is to facilitate competition in broadband access by providing alternatives to wireline broadband access. Other goals are to facilitate coexistence studies, encourage consistent worldwide allocation, and accelerate the commercialization of BWA spectrum. It will identify techniques to tolerate interference in the unlicensed bands and facilitate strategies for coexistence with other unlicensed band systems such as 802.11 and 802.15. It will also encourage consistent worldwide spectrum allocation and accelerate the commercialization of unlicensed BWA spectrum. The utilization of unlicensed frequencies will address a market that includes residences, small offices-home offices (SOHOs), telecommuters, and small and medium enterprises (SME).

The 802.16.2a[7] *project* is aimed at developing extensions and modifications to IEEE 802.16.2-2001 addressing two distinct topics. The first topic is the coexistence between multipoint systems and point-to-point systems in the frequency range 10 to 66 GHz, which were planned to focus on the range 23.5 to 43.5 GHz. Two types of point-to-point systems will be considered: those used by fixed BWA operators and those used as individually assigned links, commonly licensed on a first-come, first-served basis.

The second topic is coexistence among fixed BWA systems operating in licensed bands within the frequency range 2 to 11 GHz. The project purpose is to provide additional coexistence guidelines to license holders, service providers, deployment groups, and system integrators, covering coexistence with point-to-point systems (primarily from 23.5 to 43.5 GHz) and coexistence among licensed, fixed BWA systems operating in the 2 to 11 GHz frequency range. The equipment parameter values contained within this amended practice will benefit license holders, equipment and component vendors, and industry associations by facilitating the deployment and operation of fixed BWA systems while minimizing the need for case-by-case coordination. Another purpose is to encourage voluntary procedures that facilitate a simpler licensing process for systems operating below 11 GHz, particularly in the 2.5 GHz MMDS/ITFS bands in the United States.

The 802.16.2-2001[8] *project* covers the development of a recommended practice for the design and coordinated deployment of fixed systems operating from 10 to 66 GHz (with a focus on 23.5 to 43.5 GHz). This is done to minimize interference and to maximize system performance and/or service quality. The spec provides for coexistence using frequency and spatial separation and will cover three areas. First, it will recommend limits of in-band and out-of-band emissions from BWA transmitters through parameters, including radiated power, spectral masks, and antenna patterns. Second, it will recommend receiver-tolerance parameters, including noise floor degradation performance, for interference received from other BWA systems. Third, it will provide coordination parameters, including separation distances and power flux density limits, to enable the successful deployment of BWA systems with tolerable interference. The scope includes interference between systems deployed across geographic boundaries in the same frequency band and systems deployed in the same geographic area in different frequency bands (including different systems deployed by a single license-holder in subbands of the licensees' authorized bandwidth).

The purpose of this spec is to provide coexistence guidelines to license holders, service providers, deployment groups, and system integrators. The equipment parameters contained within this spec will benefit equipment and component vendors as well as industry associations by providing design targets. The benefits of this spec include the following:

- Coexistence of different systems with higher assurance that system performance objectives will be met.
- Minimal need for case-by-case interference studies and coordination between operators to resolve interference issues.
- Preservation of a favorable electromagnetic environment for the deployment and operation of BWA systems, including future systems compliant to IEEE 802.16 interoperability standards.
- Improved spectrum utilization.
- Cost-effective system deployment.

The 802.16.3[9] *standard* specifies the PHY and MAC layer of the air interface of interoperable, fixed point-to-multipoint BWA systems (those supporting data rates of DS1/E1 or greater). The specification enables access to data, video, and voice services with a specified QoS in licensed bands designated for public network access. It applies to systems operating between 2 and 11 GHz. The goal is to enable the rapid worldwide deployment of innovative, cost-effective, and interoperable multivendor BWA products. This is done to facilitate competition in broadband access by providing wireless alternatives to wireline broadband access and to facilitate coexistence studies, encourage worldwide allocation, and accelerate the commercialization of spectrum. The utilization of frequencies from 2 to 11 GHz will address a market that includes residences, SOHO, telecommuters, and SMEs.

The 802.16a[10] *standard* specifies the PHY and MAC layer of the air interface of interoperable fixed point-to-multipoint BWA systems (those supporting data rates of DS1/E1 or greater). The specification enables access to data, video, and voice services with a specified QoS in licensed bands designated for public network access. The spec applies to systems operating between 2 and 11 GHz. The project's purpose is to enable the rapid worldwide deployment of innovative, cost-effective, and interoperable multivendor BWA products. This is done to facilitate competition in broadband access by providing wireless alternatives to wireline broadband access and to facilitate coexistence studies, encourage consistent worldwide allocation, and accelerate the commercialization of BWA spectrum. The utilization of frequencies from 2 to 11 GHz will address a market that includes residences, SOHOs, telecommuters, and SMEs.

Finally, the 802.16b[11] *standard* specifies the MAC layer and PHYs of the air interface of interoperable, fixed, point-to-multipoint BWA systems. The specification enables the transport of data, video, and voice services. PHYs are specified for both licensed and license-exempt bands. The spec expands the scope of the original project by extending it to license-exempt bands (thereby defining the WirelessHUMAN™ standard). It specifies the PHY and MAC layer of the air interface of interoperable, fixed, broadband wireless MAN systems, including point-to-multipoint. The standard enables access to data, video, and voice services with QoS in unlicensed (license-exempt) bands designated for public network access. It focuses on the 5 to 6 GHz range and may be applied to unlicensed bands between 2 and 11 GHz. It also addresses strategies for coexistence with other unlicensed applications.

The project will utilize or modify applicable elements from the following layers: MAC, IEEE 802.16 PHY, IEEE 802.11a, and ETSI BRAN HIPER-

LAN/2. The project purpose is (i) to enable the rapid worldwide deployment of innovative, cost-effective, and interoperable multivendor BWA products; (ii) to facilitate competition in broadband access by providing alternatives to wireline broadband access; and (iii) to facilitate coexistence studies, encourage consistent worldwide allocation, and accelerate the commercialization of BWA spectrum. The undertaking enhances the original project by extending it to license-exempt bands. It identifies techniques to tolerate interference in the unlicensed bands and facilitate strategies for coexistence with other unlicensed band systems such as 802.11 and 802.15. It encourages consistent worldwide spectrum allocation and accelerates the commercialization of unlicensed BWA spectrum. The utilization of unlicensed frequencies will address a market that includes residences, SOHOs, telecommuters, and SMEs.

2G/2.5G/3G Table 2-6 depicts some of the existing WWAN 2G/2.5G/3G standards, while Table 2-7 provides a more detailed listing of key standards. Generally speaking, 3G seeks to provide up to 2 Mbps of data (for example, on the Internet) to cell phones. Multiple groups are involved in standards development: Third-Generation Partnership Project (3GPP), Third-Generation Partnership Project 2 (3GPP2), and so on. Eventually, there will be some interplay with 802.11. Two of the groups are as follows:

- **Third-Generation Partnership Project (3GPP)** This is a GSM-originated standards group. 3GPP is a collaboration agreement that was established in 1998 to bring together a number of telecommunications standards bodies such as ARIB, CWTS, ETSI, T1, TTA, and TTC in one single body.[12] ETSI, T1P1, ARIB/TTC, TTA, CWTS aim at all IP-based mobile networks (GSM focus). Originally, the 3GPP was to produce globally applicable technical specifications and technical reports for 3G based on evolved GSM core networks. This was amended to include the maintenance and development of GSM and evolved into radio access technologies (GPRS and EDGE).[13]
- **Third Generation Partnership Project 2 (3GPP2)** This is a CDMA-oriented 3G standards group. It is an American National Standards Institute (ANSI)-based effort.

3GPP supports efforts for Europe and Asia, while 3GPP2 supports efforts for North America. Table 2-8 lists some recent Internet Engineering Task Force (IETF) Request for Comments (RFCs) that focus on wireless and/or mobile systems.

Table 2-6

Key technologies for 2G, 2.5G, and 3G

Technology	Generation	Description	Notes
Time Division Multiple Access (TDMA)	2G	The standard used by AT&T Wireless services. In North America, CDMA subscribers currently outnumber TDMAs. The TDMA variant GSM is deployed in Europe and has the largest number of subscribers worldwide.	
EDGE	3G	Enhances TDMA for data rates between 384 Kbps and 2 Mbps.	North America is one of the few markets where EDGE services are likely to appear.
CDMA	2G	The leading air interface in North America, patented by Qualcomm.	Key CDMA carriers include Verizon, Sprint, and Bell Mobility.
CDMA2000 1X	2.5G	Provides CDMA users with data rates as fast as 307 Kbps. Qualcomm's pre-3G evolution of its CDMA.	Sprint PCS, BellSouth, and Verizon were planning to introduce CDMA2000 1X data service in North America.
CDMA2000 2X	3G	Provides data services to CDMA devices at bit rates as fast as 2 Mbps. Qualcomm's technology.	Has more momentum in North America than W-CDMA and could still lose the race in North America but, given the success of CDMA2000 1X, it has a lead.
W-CDMA	3G	ITU's official 3G migration path for TDMA networks, including the subscriber-rich GSM networks in Europe and Asia. It is a major competitor to CDMA2000 and is likely to become the world's leading wireless data standard.	W-CDMA, now an international standard, has a lock on most non-U.S. markets.
GSM	2G	The most widely used wireless standard in Europe, based on TDMA.	To provide global roaming for wireless data subscribers, North American carriers will need to support GSM devices. Qualcomm was reportedly planning to create a chip that enables wireless devices to communicate on both GSM and CDMA networks.
GPRS	2.5G	Supports midrange data service to TDMA subscribers, including those using GSM devices. The maximum bit rate is 115 Kbps (less than half that of CDMA2000 1X.) GRPS is a steppingstone to EDGE, a 3G alternative to W-CDMA for TDMA wireless carriers.	AT&T Wireless (while it upgrades its network for EDGE), Cingular Wireless, and Voice Stream were expected to expand their GPRS offerings in the United States.

Table 2-7

Detailed listing of key 1G/2G/3G standards

Standard's Family	Detailed Standard
TDMA Systems	EIA Standard IS-54-B, "Cellular System Dual-Mode Mobile Station—Base Station Compatibility Standard," 1992.
	EIA Interim Standard IS-136.2, "800 MHz TDMA—Radio Interface—Mobile Station—Base Station Compatibility —Traffic Channels and FSK Control Channels," 1994.
GSM	GSM Specifications 2.01, Version 4.2.0, Issued by ETSI, January 1993. Also, ETSI/GSM Specifications 2.01, "Principles of Telecommunications Services," January 1993.
	GSM Specifications 3.60, Version 6.4.1, "General Packet Radio Service (GPRS); Service Description, Stage 2," 1997.
	GSM Specifications 4.60, Version 7.2.0, "General Packet Radio Service (GPRS); Mobile Radio-Base Station Interface, Radio Link Control/Medium Access Control (RLC/MAC) Protocol," 1998.
	ETSI GSM 03.60: GPRS Service Description, Stage 2.
	ETSI GSM 03.64: Overall Description of the GPRS Radio Interface, Stage 2.
	ETSI GSM 04.60: GPRS, Mobile Station—Base Station System (BSS) Interface, Radio Link Control/Medium Access Control (RLC/MAC) Protocol.
	ETSI GSM 04.64: GPRS, Logical Link Control.
	ETSI GSM 04.65: GPRS, Subnetwork Dependent Convergence Protocol (SNDCP).
	ETSI GSM 07.60: Mobile Station (MS) Supporting GPRS.
	ETSI GSM 08.08: GPRS, Mobile Switching Center—Base Station Subsystem (MSC-BSC) Interface: Layer 3 Specification.
	ETSI GSM 08.14: Base Station Subsystem—Serving GPRS Support Node (BSS-SGSN) Interface; Gb Interface Layer 1.
	ETSI GSM 08.16: Base Station Subsystem—Serving GPRS Support Node (BSS-SGSN) Interface; Network Service.
	ETSI GSM 08.18: Base Station Subsystem—Serving GPRS Support Node (BSS-SGSN); Base Station Subsystem GPRS Protocol (BSSGP).
	ETSI GSM 09.60: GPRS Tunneling Protocol (GTP) Across the Gn and Gp Interface.
	ETSI GSM 09.61: General Requirements on Interworking Between the Public Land Mobile Network (PLMN) Supporting GPRS and Packet Data Network (PDN).
	ETSI GSM 2.01, Version 4.2.0, January 1993.
	ETSI GSM Section 4.0.2, "European Digital Cellular Telecommunication System (Phase 2); Speech Processing Functions: General Description," April 1993.

Table 2-7 cont.

Detailed listing of key 1G/2G/3G standards

Standard's Family	Detailed Standard
CDMA One	EIA Interim Standard IS-95, "Mobile Station—Base Station Compatibility Standard for Dual-Mode Wideband Spread Spectrum Cellular System," 1998.
IMT-2000	ITU-R M.1034-1, "International Mobile Telecommunications-2000 (IMT-2000)," 1997.
	ITU-R M.816-1, " Framework for Services Supported on International Mobile Telecommunications-2000 (IMT-2000)," 1997.
	ITU-R M.687-2, "International Mobile Telecommunications-2000 (IMT-2000)," 1997.
	IMT-2000: Recommendations ITU-R M.687-2, 1997.
	3G TS 22.105 Release 1999, Services and Service Capabilities.
	EIA/TIA/IS-41.1-B, Cellular Radio—Telecommunications Intersystem Operations: Functional Overview, 1991.
	EIA/TIA/IS-634-A, MSC-BS Interface (A-Interface) for Public 800 MHz, 1998.
	GSM 03.60: GPRS Service Description, Stage 2.
	GSM 03.64: Overall Description of the GPRS Radio Interface, Stage 2.
	3GPP TS 25.401: UTRAN Overall Description, 2000.
	3GPP TR 23.922: Architecture for an All IP Network, 1999.
PCS	TIA TR-45.4, Microcellular/PCS.
	TIA TR-46, Mobile and Personal Communications 1800.
	TIA TR-46.1, Services and Reference Model.
	TIA TR-46.2, Network Interfaces.
	TIA TR-46.3, Air Interfaces.
	EIA/TIA-553 Cellular System Mobile Station—Land Station Compatibility Specification.
Universal Mobile Telecommunications System (UMTS)	3G TS 22.105, Service Aspects; Services and Service Capabilities.
	3GPP TS 23.107, QoS Concept and Architecture.
	3GPP TS 25.401, UTRAN Overall Description
	3GPP TS 25.101, UE Radio Transmission and Reception.
	3GPP TS 25.104, UTRA (BS) FDD, Radio Transmission and Reception.
	3GPP TS 25.105, UTRA (BS) TDD, Radio Transmission and Reception.
	3GPP TS 25.301, Radio Interface Protocol Architecture.
	3GPP TS 25.201, Physical Layer—General Description.
	3GPP TS 25.211, Physical Channels and Mapping of Transport Channels onto Physical Channels (FDD).
	3GPP TS 25.212, Multiplexing and Channel Coding.
	3GPP TS 25.213, Spreading and Modulation (FDD).
	3GPP TS 25.214, Physical Layer Procedures.

Table 2-7 cont.

Detailed listing of key 1G/2G/3G standards

Standard's Family	Detailed Standard
	3GPP TS 25.215, Physical Layer—Measurements.
	3GPP TS 25.302, Services Provided by the Physical Layer.
	3GPP TS 25.321, MAC Protocol Specification.
	3GPP TS 25.322, RLC Protocol Specification.
	3GPP TS 25.323, Packet Data Convergence Protocol (PDCP) Specification.
	3GPP TS 25.324, Broadcast/Multicast Control (BMC) Protocol Specification.
	3G TS 25.331, RRC Protocol Specification.
	3G TS 25.303, Interlayer Procedures in Connected Mode.
	3GPP TS 25.410, UTRAN Iu Interface: General Aspects and Principles.
	3GPP TS 25.411, UTRAN Iu Interface: Layer 1.
	3GPP TS 25.412, UTRAN Iu Interface: Signaling Transport.
	3GPP TS 25.413, UTRAN Iu Interface: RANAP Signaling.
	3GPP TS 25.414, UTRAN Iu Interface: Data Transport and Transport Signaling.
	3GPP TS 25.415, UTRAN Iu Interface: CN-RAN User Plane Protocol.
	3GPP TS 25.420, UTRAN Iur Interface: General Aspects and Principles.
	3GPP TS 25.421, UTRAN Iur Interface: Layer 1.
	3GPP TS 25.422, UTRAN Iur Interface: Signaling Transport.
	3GPP TS 25.423, UTRAN Iur Interface: RNSAP Signaling.
	3GPP TS 25.424, UTRAN Iur Interface: Data Transport and Transport Signaling for CCH Data Streams.
	3GPP TS 25.425, UTRAN Iur Interface: User Plane Protocols for CCH Data Streams.
	3GPP TS 25.426, UTRAN Iur and Iub Interface Data Transport and Transport Signaling for DCH Data Streams.
	3GPP TS 25.427, UTRAN Iur and Iub Interface User Plane Protocols for DCH Data Streams.
	3GPP TS 25.430, UTRAN Iub Interface: General Aspects and Principles.
	3GPP TS 25.431, UTRAN Iub Interface: Layer 1.
	3GPP TS 25.432, UTRAN Iub Interface: Signaling Transport.
	3GPP TS 25.433, UTRAN Iub Interface: NBAP Signaling.
	3GPP TS 25.434, UTRAN Iub Interface: Data Transport and Transport Signaling for CCH Data Streams.
	3GPP TS 25.435, UTRAN Iub Interface: User Plane Protocols for CCH Data Streams.
	3G TR 23.922, Architecture of an All IP Network.
	3G TR 25.990, Vocabulary.

Table 2-8

Recent wireless/mobile IETF RFCs

RFC	Title
3141	**CDMA2000 Wireless Data Requirements for AAA.** T. Hiller, P. Walsh, X. Chen, M. Munson, G. Dommety, S. Sivalingham, B. Lim, P. McCann, H. Shiino, B. Hirschman, S. Manning, R. Hsu, H. Koo, M. Lipford, P. Calhoun, C. Lo, E. Jaques, E. Campbell, Y. Xu, S. Baba, T. Ayaki, T. Seki, A. Hameed (June 2001).
3115	**Mobile IP Vendor/Organization-Specific Extensions.** G. Dommety, K. Leung (April 2001, obsoletes RFC3025.)
3024	**Reverse Tunneling for Mobile IP, Revised.** G. Montenegro, Editor (January 2001, obsoletes RFC2344).
2977	**Mobile IP Authentication, Authorization, and Accounting Requirements.** S. Glass, T. Hiller, S. Jacobs, C. Perkins (October 2000).
2501	**Mobile Ad hoc Networking (MANET): Routing Protocol Performance Issues and Evaluation Considerations.** S. Corson, J. Macker (January 1999).
2284	**PPP Extensible Authentication Protocol (EAP).** L. Blunk, J. Vollbrecht (March 1998).
2002	**IP Mobility Support.** C. Perkins, Editor (October 1996).

IETF IP Over Bluetooth (BOF) The BOF[14] advocates the creation of an IETF Working Group to investigate the most open and efficient way to place IP over the Bluetooth Host Controller. Current work in this area within the Bluetooth Special Interest Group (SIG) concentrates on defining IP over a set of other lower-layer stacks. Currently, the Bluetooth SIG defines two options:

- Option 1: IP/PPP/RFCOM/L2CAP/Host Controller
- Option 2: IP/PAN Profile/L2CAP/Host Controller (where the PAN Profile is a Bluetooth SIG work in progress)

The IETF Working Group seeks to define a more efficient way of running IP over Bluetooth. In particular, IP would run over an IETF protocol over the Host Controller without L2CAP. This option may be adopted by the Bluetooth SIG at a later date as a profile. Since all Bluetooth SIG profiles are optional, a customer may choose any combination of profiles in a final product. Further, since Bluetooth Working Groups have it in their mandate to adopt protocols from other standards-making bodies such as the IETF, a clear path exists for IETF work to be adopted by the Bluetooth SIG. The objective of the BOF is to foster innovation and speed progress by placing IP-related protocol development within the IETF and Bluetooth-specific protocol development within the Bluetooth SIG by developing an IP over

Bluetooth IETF Working Group. This effort will define its own way of running IP over Bluetooth by carefully selecting a set of Bluetooth protocols (freely available from published specifications at www.bluetooth.com/developer/specification/specification.asp) on which to build IP.

Mobile IP

IETF RFC 2002 specifies protocol enhancements that enable the transparent routing of IP datagrams to mobile nodes on the Internet. Mobility does not necessarily imply a wireless situation, but often it does, hence our brief coverage of this technology. In a mobile environment, each mobile node is always identified by its home address, regardless of its current point of attachment to the Internet. Although situated away from its home, a mobile node is also associated with a care-of address, which provides information about its current point of attachment to the Internet. The protocol of RFC 2002 provides for registering the care-of address with a home agent. The home agent sends datagrams destined for the mobile node through a tunnel to the care-of address. After arriving at the end of the tunnel, each datagram is then delivered to the mobile node.[15]

IP[16] Version 4 assumes that a node's IP address uniquely identifies the node's point of attachment to the Internet. Therefore, a node must be located on the network indicated by its IP address in order to receive datagrams destined to it; otherwise, datagrams destined to the node would be undeliverable. For a node to change its point of attachment without losing its capability to communicate, currently one of the two following mechanisms must typically be employed:

- The node must change its IP address whenever it changes its point of attachment.
- Host-specific routes must be propagated throughout much of the Internet routing fabric.

Both of these alternatives are often unacceptable. The first makes it impossible for a node to maintain transport and higher-layer connections when the node changes location. The second has obvious and severe scaling problems, which are especially relevant considering the explosive growth in sales of notebook (mobile) computers. A new, scalable, mechanism is required for accommodating node mobility within the Internet. The RFC defines such a mechanism, which enables nodes to change their point of attachment to the Internet without changing their IP address.

Protocol Requirements

- A mobile node must be able to communicate with other nodes after changing its link-layer point of attachment to the Internet yet without changing its IP address.
- A mobile node must be able to communicate with other nodes that do not implement these mobility functions. No protocol enhancements are required in hosts or routers that are not acting as any of the new architectural entities introduced in the "New Architectural Entities" section.
- All messages used to update another node as to the location of a mobile node must be authenticated in order to protect against remote redirection attacks.

Goals The link by which a mobile node is directly attached to the Internet may often be a wireless link. This link may thus have substantially lower bandwidth and a higher error rate than traditional wired networks. Moreover, mobile nodes are likely to be battery powered, and minimizing power consumption is important. Therefore, the number of administrative messages sent over the link by which a mobile node is directly attached to the Internet should be minimized, and the size of these messages should be kept as small as reasonably possible.

Assumptions

- The protocols defined in this document place no additional constraints on the assignment of IP addresses. That is, a mobile node can be assigned an IP address by the organization that owns the machine.
- The RFC 2002 protocol assumes that mobile nodes will generally not change their point of attachment to the Internet more frequently than once per second.
- The RFC 2002 protocol assumes that IP unicast datagrams are routed based on the destination address in the datagram header (and not, for example, by the source address).

Applicability Mobile IP is intended to enable nodes to move from one IP subnet to another. It is just as suitable for mobility across homogeneous media as it is for mobility across heterogeneous media. That is, Mobile IP facilitates node movement from one Ethernet segment to another and it accommodates node movement from an Ethernet segment to a wireless

LAN as long as the mobile node's IP address remains the same after such a movement.

Think of Mobile IP as solving the macromobility management problem. It is less suited for more micromobility management applications such as handoffs amongst wireless transceivers, each of which covers only a small geographic area. As long as node movement does not occur between points of attachment on different IP subnets, link-layer mechanisms for mobility (that is, a link-layer handoff) may offer faster convergence and far less overhead than Mobile IP.

New Architectural Entities Mobile IP introduces the following new functional entities:

- **Mobile node** A host or router that changes its point of attachment from one network or subnetwork to another. A mobile node may change its location without changing its IP address; it may continue to communicate with other Internet nodes at any location using its (constant) IP address, assuming that link-layer connectivity to a point of attachment is available.
- **Home agent** A router on a mobile node's home network that tunnels datagrams for delivery to the mobile node when it is away from home. A home agent also maintains current location information for the mobile node.
- **Foreign agent** A router on a mobile node's visited network that provides routing services to the mobile node while registered. The foreign agent detunnels and delivers datagrams to the mobile node that were tunneled by the mobile node's home agent. For datagrams sent by a mobile node, the foreign agent may serve as a default router for registered mobile nodes.

A mobile node is given a long-term IP address on a home network. This home address is administered in the same way as a permanent IP address is provided to a stationary host. When away from its home network, a care-of address is associated with the mobile node and reflects the mobile node's current point of attachment. The mobile node uses its home address as the source address of all IP datagrams that it sends, except where otherwise described in this document for datagrams sent for certain mobility management functions.

Terminology The RFC frequently uses the terms shown in Table 2-9.

Table 2-9

Key mobile IP terms

Term	Definition
Agent advertisement	An advertisement message constructed by attaching a special extension to a router advertisement message.
Care-of address	The termination point of a tunnel toward a mobile node for datagrams forwarded to the mobile node while it is away from home. The protocol can use two different types of care-of addresses: a *foreign agent care-of address* is an address of a foreign agent with which the mobile node is registered, and a *co-located care-of address* is an externally obtained local address with which the mobile node has associated one of its own network interfaces.
Correspondent node	A peer with which a mobile node is communicating. A correspondent node may be either mobile or stationary.
Foreign network	Any network other than the mobile node's home network.
Home address	An IP address that is assigned for an extended period of time to a mobile node. It remains unchanged regardless of where the node is attached to the Internet.
Home network	A network, possibly virtual, having a network prefix matching that of a mobile node's home address. Note that standard IP routing mechanisms will deliver datagrams destined to a mobile node's home address to the mobile node's home network.
Link	A facility or medium over which nodes can communicate at the link layer. A link underlies the network layer.
Link-layer address	The address used to identify an endpoint of some communication over a physical link. Typically, the link-layer address is an interface's Media Access Control (MAC) address.
Mobility agent	Either a home agent or a foreign agent.
Mobility binding	The association of a home address with a care-of address, along with the remaining lifetime of that association.
Mobility security association	A collection of security contexts between a pair of nodes that may be applied to Mobile IP protocol messages exchanged between the nodes. Each context indicates an authentication algorithm and mode, a secret (a shared key or appropriate public/private key pair), and a style of replay protection in use.
Node	A host or a router.
Nonce	A randomly chosen value, different from previous choices, inserted in a message to protect against replays.
Security Parameter Index (SPI)	An index identifying a security context between a pair of nodes among the contexts available in the mobility security association. SPI values 0 through 255 are reserved and *must not* be used in any mobility security association.
Tunnel	The path followed by a datagram while it is encapsulated. The model is that, while it is encapsulated, a datagram is routed to a knowl-

Table 2-9 cont.

Key mobile IP terms

	edgeable decapsulating agent, which decapsulates the datagram and then correctly delivers it to its ultimate destination.
Virtual network	A network with no physical instantiation beyond a router (with a physical network interface on another network). The router (a home agent) generally advertises reachability to the virtual network using conventional routing protocols.
Visited network	A network other than a mobile node's home network, to which the mobile node is currently connected.
Visitor list	The list of mobile nodes visiting a foreign agent.

Protocol Overview The following support services are defined for Mobile IP:

- **Agent discovery** Home agents and foreign agents may advertise their availability on each link for which they provide service. A newly arrived mobile node can send a solicitation on the link to learn if any prospective agents are present.
- **Registration** When the mobile node is away from home, it registers its care-of address with its home agent. Depending on its method of attachment, the mobile node will register either directly with its home agent or through a foreign agent that forwards the registration to the home agent.

The following steps provide a rough outline of operation of the Mobile IP protocol:

1. Mobility agents (foreign agents and home agents) advertise their presence via agent advertisement messages. A mobile node may optionally solicit an agent advertisement message from any locally attached mobility agents through an agent solicitation message.
2. A mobile node receives these agent advertisements and determines whether it is on its home network or a foreign network.
3. When the mobile node detects that it is located on its home network, it operates without mobility services. If returning to its home network from being registered elsewhere, the mobile node deregisters with its home agent through the exchange of a registration request and a registration reply message with it.
4. When a mobile node detects that it has moved to a foreign network, it obtains a care-of address on the foreign network. The care-of address

can either be determined from a foreign agent's advertisements (a foreign agent care-of address) or by some external assignment mechanism such as the Dynamic Host Configuration Protocol (DHCP), a co-located care-of address.

5. The mobile node operating away from home then registers its new care-of address with its home agent through the exchange of a registration request and a registration reply message with it, possibly via a foreign agent.
6. Datagrams sent to the mobile node's home address are intercepted by its home agent, tunneled by the home agent to the mobile node's care-of address, received at the tunnel endpoint (either at a foreign agent or at the mobile node itself), and finally delivered to the mobile node.
7. In the reverse direction, datagrams sent by the mobile node are generally delivered to their destination using standard IP routing mechanisms, not necessarily passing through the home agent.

When away from home, Mobile IP uses protocol tunneling to hide a mobile node's home address from intervening routers between its home network and its current location. The tunnel terminates at the mobile node's care-of address. The care-of address must be an address to which datagrams can be delivered via conventional IP routing. At the care-of address, the original datagram is removed from the tunnel and delivered to the mobile node.

Mobile IP provides two alternative modes for the acquisition of a care-of address:

- **Foreign agent care-of address** A care-of address provided by a foreign agent through its agent advertisement messages. In this case, the care-of address is an IP address of the foreign agent. In this mode, the foreign agent is the endpoint of the tunnel and, upon receiving tunneled datagrams, decapsulates them and delivers the inner datagram to the mobile node. This mode of acquisition is preferred because it allows many mobile nodes to share the same care-of address and therefore does not place unnecessary demands on the already limited IPv4 address space.
- **Co-located care-of address** A care-of address acquired by the mobile node as a local IP address through some external means, which the mobile node then associates with one of its own network interfaces. The address may be dynamically acquired as a temporary address by the mobile node such as through DHCP, or it may be owned by the mobile node as a long-term address for its use only while visiting some

> foreign network. Specific external methods of acquiring a local IP address for use as a co-located care-of address are beyond the scope of this document. When using a co-located care-of address, the mobile node serves as the endpoint of the tunnel and itself performs decapsulation of the datagrams tunneled to it.

The mode of using a co-located care-of address has the advantage of allowing a mobile node to function without a foreign agent, for example, in networks that have not yet deployed a foreign agent. It does, however, place an additional burden on the IPv4 address space because it requires a pool of addresses within the foreign network to be made available to visiting mobile nodes. It is difficult to efficiently maintain pools of addresses for each subnet that may permit mobile nodes to visit.

It is important to understand the distinction between the care-of address and the foreign agent functions. The care-of address is simply the endpoint of the tunnel. It might indeed be an address of a foreign agent (a foreign agent care-of address), but it might instead be an address temporarily acquired by the mobile node (a co-located care-of address). A foreign agent, on the other hand, is a mobility agent that provides services to mobile nodes.

A home agent must be able to attract and intercept datagrams that are destined to the home address of any of its registered mobile nodes. Using the proxy and gratuitous Address Resolution Protocol (ARP) mechanisms, this requirement can be satisfied if the home agent has a network interface on the link indicated by the mobile node's home address. Other placements of the home agent relative to the mobile node's home location *may* also be possible using other mechanisms for intercepting datagrams destined to the mobile node's home address. Such placements are beyond the scope of the RFC.

Similarly, a mobile node and a prospective or current foreign agent must be able to exchange datagrams without relying on standard IP routing mechanisms, that is, those mechanisms that make forwarding decisions based upon the network prefix of the destination address in the IP header. This requirement can be satisfied if the foreign agent and the visiting mobile node have an interface on the same link. In this case, the mobile node and foreign agent simply bypass their normal IP routing mechanism when sending datagrams to each other, addressing the underlying link-layer packets to their respective link-layer addresses. Other placements of the foreign agent relative to the mobile node may also be possible using other mechanisms to exchange datagrams between these nodes, but such placements are beyond the scope of this chapter.

If a mobile node is using a co-located care-of address (as described previously), the mobile node must be located on the link identified by the network prefix of this care-of address. Otherwise, datagrams destined to the care-of address will be undeliverable.

For example, Figure 2-3 illustrates the routing of datagrams to and from a mobile node away from home once the mobile node has registered with its home agent. In the figure, the mobile node is using a foreign agent care-of address. Figure 2-4 provides another pictorial example.

Message Format and Protocol Extensibility Mobile IP defines a set of new control messages, sent with User Datagram Protocol (UDP) using well-known port number 434. Currently, the following two message types are defined:

- 1 Registration request
- 3 Registration reply

Up-to-date values for the message types for Mobile IP control messages are specified in the most recent "Assigned Numbers."

In addition, for agent discovery, Mobile IP makes use of the existing router advertisement and router solicitation messages defined for ICMP router discovery.

Mobile IP defines a general extension mechanism to enable optional information to be carried by Mobile IP control messages or by ICMP router

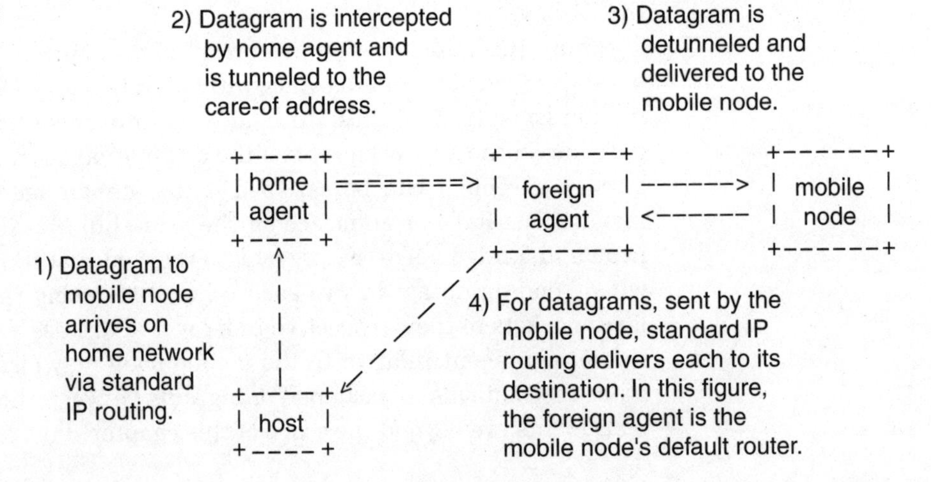

Figure 2-3 Routing datagrams

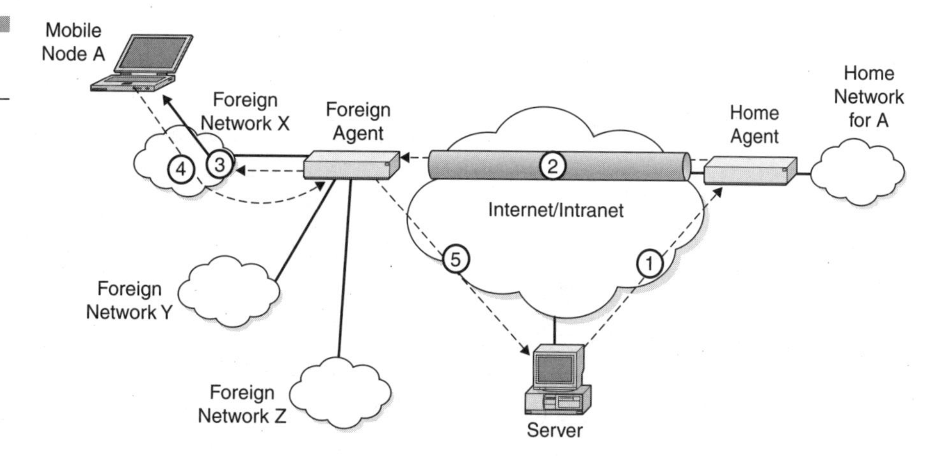

Figure 2-4
Mobile IP

To transmit IP PDUs between mobile user A and the server, the following steps take place:

1. The server transmits an IP PDU intended for mobile node A, with A's home IP address. The IP PDU is routed to A's home network.
2. The arriving IP PDU is intercepted by the home agent. The home agent encapsulates the entire PDU inside a new IP PDU, which has the A's care-of address in the header, and retransmits the PDU. This is in effect tunneling, namely, the use of an outer-layer IP PDU with a different destination IP than the one contained in the inner-layer PDU.
3. The foreign agent removes the outer IP header, encapsulates the original IP PDU in a new PDU, and forwards the original PDU to A via the foreign network.
4. When A sends traffic to the server, it uses the server's IP address. The foreign network will route the packet to the server (generally this router is also the foreign agent).
5. Past the foreign agent, the IP PDU from A to the server goes across the network or Internet to the server via normal routing procedures, utilizing the server's IP address.

discovery messages. Each of these extensions (with one exception) is encoded in the following type-length-value format:

```
 0                   1                   2
 0 1 2 3 4 5 6 7 8 9 0 1 2 3 4 5 6 7 8 9 0 1 2
+-+-+-+-+-+-+-+-+-+-+-+-+-+-+-+-+-+-+-+-+-+-+-+-
|     Type      |    Length     |    Data ...
+-+-+-+-+-+-+-+-+-+-+-+-+-+-+-+-+-+-+-+-+-+-+-+-
```

- **Type** Indicates the particular type of extension.
- **Length** Indicates the length (in bytes) of the data field within this extension. The length does not include the type and length bytes.
- **Data** The particular data associated with this extension. This field may be zero or more bytes in length. The format and length of the data field is determined by the type and length fields.

Extensions enable varying amounts of information to be carried within each datagram. The end of the list of extensions is indicated by the total length of the IP datagram.

Two separately maintained sets of numbering spaces, from which extension type values are allocated, are used in Mobile IP:

- The first set consists of those extensions that may appear only in Mobile IP control messages (those sent to and from UDP port number 434). Currently, the following types are defined for extensions appearing in Mobile IP control messages:
  - 32 mobile-home authentication
  - 33 mobile-foreign authentication
  - 34 foreign-home authentication
- The second set consists of those extensions that may appear only in ICMP router discovery messages. Currently, Mobile IP defines the following types for extensions appearing in ICMP router discovery messages:
  - 0 one-byte padding (encoded with no Length or Data field)
  - 16 mobility agent advertisement
  - 19 prefix-lengths

Each individual extension is described in detail in a separate section later in this chapter. Up-to-date values for these extension type numbers are specified in the most recent "Assigned Numbers".

Due to the separation (orthogonality) of these sets, it is conceivable that two extensions that are defined at a later date could have identical type values. This could be true as long as one of the extensions is used only in Mobile IP control messages and the other is used only in ICMP router discovery messages.

When an extension numbered in either of these sets within the range 0 through 127 is encountered but not recognized, the message containing that extension must be silently discarded. When an extension numbered in the range 128 through 255 is encountered but not recognized, that particular extension is ignored, but the rest of the extensions and message data must still be processed. The Length field of the extension is used to skip the Data field in searching for the next extension.

Protocol Details The interested reader should refer to the RFC for the detailed protocol machinery.

Quality of Service (QoS)

QoS relates to the use of design criteria, the selection of protocols, the determination of architectures, the identification of approaches, the choice of network restoration techniques, the design of node buffer management, and other network aspects. QoS ensures that end-to-end goals for congestion/availability, delay, jitter, throughput, and loss be reliably met over a specified time horizon and traffic load between any two chosen points in the network. These parameters are defined as follows:[17]

- **Congestion** A network condition where traffic bottles up in queues to the point that it noticeably and negatively impacts the operation of the application.
- **Service availability** The reliability of users' connection through the network.
- **Delay** The time taken by a packet to travel through the network from one end to another.
- **Delay jitter** The variation in the delay encountered by similar packets following the same route through the network.
- **Throughput** The rate at which packets go through the network.
- **Packet loss rate** The rate at which packets are dropped, get lost, or become corrupted (some bits are changed in the packet) while going through the network.

The industry has been working on the QoS issue for a decade now, but relatively little deployment of QoS-enabled networks has been seen on extranets, intranets, carrier networks, or on the Internet. There is no dearth of QoS literature on the topic.[18] Obviously, the protocol cornucopia leaves something to be desired; otherwise, we would have seen a statistically significant penetration of these protocols in the tens of thousands of networks that are currently deployed. This material is based on various industry sources as well as a book on Internet technologies published by Minoli and Schmidt in 1999[19] that included an extensive treatment of QoS. Additional references that covered QoS are Schmidt and Minoli, *Multiprotocol over ATM Building State of the Art ATM Intranets Utilizing RSVP, NHRP, LANE, Flow Switching, and WWW Technology*.[20] A number of analytical design techniques for broadband networks were described in Minoli, *Broadband Network Analysis and Design*.[21]

No fewer than five approaches have evolved for QoS in recent years as follows:

- Asynchronous Transfer Mode (ATM)-based QoS approaches
- Overengineering the network, without using any special QoS discipline
- Utilization of high-throughput gigarouters with advanced queue management, without using any special QoS discipline
- Per-flow QoS technology, the IETF's Integrated Services (*intserv*) Working Group recommendations
- Class-based QoS technology, the IETF's Differentiated Services (*diffserv*) Working Group recommendations

Some of these approaches reflect different philosophies regarding QoS. One school of thought believes in overprovisioning (assuming that the bandwidth exceeds demand); a second school of thought looks to traffic engineering (steering traffic away from congestion); a third school of thought looks to advanced queuing techniques where there is true contention for the resource (being considered scarce). Internet folks often take an approach of overprovisioning without much mathematically sophisticated analysis. Incumbent carriers often prefer robust (but complex) controls; however, they have focused more on Permanent Virtual Connection (PVC) networks (such as X.25 PVCs, Frame Relay PVCs, and ATM PVCs) rather than on switched/connectionless environments. In this chapter we will briefly look at the approach of advanced queue management and focus the discussion on *intserv* and *diffserv*.

QoS Basics

QoS is defined as those mechanisms that give network administrators the ability to manage traffic's bandwidth, delay, jitter, loss, and congestion throughout the network.[22] To realize true QoS, a QoS-endowed architecture must be applied end to end, not just at the edge of the network or at select network devices.[23] The solution must encompass a variety of technologies that can interoperate in such a way as to deliver scalable, feature-rich services throughout the network. The services must provide an efficient use of resources by facilitating the aggregation of large numbers of IP flows where needed while at the same time providing fine-tuned granularity to those premium services defined by service level agreements (SLAs) in general and real-time requirements in particular. The architecture must also provide the mechanisms and capabilities to monitor, analyze, and report detailed network status, since the need to continuously undertake traffic engineering, network tuning, and provisioning of new facilities is not going to go away given that the growth of the demand on the network will continue to be in the double-digit percentage points for years to come. Armed with this knowledge, network administrators or network monitoring software can react quickly to changing conditions, ensuring the enforcement of QoS policies. Finally, the architecture must also provide mechanisms to defend against the possibility of theft, to prevent denial of service, and to anticipate equipment failure.[24]

In general terms, QoS services in packet-based networks can be achieved in two possible ways:

- Using out-of-band signaling mechanisms to secure allocations of shared network resources. This includes signaling for different classes of services in ATM and Resource Reservation Protocol (RSVP). It should be immediately noted, however, that RSVP only reserves, and does not provide, bandwidth. As such, it augments existing unicast/multicast routing protocols, IP in particular. In turn, IP will have to rely on Packet over SONET (POS), ATM (say via Classical IP Over ATM [CIOA]), or Generalized Multiprotocol Label Switching (GMPLS, optical switch control) to obtain bandwidth. This approach is used in the *intserv* model described later.
- Using in-band signaling mechanisms where carriers and ISPs can provide a priority treatment to packets of a certain type. This could be done, for example, with the Type of Service (TOS) field in the IPv4 header, the Priority field in the IPv6 header, or the Priority field in the Virtual LAN (VLAN) IEEE 802.1Q/1p header. The MPLS label is another way to identify to the router/IP switch that special treatment is required. If routers, switches, and end systems all used or recognized the appropriate fields, if the queues in the routers or switches were effectively managed according to the priorities, and if adequate resources (buffers, links, backup routes, and so on) were provided in the network, then this method of providing QoS guarantees could be called the simplest. This is because no new protocols would be needed, the carrier's router can be configured in advance to recognize labels of different types of information flows, and relatively little state needs to the kept in the network. This approach is used in the *diffserv* model described later.

Specific tools available to the designer of an IP/MPLS network that is intended to support VoIP include the following:

- ***intserv*/RSVP** A bandwidth reservation mechanism targeted to enterprise networks (because of size considerations). Also, it is being targeted to MPLS label distribution and MPLS QoS.
- ***diffserv*** This associates a DSCP (*diffserv* code point) for every packet and defines per hop behaviors (PHBs).
- **MPLS** This defines label-switched paths (especially in the core of the network for aggregating traffic flows) that have different characteristics (link utilization, link capacity, the number of link hops, and so on). It utilizes the approach of mapping *diffserv* PHB in an access network to MPLS flows in a core network.
- **Traffic management mechanisms** This includes traffic shaping, marking, dropping, and queue handling. It also includes priority- and class-based queuing with disciplines such as Random Early Detection (RED) and other methods.

As noted earlier, two philosophical approaches satisfy the service requirements of applications:

- Overprovisioning or overallocation of resources that meet or exceed peak load requirements.
- Managing and controlling the allocation of network and computing resources.

Depending on the deployment, overprovisioning can be viable if it is a simple matter of upgrading to faster LAN switches and network interface cards (NICs), adding memory, adding a central processing unit (CPU) or disk, and so on. However, it may not be viable or cost-effective in many other cases, such as when dealing with relatively expensive long-haul WAN links. Overprovisioned resources remain underused and are utilized only during short peak periods. Better management consists of optimizing existing resources such as limited bandwidth, CPU cycles, and so on. VoIP stakeholders (carriers and intranet planners) have an economic incentive to deploy viable QoS capabilities so that an acceptable grade of service can be provided to the end users.[25,26]

QoS Approaches

Per-flow QoS The IETF Integrated Services (*intserv*) Working Group has developed the mechanisms with link-level, per-flow QoS control, while

RSVP is used for signaling. *intserv* services are guaranteed and controlled load services; these have been renamed by the International Telecommunication Union-Telecommunications (ITU-T) IP traffic control (Y.iptc) to "delay sensitive statistical bandwidth capability" and "delay insensitive statistical bandwidth capability," respectively. (ITU Y.itcp effort uses *intserv* services and *diffserv* expedited forwarding.)

The *intserv* architecture[27] defines QoS services and reservation parameters to be used to obtain the required QoS for an Internet flow. RSVP[28] is the signaling protocol used to convey these parameters from one or multiple senders towards a unicast or multicast destination. RSVP assigns QoS with the granularity of a single application's flows.[29] The work group is now also looking at new RSVP extensions.

Signaling traffic is exchanged between routers belonging to a core area. After a reservation has been established, each router must classify each incoming IP packet to determine whether it belongs to a QoS flow or not, and, in the former case, to assign the needed resources to the flow. The *intserv* classifier is based on a MultiField classification, because it checks five parameters in each IP packet, namely the source IP address, destination IP address, protocol ID, source transport port, and destination transport port. The classifier function generates a FLOWSPEC object.

intserv addresses the following categories of applications:

- **Elastic applications** No constraints for delivery are used as long as packets reach their destination. There is no specific demand on the delay bounds or bandwidth requirements. Examples are web browsing and e-mail.
- **Real-Time Tolerant (RTT) applications** These applications demand weak bounds on the maximum delay over the network. Occasional packet loss is acceptable. An example is Internet radio applications; these use buffering, hiding the packet losses from the application.
- **Real-Time Intolerant (RTI) applications** This class of applications demands tight bounds on latency and jitter. An example is a VoIP application; here excessive delay and jitter are hardly acceptable.

To service these classes, *intserv*, utilizing the various mechanisms at the routers, supports the following classes of service:

- **Guaranteed service** This service is meant for RTI applications. This service "guarantees"

- Bandwidth for the application traffic
- Deterministic upper bound on delay

- **Controlled load service** This is intended to service the RTT traffic. The average delay is guaranteed, but the end-to-end delay experienced by some arbitrary packet cannot be determined deterministically, such as H.323 traffic.

RSVP can support an *intserv* view of QoS; it can also be used as a signaling protocol for MPLS for distributing labels (although a distinct label distribution protocol is also available to MPLS). In the mid-1990s, RSVP was developed to address network congestion by enabling routers to decide in advance whether they could meet the requirements of an application flow and then reserve the desired resources if they were available. RSVP was originally designed to install the forwarding state associated with resource reservations for individual traffic flows between hosts.[30] The physical path of the flow across a service provider's network was determined by conventional destination-based routing (such as the Routing Information Protocol [RIP], Open Shortest Path First [OSPF], or the Interior Gateway Protocol [IGP]). By the late 1990s, RSVP became a proposed standard and has since been implemented in a variety of IP networking equipment. However, RSVP has not been widely used in service provider/carrier networks because of operator concerns about its scalability and the overhead required to support potentially millions of host-to-host flows.

An informational IETF document[31] discusses issues related to the scalability posed by the signaling, classification, and scheduling mechanisms. An important consequence of this problem is that *intserv*-level QoS can be provided only within peripheral areas of a large network, preventing its extension inside core areas and the implementation of end-to-end QoS. IETF RSVP-related Work Groups have undertaken some work to overcome these problems. The RSVP Work Group has recently published the RFC2961 that describes a set of techniques to reduce the overhead of RSVP signaling; however, this RFC does not deal with the classification problem still to be addressed. The Baker, Iturralde, Le Faucheur, and Davie[32] draft discusses the possibility of aggregating RSVP sessions into a larger one. The aggregated RSVP session uses a *diffserv* code point (DSCP) for its traffic.[29]

Class-based QoS The IETF Differentiated Services (*diffserv*) Working Group has developed a class-based QoS. Packets are marked at the network "edge." Routers use markings to decide how to handle packets. There are four services:

- **Best efforts** Normal Internet traffic
- **Seven precedence levels** Prioritized classes of traffic
- **Expedited forwarding (EF)** Leased-line-like service
- **Assured forwarding (AF)** Four queues with three drop classes

This approach requires edge policing, but this technology is not yet defined.

In a *diffserv* domain (RFC-2475), all the IP packets crossing a link and requiring the same *diffserv* behavior are said to constitute a behavior aggregate (BA). At the ingress node of the *diffserv* domain, the packets are classified and marked with a *diffserv* code point (DSCP) that corresponds to their BA. At each transit node, the DSCP is used to select the per hop behavior (PHB) that determines the scheduling treatment and, in some cases, the drop probability for each packet.

At face value, *diffserv* appears to be able to scale more easily than *intserv*; also it is simpler. Packet purists (will probably) argue that *diffserv* is the best approach because there is very little if any state information kept along the route, while folks more in the carriers' camp (will probably) argue that *intserv* is a better approach because resource reservations and allocations can be better managed in the network in terms of being able to engineer networks and maintain SLAs. It is within reason to assume that if the design is properly supported by statistically valid and up-to-date demand information,[33] and resources are quickly added when needed, either approach would probably provide reasonable results.

One is not able to generalize as to which of these techniques is better for delay-sensitive traffic such as VoIP, because the decision will have to be based on the type of network architecture one chooses to implement and the size of the network both in terms of network elements (NEs) and lines supported. One cannot argue that a metric wrench is better than a regular wrench. If one is working on a European-made engine, then the metric wrench is obviously best; if one is working on a U.S.-built engine, then regular wrenches are the answer.

For example, in a small network where the end-to-end hop diameter is around three to seven hops, a reservation scheme (specifically *intserv*) would seem fine (the U.S. voice network kind of fits this range). A network with a large diameter where paths may be 8 to 15 hops may find a reservation scheme too burdensome and a node-by-node distributed approach (specifically, *diffserv*) may be better (the Internet kind of fits this range). The same kind of argument also applies when looking at the total number of nodes (separate and distinct from the network diameter). If the network in question is the national core network with 10 to 20 core nodes, the reservation/*intserv* may be fine; if the network in question covers all the tiers of a voice network with around 400 to 500 interacting nodes, the *diffserv* approach may be better. These are just general observations: the decision regarding the best method must be made based on careful network-specific analysis and product availability.

MPLS-based QoS Prima facie, the use of MultiProtocol Label Switching (MPLS) affords a packet network the possibility for an improved level of QoS control compared with pure IP. MPLS developers have proposed both a *diffserv*-style and an *interv*-style approach to QoS in MPLS. QoS controls are critical for multimedia application in intranets, dedicated (WAN) IP networks, virtual private networks (VPNs), and a converged Internet. Services such as VoIPoMPLS, VoMPLS, MPLS VPNs, Layer 2 VPN (L2VPN), Differentiated Services Traffic Engineering (DS-TE), and draft-martini typically require service differentiation in particular and QoS support in general. It is important to realize, however, that MPLS per se is not a QoS solution: it still needs a distinct mechanism to support QoS. The issue of QoS in an MPLS network was treated at length in Minoli, *Delivering Voice over MPLS Networks*.[34]

In the *diffserv*-style case, the EXPerimental (EXP) bits of the header are used to trigger scheduling and/or drop behavior at each label-switching router (LSR). This solution, based on Francois Le Faucheur's, "MPLS Support of Differentiated Services,"[35] enables the MPLS network administrator to select how *diffserv* BAs are mapped onto label-switched paths (LSPs) so that he or she can best match the *diffserv*, traffic engineering, and protection objectives within his or her particular network. The proposed solution enables the network administrator to decide whether different sets of BAs are to be mapped onto the same LSP or mapped onto separate LSPs. The MPLS solution relies on the combined use of two types of LSPs:

- LSPs that can transport multiple ordered aggregates, so that the EXP field of the MPLS shim header conveys to the LSR the PHB to be applied to the packet (covering both information about the packet's scheduling treatment and its drop precedence).
- LSPs that only transport a single ordered aggregate, so that the packet's scheduling treatment is inferred by the LSR exclusively from the packet's label value while the packet's drop precedence is conveyed in the EXP field of the MPLS shim header or in the encapsulating link-layer-specific selective drop mechanism (ATM, Frame Relay, or 802.1).

Some developers[29] have proposed a solution that efficiently combines the application-oriented intserv QoS with the power of MPLS label switching. The proposal is contained in Tommasi, Molendini, and Tricco's "Integrated Services Across MPLS Domains Using CR-LDP Signaling."[36] The cited document defines the following intserv-like QoS services in MPLS domains targeting certain problems:

- Providing a user-driven MPLS QoS path setup. An application uses the standard *intserv* reservation application programming interface (API) to allocate network resources. *intserv* reservation (signaled using RSVP) are then mapped at the Ingress LSR of the MPLS domain into proper constraint-based routed LSPs (CR-LSPs).
- Reducing the constraint-based routing Label Distribution Protocol (CR-LDP) signaling overhead providing caching and aggregation of CR-LSPs. Both manual configuration of the bandwidth/signaling trade-off as well as automatic load discovery mechanisms are allowed.

The key element of this solution is the MPLS Ingress LSR that acts like an MPLS/*intserv* QoS gateway. The CR-LDP protocol enables the definition of a LSP with QoS constraints[37] that is to perform QoS classification *using a single valued label* (not a MultiField one). The main limitation of this current approach is that end hosts cannot use it because they cannot support CR-LDP signaling. On the other hand, *intserv* has been designed to enable applications to signal QoS requirements on their own (reservation APIs are available and many operating systems enable applications to use them.)

The basic idea of the "Integrated Services Across MPLS Domains Using CR-LDP Signaling, Internet Draft" is to combine the application-oriented *intserv* QoS with the power of MPLS label switching, that is, to define *intserv*-like QoS services in MPLS domains. Using these mechanisms, end-to-end QoS is reached without service disruptions between MPLS domains and *intserv* areas. Here the MPLS Ingress LSR acts like an MPLS/*intserv* QoS gateway. At the same time, the number and the effects of the changes to the current CR-LDP specifications are minimal. Most of the integration work is included in the Ingress LSR at the sender side of the MPLS domain's border.

Traffic Management/Queue Management As noted earlier, two approaches have been used over time to allocate resources. The first is the out-of-band reservation model (*intserv*/RSVP and ATM), requiring applications to signal their traffic requirements to the serving switch. This in turn sets up a path from the source to the destination with reserved resources such as bandwidth and buffer space that either guarantees the desired QoS service or assures with reasonable certainty that the desired service will be provided. The second approach is in-band precedence priority. Here packets are marked or tagged according to priority, such as *diffserv* DSCP, IP Precedence TOS, and IEEE 802.1Q/1p. A router takes aggregated traffic, segregates the traffic flows into classes, and provides preferential treatment of classes. Routers then read these markings and treat the packets accordingly. Both of these approaches require advanced traffic and queue management, especially the in-band priority.

Typically, delays and QoS degradation are accumulated at points in the network where queues exist. Queues arise when the "server" capacity (an outgoing link or a CPU undertaking a task such as a sort or table lookup) is less than the aggregated demand for service "brought along" by the incoming "jobs" (packets). Because of the way that internetworking technology has been developed in the past 15 years, queues are typically found at routing points rather than at switching points. Furthermore, the distribution of the delay (and, hence, jitter) increases as the number of queues that have to be traversed increases, as shown in Table 2-10.

Managing resources and supporting QoS routers require sophisticated queue management. QoS mechanisms for controlling resources that achieve more predictable delays include

- Classification
- Conditioning, specifically policing/shaping traffic (such as Token Bucket)
- Queuing management (such as Random Early Detection [RED])
- Queue/packet scheduling (such as Weighted Fair Queuing [WFQ])
- Bandwidth reservation via signaling and path establishment (such as RSVP, H.225, MPLS CR-LDP)

Routers can implement the following mechanisms to deal with QoS:[25]

- **Admission control** Accepting or rejecting access to a shared resource. This a key component for *intserv* and ATM networks. This ensures that resources are not oversubscribed and hence are more expensive and less scalable.
- **Congestion management** Prioritizing and queuing traffic access to a shared resource during congestion periods (as done in *diffserv)*.
- **Congestion avoidance** Instead of waiting for congestion to occur, use measures to prevent it. Algorithms such as Weighted Random Early Detection (WRED) make use of the Transmission Control

Table 2-10

Increasing variance as the number of queues increases

Figures of merit	Values	Random variable	Sample space	Probability values
		X	1	0.333333
			3	0.333333
			5	0.333333
$E(X) =$	3			
$E(X^2) =$	11.66667			
$V(X) = E(X^2) - E(X)^2 =$	**2.666667**			
		Y	1	0.333333
			3	0.333333
			5	0.333333
		A=X+Y	2	0.111111
			4	0.222222
			6	0.333333
			8	0.222222
			10	0.111111
$E(A) =$	6			
$E(A^2) =$	41.33333			
$V(A) = E(A^2) - E(A)^2 =$	**5.333333**			
		Z	1	0.333333
			3	0.333333
			5	0.333333
		B=X+Y+Z	3	0.037037
			5	0.111111
			7	0.222222
			9	0.259259
			11	0.222222
			13	0.111111
			15	0.037037
$E(B) =$	9			
$E(B^2) =$	89			
$V(B) = E(B^2) - E(B)^2 =$	**8**			

Protocol's (TCP) congestion-avoidance algorithms to reduce traffic injected into the network and prevent congestion.

- **Traffic shaping** Reducing the burstiness of ingress network traffic by smoothing the traffic and then forwarding it to the egress link.

Basic elements of a router include some or all of the following:[25]

- **Packet classifier** This functional component is responsible for identifying flows and matching them with a filter. The filter is composed of parameters such as source and destination, IP address, port, protocol, and TOS field. The filter is also associated with information that describes the treatment of this packet. Aggregate ingress traffic flows are compared against these filters. Once a packet header is matched with a filter, the QoS profile is used by the meter, marker, and policing/shaping functions.

- **Metering** The metering function compares the actual traffic flow against the QoS profile definition.
- **Marking** Marking is related with metering in that when the metering function compares the actual measured traffic against the agreed QoS profile, the traffic is handled appropriately.
- **Policing/shaping** The policing functional component uses the metering information to determine if the ingress traffic should be buffered or dropped. Shaping means packets are dispensed at a constant rate, buffering packets in order to achieve a constant output rate. A common algorithm used here is the token bucket algorithm to shape the egress traffic as well as police ingress traffic.
- **Queue manager/scheduler** This is a capability that handles the packets that are in the router's (set of) queue(s), based on the priority management and traffic-handling machinery just described.

Conclusion

This chapter identifies the various standardization efforts that are either under way or have just completed in support of WLANs, WPANs, and WWANs. It is important to note, however, that at the time of this writing (and in the foreseeable future) the commercial emphasis is squarely in favor of IEEE 802.11b technology.

IEEE 802.11b works well for corporate WLAN applications in indoor environments where external interference is at a minimum. This is because the enterprise user controls who is in the building or on the floor (namely, who can deploy other antennas), and the walls attenuate potential interference from the outside. Hotspot services in confined environments (for example, airports) also work well on IEEE 802.11b technology. Hotspot services for outdoor environments (for example, malls, stadiums, and so on) have to deal with the issue of potential interference from other ISM sources.

One other issue of consideration is the overhead added by the use of very tight security. IEEE 802.11b has experienced three overlays of security:

WEP, dynamic WEP, and IEEE 802.11x (EAP) and 802.11i overlays. These build on one another, so that there is not a totally cohesive apparatus; instead, it forms an incrustation of protocols. Several levels of security can be implemented, as shown in Figure 2-5 (from Cisco sources, describing their Lightweight EAP [LEAP]). The use of the tightest form of security can make a hotspot service more difficult to sustain because near-constant reauthentication occurs for users who have mid to low levels of signal.

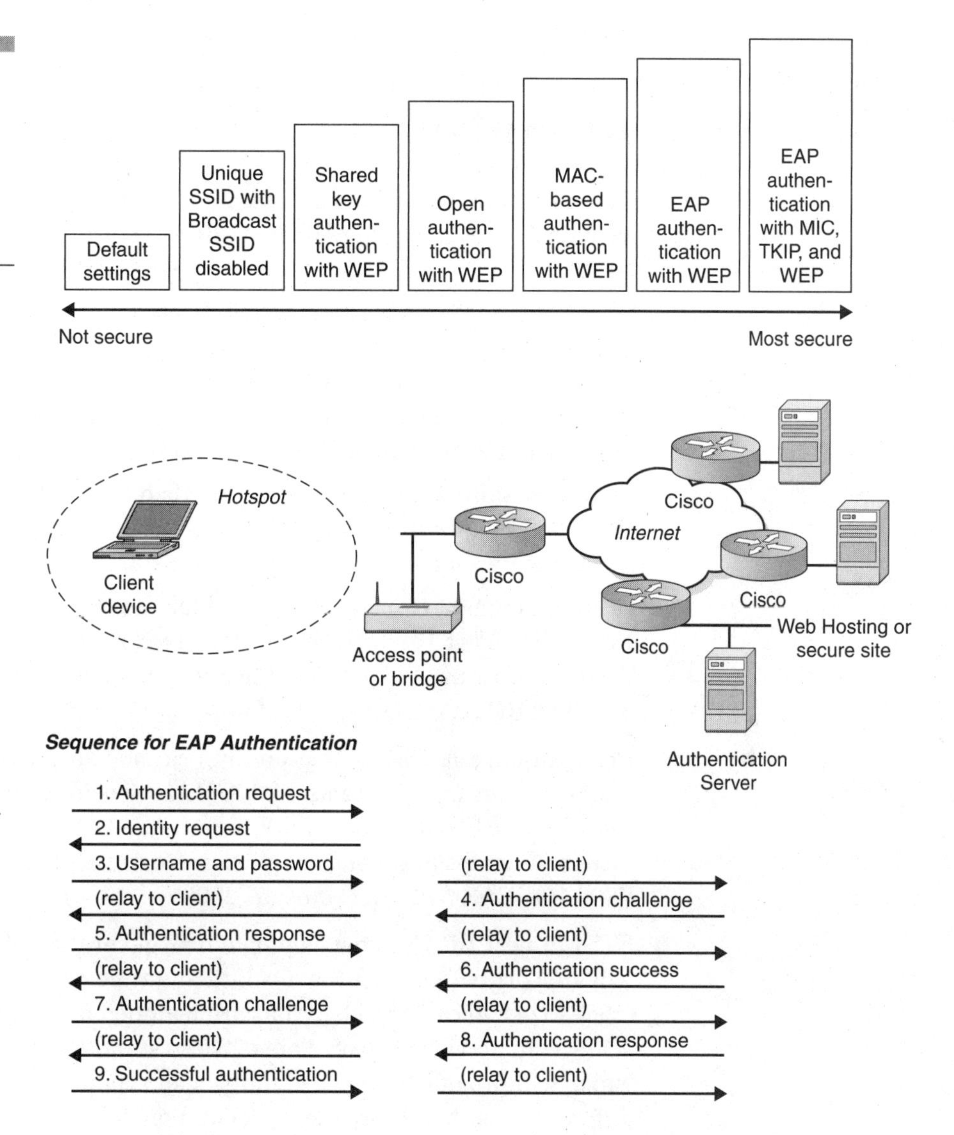

Figure 2-5 Levels of security in practical implementations of IEEE 802.11b technology (Source: Cisco Systems)

While this handshake with the authentication server is occurring, the user is locked out of the service, although this is all supposed to be automatic and transparent to the user. For Internet-access services, a mid-range security discipline (such as MAC-based authentication with WEP) should be adequate.

In summary, in spite of some ultimate limitations hinted at previously, IEEE 802.11b remains the leading hotspot technology for the foreseeable future.

End Notes

1. Glenn Fleishman, "New Wireless Standards Challenge 802.11b," www.oreillynet.com/pub/a/wireless/2001/05/08/standards.html, June 8, 2001.
2. Subsequently, Texas Instruments indicated it would likely pursue the technology and sell it as a wireless networking chipset compatible with 802.11b up to 11 Mbps. PBCC was already approved for use by the IEEE with 802.11b at 11 Mbps (however, manufacturers have not implemented the technology).
3. Figure partially based on IBM materials by V. Malhotra, "Checking on IEEE 802.15," www-106.ibm/developerworks/library/wi-checking/?dwzone=wireless, September 2001.
4. Designation: 802.16-2001. "Standard for Local and Metropolitan Area Networks—Part 16: Air Interface for Fixed Broadband Wireless Access Systems." Status: Approved Publication of IEEE. PAR APP: January 30, 2000, BD APP: December 6, 2001.
5. Designation: 802.16.1b. "Telecommunications and Information Exchange Between Systems—LAN/MAN Specific Requirements—Air Interface for Fixed Broadband Wireless Access Systems Including License-Exempt Frequencies." Status: Changed Designation to P802.16b. PAR APP: December 7, 2000.
6. ETSI Project BRAN is developing a new generation of standards that will support both asynchronous data and time-critical services (packetized voice and video) that are bounded by specific time delays to achieve an acceptable QoS. One of these standards is HIPERLAN/2, which will provide high-speed multimedia communications between different broadband core networks and mobile terminals. HIPERLAN/2

provides a platform for a variety of business and home multimedia applications that can support a set of bit rates up to 54 Mbps. In a typical business application scenario, a mobile terminal gets services over a fixed corporate/public network infrastructure. In addition to QoS, the network will provide mobile terminals with security and mobility management services when moving. In an exemplary home application scenario, a low-cost and flexible networking system is supported to interconnect wireless digital consumer devices (www.etsi.org/frameset/home.htm?/technicalactiv/Hiperlan/hiperlan2tech.htm).

7. Designation: 802.16.2a. "Local and Metropolitan Area Networks—Amendment to Recommended Practice for Coexistence of Fixed Broadband Wireless Access Systems." Status: Amendment. PAR APP: August 17, 2001.
8. Designation: 802.16.2-2001. "Local and Metropolitan Area Networks—IEEE Recommended Practice for Coexistence of Fixed Broadband Wireless Access Systems." Status: Approved Publication of IEEE. PAR APP: September 16, 1999, BD APP: July 6, 2001, ANSI APP: November 1, 2001.
9. Designation: 802.16.3. "Telecommunications and Information Exchange Between Systems—LAN/MAN Specific Requirements—Air Interface for Fixed Broadband Wireless Access Systems in Licensed Bands from 2 to 11 GHz." Status: Changed Designation to P802.16a. PAR APP: March 30, 2000.
10. Designation: 802.16a. "Local and Metropolitan Area Networks—Amendment to Standard Air Interface for Fixed Broadband Wireless Access Systems—Media Access Control Modifications and Additional Physical Layer for 2-11 GHz RP." History: PAR APP: March 30, 2000.
11. Designation: 802.16b. "Local and Metropolitan Area Networks—Amendment to Standard Air Interface for Fixed Broadband Wireless Access Systems—Media Access Control Modifications and Additional Physical Layer for License-Exempt Frequencies RP." History: PAR APP: December 7, 2000.
12. www.3gpp.org.
13. Oliver Thylmann, www.infosync.no/show.php?id=1246, February 2, 2002.
14. Kulwinder Atwal, www.imc.org/ietf-sacred/mail-archive/msg00152.html, February 9, 2001.

15. C. Perkins (ed.), Request for Comments 2002, "IP Mobility Support," IETF, October 1996.

16. This section is based on RFC 2002. Copyright© The Internet Society (1996). This document and translations of it may be copied and furnished to others, and derivative works that comment on or otherwise explain it or assist in its implementation may be prepared, copied, published, and distributed, in whole or in part, without restriction of any kind, provided that the above copyright notice and this paragraph are included on all such copies and derivative works.

17. Paul Arindam, "QoS in Data Networks: Protocols and Standards," www.cis.ohio-state.edu/~jain/cis788-99/qos_protocols/index.html.

18. For example, see www.cis.ohio-state.edu/~jain/refs/ipq_book.htm.

19. Daniel Minoli and Andrew Schmidt, *Internet Architectures* (New York: Wiley, 1999).

20. Andrew Schmidt and Daniel Minoli, *Multiprotocol over ATM Building State of the Art ATM Intranets Utilizing RSVP, NHRP, LANE, Flow Switching, and WWW Technology* (New York: Prentice Hall, 1998). Dan Minoli and Andrew Schmidt, *Network Layer Switched Services* (New York: Wiley, 1998) (includes LANE, MPOA, IP switching, tag switching).

21. Daniel Minoli, *Broadband Network Analysis and Design* (Norwood, MA: Artech House, 1993).

22. J. Zeitlin, "Voice QoS in Access Networks—Tools, Monitoring, and Troubleshooting", Next-Generation Networks Conference (NGN) Proceedings, Boston, MA, November 2001.

23. However, if IP is actually deployed at the core of the network in support of VoIP, as discussed in Chapter 11, the QoS can also initially be targeted for the core.

24. Robert Pulley and Peter Christensen, "A Comparison of MPLS Traffic Engineering Initiatives," NetPlane Systems, Inc., www.netplane.com.

25. Deepak Kakadia, "Tech Concepts: Enterprise QoS Policy-Based Systems and Network Management," Sun Microsystems, www.sun.com/software/bandwidth/wp-policy.

26. Grade of service relates to an overall level of service delivery (similar to an SLA-oriented view), while QoS refers to the achievement of specific network parameters within defined ranges (for example, $0.100 < \text{delay} < 0.200$ seconds).

27. R. Braden, D. Clark, and S. Shenker, "Integrated Services in the

Internet Architecture: An Overview," IETF RFC 1633, June 1994.

28. R. Braden, (ed.), L. Zhang, S. Berson, S. Herzog, and S. Jamin, "Resource ReSerVation Protocol (RSVP)—Version 1 Functional Specification," IETF RFC 2205, September 1997.

29. F. Tommasi, S. Molendini, and A. Tricco, University of Lecce, "Integrated Services Across MPLS Domains Using CR-LDP Signaling," Internet Draft, http://search.ietf.org/ . . . /draft-tommasi-mpls-intserv-01.txt, May 2001.

30. Chuck Semeria, "RSVP Signaling Extensions for MPLS Traffic Engineering," White Paper, Juniper Networks, Inc., www.juniper.net, 2000.

31. A. Mankin, (ed.), F. Baker, B. Braden, S. Bradner, M. O'Dell, A. Romanow, A. Weinrib, and L. Zhang, "Resource ReSerVation Protocol (RSVP)—Version 1 Applicability Statement Some Guidelines on Deployment," September 1997.

32. F. Baker, C. Iturralde, F. Le Faucheur, and B. Davie, "Aggregation of RSVP forIPv4 and IPv6 Reservations," work in progress, draft-ietf-issll-rsvp-aggr-04, April 2001.

33. The $700 billion debt created by the telecom industry in 2000 and the abundance of carrier failures in 2001 strongly argue for mathematically sound, statistically significant primary market research; it also argues for mathematically sound forecasting of demand and analytical decision-making regarding engineering and deployment.

34. Daniel Minoli, *Delivering Voice over MPLS Networks* (New York: McGraw-Hill, 2002).

35. http://search.ietf.org/internet-drafts/draft-ietf-mpls-diff-ext-09.txt.

36. CR stands for constraint-based routing.

37. B. Jamoussi et al., "Constraint-Based LSP Setup Using LDP," work in progress, draft-ietf-mpls-cr-ldp-05, August 2001.

CHAPTER 3

Technologies for Hotspots

This chapter expands on the discussions of the wireless personal area network (WPAN), wireless local area network (WLAN), and wireless wide area network (WWAN) technology that appeared in Chapter 1, "Introduction to Wireless Personal Area Networks (WPANs), Public Access Locations (PALs), and Hotspot Services." This chapter drills down to another level of detail for the major systems that play a role in hotspot networks. It closes with an introduction to antenna technology for hotspot applications.

Wireless Local Area Networks (WLANs)

WLANs can be utilized to replace wired LANs or act as extensions to the wired LAN. 802.11 WLANs support communication between stations and access points (APs) using radio technology that does not require line of sight (LOS) between the AP and station. This technology works well for indoors applications and works fine for controlled outdoors applications. Currently, the Federal Communications Commission (FCC) (and, by extension, most of the regulatory bodies from other countries) allows only two kinds of spread spectrum use: frequency-hopping spread spectrum (FHSS), which is employed by Home Radio Frequency (HomeRF) and Bluetooth, and direct sequence spread spectrum (DSSS), which is used by 802.11b.[1] The radio signal is confined to the 2.4 GHz Industry, Scientific, and Medical (ISM) band.

IEEE 802.11 defines data rates of 1 and 2 Mbps via radio waves using FHSS or DSSS. 802.11b is an enhancement of 802.11 that employs DSSS to achieve a maximum throughput of 11 Mbps.[2] Independent of the data rate (whether it is 1, 2, 5.5, or 11 Mbps), the channel bandwidth for a DSSS system is about 20 MHz; therefore, the ISM band accommodates up to three nonoverlapping channels (see Figure 3-1). The recently defined complimentary code keying (CCK) modulation scheme (IEEE 802.11b) allows speeds of 5.5 and 11 Mbps in the same bandwidth as the original 1 and 2 Mbps DSSS radios. CCK is backward compatible. Table 3-1, which is based directly on the IEEE 802.11 standard, describes key concepts that are applicable to WLANs.

Figure 3-2 depicts the basic topology of an IEEE 802.11 environment. A basic service set (BSS) consists of two or more wireless stations that have recognized each other and have established communications. Within a given cell coverage area, stations communicate directly with each other on a peer-to-peer basis; this type of network is usually established on a temporary timeframe and is commonly referred to as an *ad hoc* network or an inde-

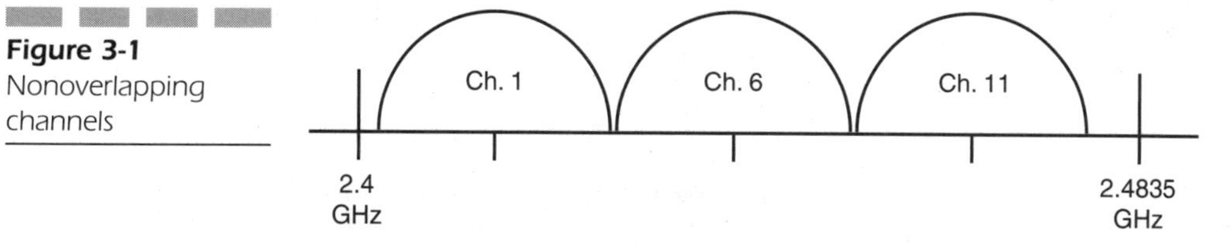

Figure 3-1 Nonoverlapping channels

Table 3-1 Key IEEE 802.11 definitions

Term	Definition
Access control	The prevention of unauthorized usage of resources.
Access point (AP)	Any entity that has station functionality and provides access to the distribution services via the wireless medium (WM) for associated stations.
Ad hoc network	A network composed solely of stations within the mutual communication range of each other via the WM. An ad hoc network is typically created in a spontaneous manner. The principal distinguishing characteristic of an ad hoc network is its limited temporal and spatial extent. These limitations enable the act of creating and dissolving the ad hoc network to be sufficiently straightforward and convenient so as to be achievable by nontechnical users of the network facilities; that is, no specialized technical skills are required and little or no time or additional resources are required beyond the stations that are to participate in the ad hoc network. The term *ad hoc* is often used to refer to an IBSS.
Association	The service used to establish AP/station mapping and enable the station invocation of the distribution system services (DSSs).
Authentication	The service used to establish the identity of one station as a member of the set of stations authorized to associate with another station.
Basic service area (BSA)	The conceptual area within which members of a BSS may communicate.
Basic service set (BSS)	A set of stations controlled by a single coordination function.
BSS basic rate set	The set of data transfer rates that all the stations in a BSS will be capable of using to receive frames from the WM. The BSS basic rate set data rates are preset for all stations in the BSS.
Broadcast address	A unique multicast address that specifies all stations.
Channel	An instance of medium use for the purpose of passing protocol data units (PDUs) that may be used simultaneously in the same volume of space with other instances of medium use (on other channels) by other instances of the same physical layer (PHY). It has an acceptably low frame error ratio due to mutual interference.
Clear channel assessment (CCA) function	The logical function in PHY that determines the current state of the WM.

Table 3-1 cont.

Key IEEE 802.11 definitions

Term	Definition
Confidentiality	The property of information that is not made available or disclosed to unauthorized individuals, entities, or processes.
Coordination function	The logical function that determines when a station operating within a BSS is permitted to transmit and receive PDUs via the WM. The coordination function within a BSS may have one point coordination function (PCF) and will have one distributed coordination function (DCF).
Coordination function pollable	A station able to respond to a coordination function poll with a data frame, if such a frame is queued and able to be generated and interpret acknowledgments in frames sent to or from the point coordinator.
Deauthentication	The service that voids an existing authentication relationship.
Disassociation	The service that removes an existing association.
Distributed coordination function (DCF)	A class of coordination function where the same coordination function logic is active in every station in the BSS whenever the network is in operation.
Distribution	The service that, by using association information, delivers Medium Access Control (MAC) service data units (MSDUs) within the DS.
Distribution system (DS)	A system used to interconnect a set of BSSs and integrated LANs to create an ESS.
Distribution system medium (DSM)	The medium or set of media used by a DS for communications between APs and portals of an ESS.
Distribution system service (DSS)	The set of services provided by the DS that enable the MAC to transport MSDUs between stations that are not in direct communication with each other over a single instance of the WM. These services include the transport of MSDUs between the APs of BSSs within an ESS, the transport of MSDUs between portals and BSSs within an ESS, and the transport of MSDUs between stations in the same BSS in cases where the MSDU has a multicast or broadcast destination address or where the destination is an individual address, but the station sending the MSDU chooses to involve DSS. DSSs are provided between pairs of IEEE 802.11 MACs.
Extended rate set (ERS)	The set of data transfer rates supported by a station (if any) beyond the ESS) basic rate set. This set may include data transfer rates that will be defined in future PHY standards.
Extended service area (ESA)	The conceptual area within which members of an ESS may communicate. An ESA is larger than or equal to a BSA and may involve several BSSs in overlapping, disjointed, or both configurations.
Extended service set (ESS)	A set of one or more interconnected BSSs and integrated LANs that appears as a single BSS to the Logical Link Control (LLC) layer at any station associated with one of those BSSs.

Table 3-1 cont.

Key IEEE 802.11 definitions

Term	Definition
Gaussian frequency shift keying (GFSK)	A modulation scheme in which the data is first filtered by a Gaussian filter in the baseband and then modulated with a simple frequency modulation.
Independent basic service set (IBSS)	A BSS that forms a self-contained network and in which no access to a DS is available.
Infrastructure	The infrastructure includes the DSM, AP, and portal entities. It is also the logical location of distribution and integration service functions of an ESS. An infrastructure contains one or more APs and zero or more portals in addition to the DS.
Integration	The service that enables the delivery of MSDUs between the DS and an existing, non-IEEE 802.11 LAN (via a portal).
MAC management protocol data unit (MMPDU)	The unit of data exchanged between two peer MAC entities to implement the MAC management protocol.
MAC protocol data unit (MPDU)	The unit of data exchanged between two peer MAC entities using the services of the PHY.
MAC service data unit (MSDU)	Information that is delivered as a unit between MAC service access points (SAPs).
Minimally conformant network	An IEEE 802.11 network in which two stations in a single BSA are conformant with ISO/IEC 8802-11: 1999.
Mobile station	A type of station that uses network communications while in motion.
Multicast	A MAC address that has the group bit set. A multicast MSDU has a multicast destination address. A multicast MPDU or control frame has a multicast receiver address.
Network allocation vector (NAV)	An indicator, maintained by each station, of time periods when transmission onto the WM will not be initiated by the station whether or not the station's CCA function senses that the WM is busy.
Point coordination function (PCF)	A class of possible coordination functions in which the coordination function logic is active in only one station in a BSS at any given time that the network is in operation. PCF is an optional extension to DCF. For example, it can provide a time-division capability to accommodate time-bounded, connection-oriented services such as cordless telephony.
Portable station	A type of station that may be moved from location to location, but that only uses network communications while at a fixed location.
Portal	The logical point at which MSDUs from a non-IEEE 802.11 LAN enter the DS of an ESS.

Table 3-1 cont.

Key IEEE 802.11 definitions

Term	Definition
Privacy	The service used to prevent the content of messages from being read by someone other than the intended recipient.
Reassociation	The service that enables an established association (between the AP and station) to be transferred from one AP to another (or the same) AP.
Station	Any device that contains an IEEE 802.11 conformant MAC and PHY interface to the WM.
Station basic rate	A data transfer rate belonging to the ESS basic rate set that is used by a station for specific transmissions. The station basic rate may change dynamically as frequently as a search MPDU transmission attempt based on local considerations at that station.
Station service (SS)	The set of services that support the transport of MSDUs between stations within a BSS.
Time unit (TU)	A measurement of time equal to 1,024 ms.
Unauthorized disclosure	The process of making information available to unauthorized individuals, entities, or processes.
Unauthorized resource use	The use of a resource that is not consistent with the defined security policy.
Unicast frame	A frame that is addressed to a single recipient, not a broadcast or multicast frame.
Wired Equivalent Privacy (WEP)	The optional cryptographic confidentiality algorithm specified by IEEE 802.11 used to provide data confidentiality that is subjectively equivalent to the confidentiality of a wired LAN medium that does not employ cryptographic techniques to enhance privacy.
Wireless medium (WM)	The medium used to implement the transfer of PDUs between peer PHY entities of a WLAN.

Figure 3-2
IBSS

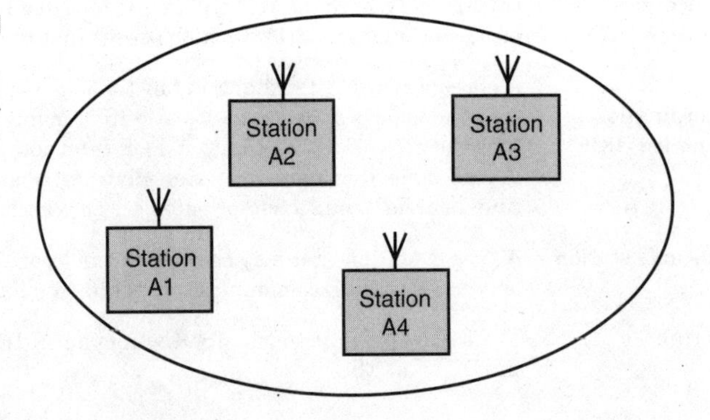

pendent basic service set (IBSS). More typically, the BSS contains an AP. The AP operates as a bridge between the wireless and wired LAN. APs are fixed-location network elements and are considered part of the wired network infrastructure. When an AP is present, the stations do not communicate on a peer-to-peer basis, but communicate through the AP instead. A BSS with APs is said to be operating in the *infrastructure mode*. All station clocks within a BSS are synchronized by the periodic transmission of time-stamped beacons. Beacons are used to convey network parameters such as hop sequence. Probe requests and responses are utilized to join a network.

There are two power-saving modes defined for stations: *awake* and *doze*. In the awake mode, stations can receive packets at any time. In the doze mode, stations must "wake up" periodically to listen for beacons that indicate that the AP has queued messages. Stations must inform the AP before entering the doze state.

Figure 3-3 depicts an extended service set (ESS). The ESS consists of a group of overlapping BSSs (each containing an AP) connected together via a distribution system (DS). Although the DS can be any type of network, it is often an Ethernet LAN. Mobile stations can roam between APs and seamless coverage is maintained. The IEEE standard identifies the basic message formats to support roaming, but the details are left up to network vendors.

Figure 3-3
ESS

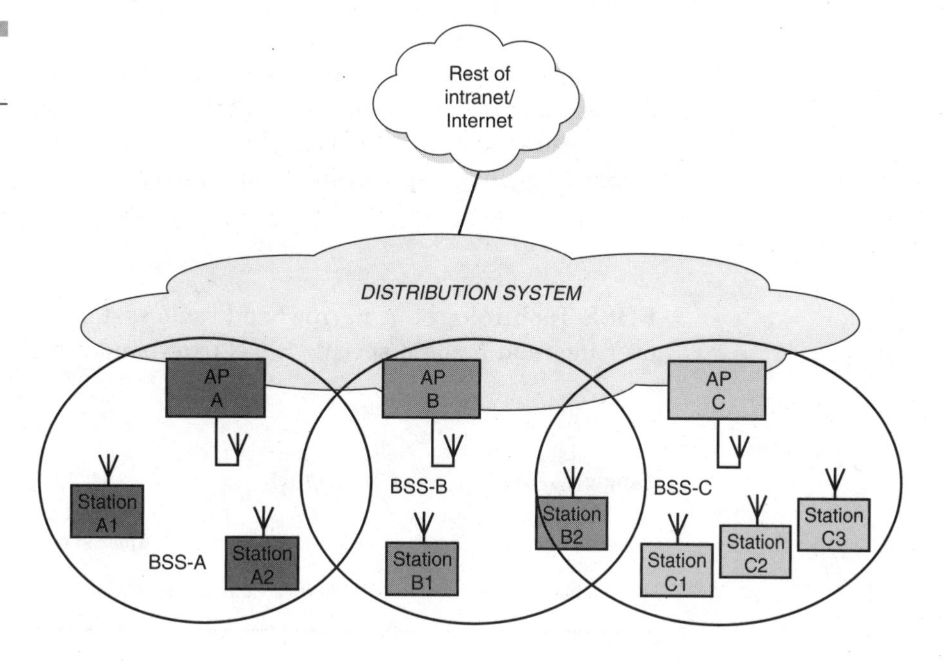

802.11 uses carrier sense multiple access/collision avoidance (CSMA/CA). Although 802.3 uses carrier sense multiple access/collision detection (CSMA/CD), which has a 100 percent collision-detect mechanism to support reliable data transfer, the CA apparatus is better optimized to an environment where there are large differences in signal strengths. With CA, collisions can only be inferred after the fact. The transmitter fails to get a response and the receiver sees corrupted information via the cyclic redundancy check (CRC). This topic is revisited later in the section "Multiple Access."

Multipath fading can inhibit signal reception. Multiple antennas can be used to minimize this problem via spatial separation of orthogonality. This is called *antenna diversity*.

Wireless PHYs

Spread Spectrum Most WLAN systems[3] use spread spectrum technology, a wideband radio frequency (RF) technique developed by the military for use in reliable, secure, mission-critical communications systems. Some PHYs provide only one channel, whereas others provide multiple channels; examples of channel types are shown in Table 3-2.

Spread spectrum is designed to trade off bandwidth efficiency for reliability, integrity, and security. That is, more bandwidth is consumed than in the case of narrowband transmission, but the trade-off produces a signal that is easier to detect, provided that the receiver knows the parameters of the spread spectrum signal being broadcast. If a receiver is not tuned to the correct frequency, a spread spectrum signal looks like background noise. As noted previously, there are two types of spread spectrum radio: FHSS and DSSS.

FHSS Technology A narrowband radio system transmits and receives user information on a specific RF. Narrowband radio keeps the radio sig-

Table 3-2
Channel types

Single Channel	n-channel
Narrowband RF channel	Frequency division multiplexed channels
Baseband infrared (IR)	DSSS with Code Division Multiple Access (CDMA)

nal frequency as narrow as possible just to pass the information. Undesirable crosstalk between communications channels is avoided by coordinating different users on different channel frequencies. In a radio system, privacy and noninterference are accomplished by the use of separate radio frequencies. The radio receiver filters out all radio signals except the ones on its designated frequency. FHSS uses a narrowband carrier that changes frequency in a pattern known to both the transmitter and receiver. Properly synchronized, the net effect is to maintain a single logical channel. To an unintended receiver, FHSS appears to be short-duration impulse noise (see Figure 3-4).

DHSS Technology DHSS generates a redundant bit pattern for each bit to be transmitted. This bit pattern is called a *chip* (or *chipping code*). The longer the chip, the greater the probability that the original data can be recovered (and, of course, the more bandwidth is required). Even if one or more bits in the chip are damaged during transmission, statistical techniques embedded in the radio can recover the original data without the need for retransmission. To an unintended receiver, DSSS appears as low-power wideband noise and is ignored by most narrowband receivers (see Figure 3-5).

In a DSSS system, each information bit is combined via an exclusive OR (XOR) function with a longer pseudorandom numerical sequence. The result is a high-speed digital stream that is then modulated onto a carrier frequency using differential phase-shift keying (DPSK). Receive-end

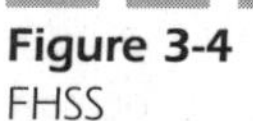

Figure 3-4
FHSS

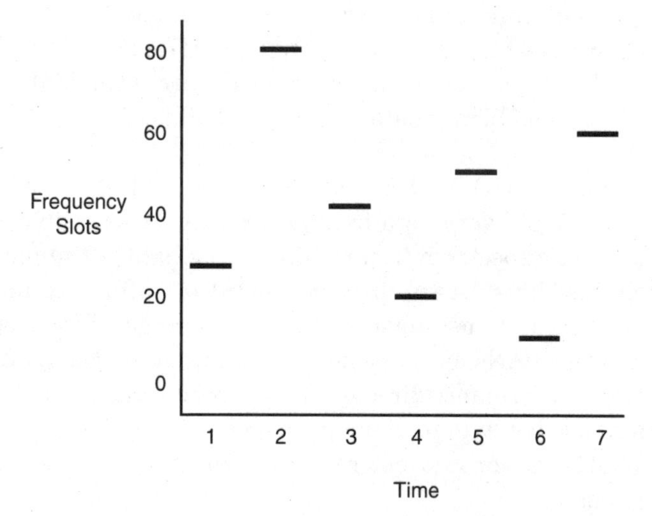

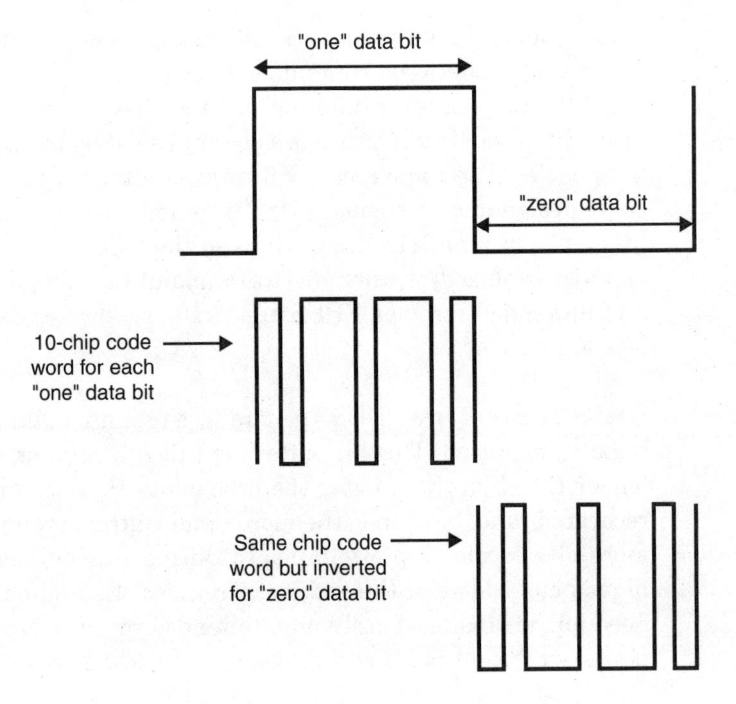

Figure 3-5
DSSS

electronics remove the PN sequence and recover the original data stream. The high-rate modulation method is called CCK. The effects of using pseudorandom numerical sequence codes to generate the spread spectrum signal are shown in Figure 3-6.[4] As illustrated in Figure 3-6a, the pseudorandom numerical sequence spreads the transmitted bandwidth of the signal and reduces the *peak* power. (Note, however, that the *total* power is unchanged.) Upon reception, the signal is correlated with the same pseudorandom numerical sequence to reject narrowband interference and recover the original binary data (see Figure 3-6b).

Infrared (IR) Technology IEEE 802.11 also specifies an IR PHY. IR systems use very high frequencies just below visible light in the electromagnetic spectrum to carry data. Like light, IR cannot penetrate opaque objects; it is either directed (LOS) or diffuse technology. Inexpensive directed systems provide very limited range (three feet) and are typically used for PANs, but occasionally are used in specific WLAN applications. High-performance directed IR is impractical for mobile users and is therefore used only to implement fixed subnetworks. Diffuse (or reflective) IR WLAN systems do not require LOS, but cells are limited to individual rooms.

Figure 3-6
DSSS signal operation

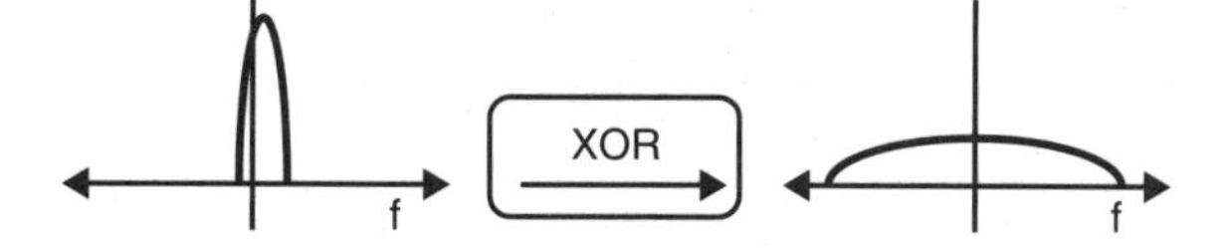

(a) Effect of PN sequence on transmit spectrum

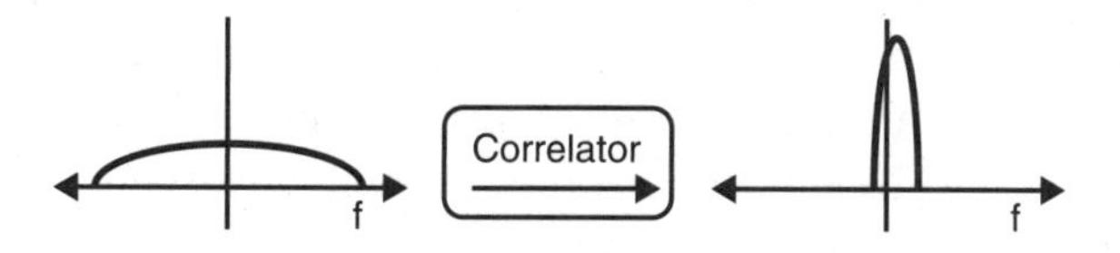

(b) Received Signal is correlated with PN to recover data

Multiple Access

The 802.11 MAC layer is based on two methods: the DCF for asynchronous contention-based access and the PCF for centralized contention-free access. The former is considered to be the default access mechanism; the latter is an optional function envisaged to support collision-free and time-bounded services.[5] DCF permits automatic medium sharing between compatible PHYs through the use of CSMA/CA and a random backoff time following a busy medium condition. In addition, all directed traffic uses an immediate positive acknowledgment (ACK) frame where retransmission is scheduled by the sender if no ACK is received.[6]

The CSMA/CA protocol is designed to reduce the collision probability between multiple stations accessing a medium at the point where collisions would most likely occur. Just after the medium becomes idle following a busy medium (as indicated by the Carrier Sense [CS] function) is when the highest probability of a collision exists. This is because multiple stations could have been waiting for the medium to become available again. This situation necessitates a random backoff procedure to resolve medium contention conflicts.

CSMA/CA mechanisms rely on *physical carrier sense*; namely, the underlying assumption is that each station can hear all other stations. CSMA/CA requires each station to listen for other users. If the channel is idle, the station is allowed to transmit; if the channel is busy, each station must wait until transmission stops and then can enter into a *backoff* procedure. The backoff randomization procedure prevents multiple stations from seizing the medium immediately after completion of the preceding transmission. Packet reception requires an acknowledgement procedure that entails the transmission of ACK frames (these frames have a higher priority than other traffic). The time interval between the completion of packet transmission and the transmission of the ACK frame is one short interframe space (SIFS). See Figure 3-7.

Data frame transmissions (other than ACKs) can occur only after at least one DCF interframe space (DIFS). More specifically, if a transmitting station senses a busy channel, it determines a random backoff period. The timer is an integer number of slot times beyond the expiration of a DIFS; that is, the timer begins to decrement after one DIFS. When the timer reaches zero, the station may begin transmission. Note that if the channel is seized by another station before the timer at the station in question reaches zero, the timer is retained at the current decremented value for a subsequent transmission.

There are situations where every station cannot hear all other stations. In this *hidden node* situation, the probability of collision has significantly increased. To address this issue, an additional carrier sense mechanism is available. The *virtual carrier sense* mechanism enables a station to reserve the channel for a specified period of time through the use of Request to Send/Clear to Send (RTS/CTS) frames. Here, Station 1 (STA-1) sends an RTS frame to the AP, but the RTS is not heard by Station 2 (STA-2). The RTS frame contains a Duration/ID field that specifies the period of time that the channel is reserved for a follow-on transmission. The reservation information is maintained in the NAV of all stations receiving the RTS frame. When the AP receives an RTS, it responds with a CTS frame that also contains a Duration/ID field that specifies the period of time for which

Figure 3-7 DCF procedure

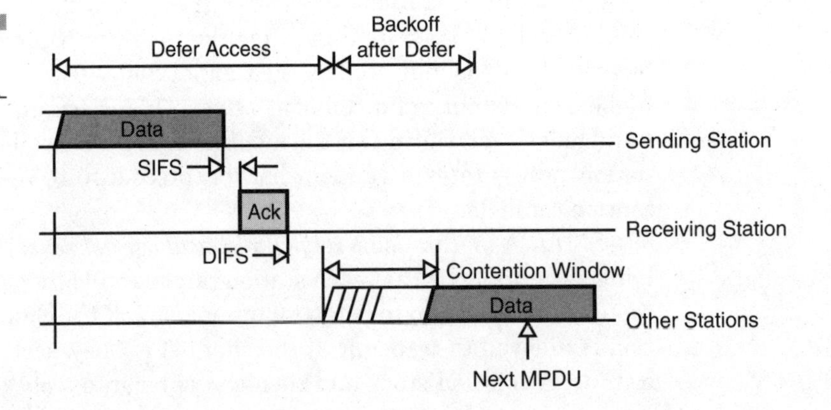

the channel is reserved. Although STA-2 did not receive the original RTS, it will be able to receive the CTS and update its NAV.

The RTS/CTS procedure is invoked based on a user-specified parameter. The procedure can always be used, never be used, or be used for packets that exceed a defined length. The RTS/CTS mechanism need not be used for every data frame transmission. Because the additional RTS and CTS frames add overhead inefficiency, the mechanism is not always justified, especially for short data frames.

Therefore, the virtual carrier sense mechanism is achieved by distributing reservation information announcing the impending use of the medium. The exchange of RTS and CTS frames prior to the actual data frame is one means of distributing this medium reservation information. The RTS and CTS frames contain a Duration/ID field, which defines the period of time that the medium is to be reserved to transmit the actual data frame and returning ACK frame. All stations within the reception range of either the originating station (which transmits the RTS) or the destination station (which transmits the CTS) can learn about the medium reservation. Thus, a station may be unable to receive from the originating station, yet it still knows about the impending use of the medium to transmit a data frame.

Another means of distributing the medium reservation information is the Duration/ID field in directed frames. This field gives the time that the medium is reserved, either to the end of the immediately following ACK or, in the case of a fragment sequence, to the end of the ACK following the next fragment.[6]

The RTS/CTS exchange also performs a type of fast collision inference and transmission path check. If the return CTS is not detected by the stations originating the RTS, the originating station may repeat the process (after observing the other medium-use rules) more quickly than if the long data frame had been transmitted and a return ACK frame had not been detected. Another advantage of the RTS/CTS mechanism can be found where multiple BSSs utilize the same channel overlap. The medium reservation mechanism works across the BSA boundaries. The RTS/CTS mechanism may also improve operation in a typical situation where all stations can receive from the AP, but cannot receive from all other stations in the BSA. The RTS/CTS mechanism cannot be used for MPDUs with broadcast and multicast immediate addresses because there are multiple destinations for the RTS and thus potentially multiple concurrent senders of the CTS in response.

The nominal peak throughput offered to the Internet Protocol (IP) layer for a maximum transmission unit (MTU) of 1,500 bytes is shown in Table 3-3 for four IEEE 802.11/11b rates.5 As the table indicates, 44 percent

Table 3-3

Throughput rates

Bit Rate (Mbps)	Nominal Throughput (Mbps)	Bit Rate (%)
11	6.2	56
5.5	3.9	71
2	1.7	85
1	0.9	90

signaling rate consumption is encountered when 11 Mbps is used, mainly due to the overhead of the preamble and physical header in IEEE 802.11, which appears quite large in relation to the payload transmission time. In order to guarantee compatibility with lower rates, these 24 bytes of PHY preamble and header overhead have to be transmitted at 1 Mbps for the four supported rates, which consumes 192 ms for each data or acknowledgment frame. Because this overhead is constant for every data packet transmitted, the loss in throughput efficiency becomes more pronounced with shorter data packets. It has been recommended to implement the short preamble and header (which lasts only half as long as the long one) defined as an optional feature within the standard.[7]

Since the IEEE 802.11b specification was finalized in 1999, the 802.11 Work Group is also developing other specifications such as 802.11a for data rates up to 54 Mbps (using orthogonal frequency division multiplexing [OFDM] at the 5 GHz band) and 802.11e for quality of service (QoS) and multimedia traffic support. In particular, the aim of these last specifications is to add some new traffic management policies and error control mechanisms (such as forward error correction [FEC] and selective retransmission) to the high-rate extensions of the 802.11 standard.

IEEE 802.11 provides for security via two methods: *authentication* and *encryption*. Authentication is the mechanism by which one station is verified to have authorization to communicate with a second station in a given coverage area; in the infrastructure mode, authentication is established between an AP and each station. IEEE 802.11 defines two subtypes of authentication service: *open system* and *shared key*. The subtype invoked is indicated in the body of authentication management frames. Therefore, authentication frames are self-identifying with respect to the authentication algorithm. All management frames of subtype authentication are unicast frames as authentication is performed between pairs of stations (that is, multicast authentication is not allowed). Management frames of subtype deauthentication are advisory and may, therefore, be sent as group-addressed frames.

A mutual authentication relationship exists between two stations following a successful authentication exchange. Authentication is used between stations and the AP in an infrastructure BSS. Authentication may be used between two stations in an IBSS. In an open system, any station may request authentication. The station receiving the request may grant authentication to any request or only those from stations on a user-defined list. In a shared key system, only stations that possess a secret encrypted key can be authenticated. Shared key authentication is only available to systems having the optional encryption capability.

Next we discuss encryption. Eavesdropping is a familiar problem that users of other types of wireless technology face. IEEE 802.11 specifies a WEP data confidentiality algorithm. WEP was designed to protect authorized users of a WLAN from casual eavesdropping. This service is intended to provide functionality for the WLAN that is equivalent to that provided by the physical security attributes inherent to a wired medium. Data confidentiality depends on an external key management service to distribute data enciphering/deciphering keys. The IEEE 802.11 standards committee specifically discourages running an IEEE 802.11 LAN with privacy but without authentication. Although this combination is possible, it leaves the system open to significant security threats. Encryption (see Figure 3-8) is intended to provide integrity and confidentiality.

The following section is reprinted material from the ANSI/IEEE 802.11 standard, 1999 Edition.[6]

The WEP feature uses the Ron's Code 4 Pseudorandom Number Generator (RC4 PRNG) algorithm from RSA Data Security, Inc. The description that follows is based directly on the IEEE 802.11 specification. Viewing from left to right in Figure 3-9, enciphering begins with a secret key that has been distributed to cooperating stations by an external key management service. WEP is a symmetric algorithm in which the same key is used for coding and decoding. The secret key is concatenated with an initialization vector (IV) and the resulting *seed* is input to a PRNG. The PRNG outputs a key sequence k of pseudorandom octets equal in length to the number of data octets that are to be transmitted in the expanded MPDU plus four (because the key sequence is used to protect the integrity check value [ICV] as well as the data). Two processes are applied to the plaintext MPDU. To protect against unauthorized data modification, an integrity algorithm operates on the plaintext P to produce an ICV.

Enciphering is then accomplished by mathematically combining the key sequence with the plaintext concatenated with the ICV. The output of the

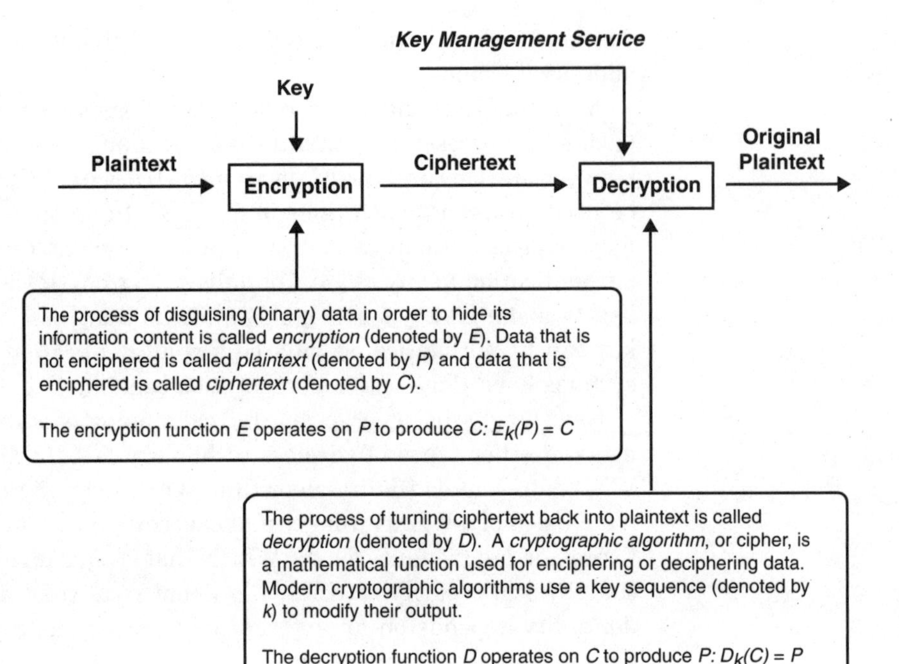

Figure 3-8 Basics of encryption

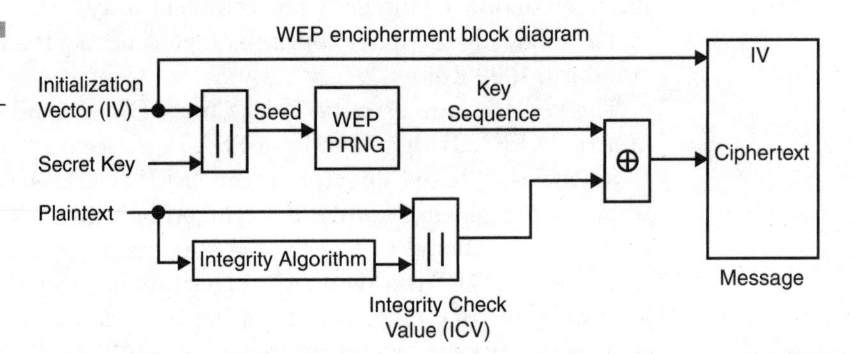

Figure 3-9 WEP

process is a message containing the IV and ciphertext. The WEP PRNG is the key component of this process because it transforms a relatively short secret key into an arbitrarily long key sequence. This simplifies the task of key distribution as only the secret key needs to be communicated between stations. The IV extends the useful lifetime of the secret key and provides the self-synchronous property of the algorithm. The secret key remains constant while the IV changes periodically. Each new IV results in a new seed and key sequence; thus, there is a one-to-one correspondence between the IV and *k*. The IV may be changed as frequently as every MPDU and because it travels with the message, the receiver will always be able to decipher any message. The IV is transmitted in the clear because it does not provide an attacker with any information about the secret key and its value must be known by the recipient in order to perform the decryption.

When choosing how often to change IV values, implementers should consider that the contents of some fields in higher-layer protocol headers as well as certain other higher-layer information are constant or highly predictable. When such information is transmitted while encrypting with a particular key and IV, an eavesdropper can readily determine portions of the key sequence generated by that (key, IV) pair. If the same (key, IV) pair is used for successive MPDUs, it may substantially reduce the degree of privacy conferred by the WEP algorithm, enabling an eavesdropper to recover a subset of the user data without any knowledge of the secret key.

Changing the IV after each MPDU is a simple method of preserving the effectiveness of WEP in this situation. The WEP algorithm is applied to the frame body of an MPDU. The (IV, frame body, ICV) triplet forms the actual data to be sent in the data frame. For WEP-protected frames, the first four octets of the frame body contain the IV field for the MPDU. The PRNG seed is 64 bits. Bits 0 through 23 of the IV correspond to bits 0 through 23 of the PRNG seed, respectively. Bits 0 through 39 of the secret key correspond to bits 24 through 63 of the PRNG seed, respectively. The numbering of the octets of the PRNG seed corresponds to that of the RC4 key. The IV is followed by the MPDU, which is followed by the ICV. The WEP ICV is 32 bits. The WEP integrity check algorithm is CRC-32. As stated previously, WEP combines *k* with *P* using bitwise XOR.

As noted in Chapter 1, WEP has been found to have holes. Chapter 4, "Security Considerations for Hotspot Services," will explore the solutions available to control them.

WPANs: A Capsule View of Bluetooth

This section provides an overview of Bluetooth and is based on openly available information from the Bluetooth Special Interest Group (SIG). The Bluetooth SIG (www.bluetooth.com) is an industry group consisting of leaders in the telecommunications, computing, and networking industries that are driving the development of the technology and bringing it to market.[9]

Overview

Bluetooth wireless technology is a low-cost, low-power, short-range radio link for mobile devices and WAN/LAN APs. It offers fast and reliable digital transmissions of both voice and data over the globally available 2.4 GHz ISM band. The Bluetooth wireless technology is set to advance the personal connectivity market by providing freedom from wired connections (see Table 3-4 for a comparison). It is a specification for a small-form-factor, low-cost radio solution providing links between mobile computers, mobile phones, and other handheld devices, and connectivity to the Internet. The Bluetooth Specification defines a short (around 10 meters) or optionally a medium-range (around 100 meters) radio link that is capable of voice or data transmission up to a maximum capacity of 720 Kbps per channel.

Products are now beginning to be delivered and Ericsson anticipated that by the 2002, tens of millions of Bluetooth wireless technology devices

Table 3-4

Comparison of Bluetooth and cabled connections

	Bluetooth	Regular Cable
Topology	Supports up to seven simultaneous links	Each link requires another cable.
Flexibility	Goes through walls, bodies, cloths, and so on	LOS or modified environment.
Data rate	1 megasamples per second (MSPS), 720 Kbps	Varies with use and cost.
Power	0.1 watts active power	0.05 watts active power or higher.
Size and weight	25 mm×13 mm×2 mm, several grams	Size is equal to range. Typically 1–2 meters. Weight varies with length (ounces to pounds).
Cost	Long-term $5 per endpoint	~$3–100 per meter (end-user cost).
Range	10 meters or less Up to 100 meters with PA	Range is equal to size. Typically 1–2 meters.
Universal	Intended to work anywhere in the world	Cables vary with local customs.
Security	Link layer security, SS radio	Secure (a cable).

will be around. Cahners-Instat projects that close to 700 million Bluetooth devices will ship annually in 2005.

The raw throughput of Bluetooth wireless technology is 1 Mbps with an actual data rate of 728 Kbps. One of the key requirements in the design of the air interface for Bluetooth wireless technology is that it must be able to operate worldwide. The only frequency band that satisfies these requirements is at 2.4 GHz at the ISM band. The ISM band is also license free and open to any radio system. To avoid interference from other devices operating in the ISM band, Bluetooth wireless technology uses FHSS to make the link robust. The radio transceiver hops to a different frequency after each transmission and reception (1,600 hops per second). The system hops among 79 frequencies at 1 MHz intervals. Up to seven simultaneous connections can be established and maintained. Devices can form piconets with up to 256 units (one master and seven slaves with the balance in standby modes). Devices automatically modify transmitting power in relation to distance and activity.

Bluetooth wireless technology is used to connect computing and telecommunication devices without the need of cables. It delivers opportunities for rapid ad hoc connections and the possibility of automatic connections between devices for synchronizing calendars and so on. It creates the possibility of using mobile data in a different way for different applications such as a cordless connection between a headset and the mobile phone, and the cordless transfer of files between two laptops. It also enables you to use the mobile phone for three functions: intercom, portable, and cellular. See Figure 3-10 for an example.

An important aspect is that Bluetooth products do not have to incorporate the entire Bluetooth specification. To get compatible products out more rapidly, the Bluetooth SIG has developed the concept of *profiles*. Profiles describe how the implementation of user models can be supported. As long as they share the same Bluetooth profiles, devices from different manufacturers can communicate anywhere in the world. A profile is defined as a combination of protocols and procedures that are used by devices to implement specific services. The profiles are described in the Bluetooth Specification. A vendor that wants its chip to interact with headsets can just support the Headset Profile; it will not need to look into compatibility tests for other profiles.[8] There are four fundamental profiles:

- Generic Access Profile (GAP Profile) for discovery and link management
- Service Discovery Application Profile (SDAP Profile) for discovering services and retrieving information

Figure 3-10 Bluetooth applications Source: Tom Siep, TI, and Chatschik Bisdikian, IBM

What does Bluetooth do for me?

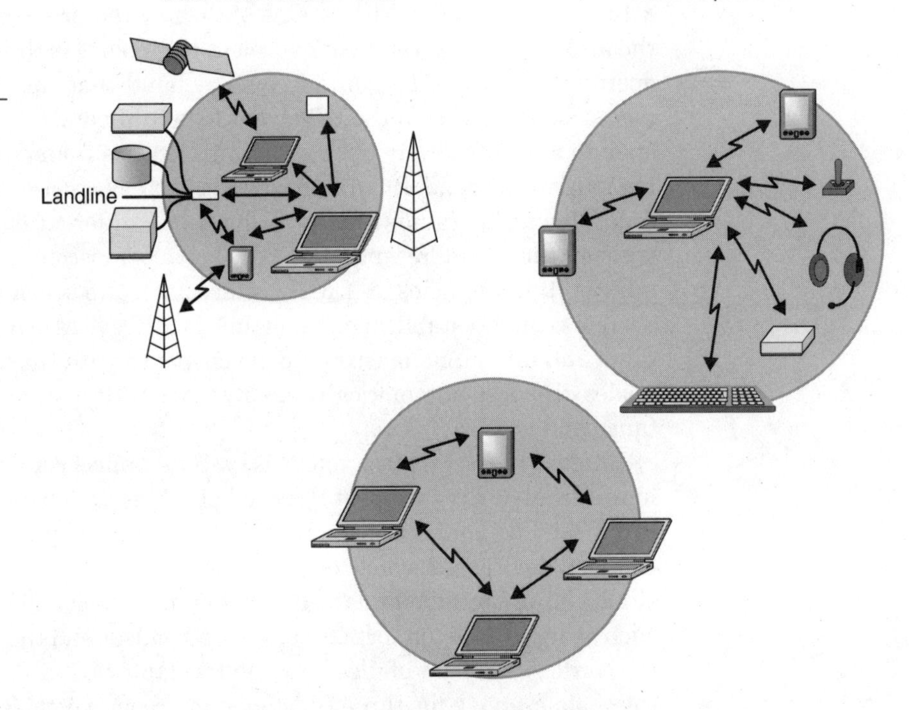

- The Bluetooth SIG (Special Interest Group) was formed in Feb. 1998 by 5 promoter companies
  - —Ericsson, IBM, Intel, Nokia, Toshiba
- The Bluetooth SIG went "public" in May 1998
- The Bluetooth SIG work (the spec: 1,600+ pages) became public in July 1999 (ver. 1.0A)
  - —Ver. 1.0B was released on December 6, 1999
  - —Ver. 1.1 was released on March 1, 2001
- The promoter group increased in December 1999 to nine
  - —Added: 3Com, Lucent, Microsoft, Motorola
- There are 2,164 adopters (as of 3/15/2001)
  - —Adopters enjoy royalty free use of the Bluetooth technology

- Serial Port Profile (SPP Profile) for emulating serial cable connections
- Generic Object Exchange Profile (GOEP Profile) for synchronization, file transfer, or object push

There are nine usage profiles:

- Cordless Telephone Profile (CTP Profile) for telephony features
- Intercom Profile (IP Profile) for intercom functionality (also referred to as the *walkie-talkie usage*)
- Headset Profile (HS Profile)
- Dial-up Networking Profile (DNP Profile) for modems, mobile phones, and so on
- Fax Profile (FP Profile)
- LAN Access Profile (LAP Profile) for LAN access using the Point-to-Point Protocol (PPP)
- Object Push Profile (OPP Profile) for the push/pull of data objects between devices
- File Transfer Profile (FTP Profile)
- Synchronization Profile (SP Profile)

Bluetooth wireless technology has built-in encryption and authentication and thus is seen as being secure in any environment. In addition, as noted, a frequency-hopping scheme with 1,600 hops per second is employed. All of this, together with an automatic output power adaptation to reduce the range exactly to the requirement, makes the system difficult to eavesdrop, according to proponents.

Bluetooth wireless technology is a layer 1 and 2 standard for wireless communication between devices. However, this should not be confused with Wireless Application Protocol (WAP). WAP is an application-oriented standard for communication. WAP does not deal with the lower layers of the Open Systems Interconnection (OSI) model like Bluetooth wireless technology does. WAP can be successfully used in combination with Bluetooth wireless technology in e-commerce applications, for example. WAP is discussed in detail in Chapter 7, "Wireless Application Protocol (WAP)."

The following sections from "Bluetooth Technical Summary" through "Licensing Technologies" (except for "Constituent Products") have been reprinted from the "Beginner's Guide" from Ericsson's web site at www.ericsson.com/bluetooth/training.[9]

Bluetooth Technical Summary

Harald Bluetooth was a Viking and the king of Denmark between 940 and 981. One of his skills was to make people talk to one another, and during his rule, Denmark and Norway were united. The idea of easy proximate communication is the reason why proponents use the term Bluetooth. Today Bluetooth wireless technology enables people to talk to each other; this time by means of a low-cost, short-range radio link.

As a global standard, Bluetooth can

- Eliminate wires and cables between both stationary and mobile devices.
- Facilitate both data and voice communication.
- Support ad hoc networks and synchronize personal devices.

The Bluetooth wireless technology comprises hardware, software, and interoperability requirements. It has been adopted by major players not only in the telecom, computer, and home entertainment industry, but also in such diverse areas as the automotive industry, healthcare, automation, toys, and so on. The idea that resulted in the Bluetooth wireless technology was born in 1994 when Ericsson Mobile Communications decided to investigate the feasibility of a low-power, low-cost radio interface between mobile phones and their accessories. The idea was that a small radio built into both the cellular telephone and the laptop would replace the cumbersome cable used today to connect the two devices.

A year later the engineering work began and the potential of the technology began to crystallize. However, in addition to untethering devices by replacing cables, the radio technology created the possibility of becoming a bridge to existing data networks, a peripheral interface, and a mechanism to form small private ad hoc groupings of connected devices away from fixed network infrastructures. In other words, it supports WPAN connectivity.

The Bluetooth SIG was formed in early 1988. Today the Bluetooth SIG includes promoter companies such as 3Com, Ericsson, IBM, Intel, Lucent, Microsoft, Motorola, Nokia, Toshiba, and thousands of Adopter/Associate member companies. The assignment of the SIG originally was to monitor the technical development of short-range radio and create an open global standard, thus preventing the technology from becoming the property of a single company. This work resulted in the release of the first Bluetooth Specification in mid-1999. The further development of the specification still is one of the main tasks for the SIG. Other important tasks are meeting interoperability requirements, harmonizing the frequency band, and promoting the technology. As noted earlier in the book, the IEEE and Internet Engineering Task Force (IETF) are also becoming involved.

From the very start, one of the main goals for the SIG has been to include a regulatory framework in the specification that will guarantee interoperability between different devices from various manufacturers—as long as they share the same profile. Although the usage models describe applications and intended devices, the profiles specify how to use the Bluetooth protocol stack for an interoperable solution. Each profile states how to reduce options and set parameters in the base standard and how to use procedures from several base standards. A common user experience is also defined. For example, a computer mouse does not need to communicate with a headset, so they are built to comply with different profiles.

Compliance

The Bluetooth Qualification Program guarantees global interoperability between devices regardless of the vendor and the country in which they are used. During the test procedure that all devices must pass, it must be verified that the devices meet all of the requirements regarding radio link quality, lower-layer protocols, profiles, and information to end users. The profiles are a part of the Bluetooth Specification, and all devices must be tested against one or more of the profiles in order to fulfill the Bluetooth certification requirements. The number of profiles will continue to grow as new Bluetooth applications arise. All qualified devices are listed at the SIG official web site at www.bluetooth.com. The profiles defined in version 1 of the Bluetooth Specification mainly address usage models concerning the telecom and computing industries. Three examples are the Internet Bridge, the Ultimate Headset, and the Automatic Synchronizer profiles.

An Internet Bridge profile providing constant access to the Internet is a useful and time-saving feature, especially when the bandwidth of mobile phones is increasing rapidly. Bluetooth wireless technology lets one access the Internet without any cable connections wherever one is, either by using a computer or the phone itself. When close to a wire-bound connection point, the mobile computer or handheld device can also connect directly to the landline, but it still connects without cables. This is the localized hotspot described in Chapter 1 (larger footprint and/or outdoor hotspots must rely on other technologies, such as, IEEE 802.11b).

The Ultimate Headset profile lets users operate their mobile phones even if they're in a briefcase, thereby freeing the user's hands for more important tasks when at the office or in the car.

The Automatic Synchronization profile, which synchronizes calendars, address books, and so on, is a feature that was long awaited by many professionals. Simply by entering a user's office, the calendar in the phone or PDA will be automatically updated to agree with the one in the desktop PC, or vice versa. Phone numbers and addresses will always be correct in all portable devices without docking through cables or IR.

Constituent Products

Many companies have declared that Bluetooth wireless technology will be incorporated into their products, especially when components become cheaper. In a forecast made by Cahners In-Stat Group in July 2000, the product availability was defined as appearing in three waves.

The first wave was planned to occur around the turn of 2001 and was expected to include products like the following:

- Adapters for mobile phones and adapters (dongles) and PC Cards for notebooks and PCs.
- High-end mobile phones and notebook PCs with integrated Bluetooth communication for the business users.
- Bluetooth headsets were expected to enter the market by the first half of 2002.
- Cordless phones, handheld PCs, and PDAs will also be included in this first wave. The first handheld PCs and PDAs were expected to enter the market during 2002.

The second wave will in many respects overlap the first:

- PCs with Bluetooth circuitry on the motherboard will appear.
- Printers, fax machines, digital still cameras, and products for industrial/medical and vertical industries will also begin to move in the second wave.
- Some industrial solutions were planned to become available in the first quarter of 2002.
- In the automotive sector, the first Bluetooth options were expected to appear for the 2003 model year (hands-free mobile phone usage with the regular mobile phone).

The third wave will add the following:

- Low-cost mobile phones
- Lower-cost portable devices and desktop PCs

Countless electronic devices for home, personal, and business use have been brought to the market during recent years, but there has been no widespread technology to address the needs of connecting personal devices in PANs. The demand for a system that could easily connect devices for transferring data and voice over short distances without cables has grown stronger. Bluetooth wireless technology aims to fill this important communication need with its capability to communicate both voice and data wirelessly using a standard low-power, low-cost technology that can be integrated in all devices to enable complete mobility. The price is expected to be low and result in mass production.

The Technology

The Bluetooth Specification defines a short (around 10 meters) or optionally a medium-range (around 100 meters) radio link that is capable of voice or data transmission up to a maximum capacity of 720 Kbps per channel. RF operation is in the unlicensed ISM band at 2.4 to 2.48 GHz and uses a spread spectrum, frequency-hopping, full-duplex signal at up to 1,600 hops per second. The signal hops among 79 frequencies at 1 MHz intervals to give a high degree of interference immunity. RF output is specified as 0 dBm (1 mW) in the 10-meter-range version and −30 to +20 dBm (100 mW) in the longer-range version. When producing the radio specification, a high emphasis was put on making a design enabling single-chip implementation in complementary metal oxide semiconductor (CMOS) circuits, thereby reducing cost, power consumption, and the chip size required for implementation in mobile devices.

Voice

Up to three simultaneous synchronous voice channels are used, or a channel that simultaneously supports asynchronous data and synchronous voice. Each voice channel supports a 64 Kbps synchronous (voice) channel in each direction.

Data

The asynchronous data channel can support maximal 723.2 Kbps asymmetric (and still up to 57.6 Kbps in the return direction) or 433.9 Kbps symmetric.

- A master can share an asynchronous channel with up to seven simultaneously active slaves in a piconet.
- By swapping active and parked slaves out respectively in the piconet, 255 slaves can be virtually connected using the PM_ADDR. A device can participate again within 2 milliseconds.
- To park even more slaves, the BD_ADDR can be used. There is no limitation to the number of slaves that can be parked.

Slaves can participate in different piconets and a master of one piconet can be the slave in another. This is known as a *scatternet*. Up to 10 piconets within range can form a scatternet with few collisions.

Network Topology

Bluetooth units that come within range of each other can set up ad hoc point-to-point and/or point-to-multipoint connections. Units can dynamically be added or disconnected to the network. Two or more Bluetooth units that share a channel form a piconet. Several piconets can be established and linked together in ad hoc scatternets to enable communication and data exchange in flexible configurations. If several other piconets are within range, they each work independently and each have access to full bandwidth. Each piconet is established by a different frequency-hopping channel. All users participating on the same piconet are synchronized to this channel. Unlike IR devices, Bluetooth units are not limited to LOS communication.

To regulate traffic on the channel, one of the participating units becomes a master of the piconet, whereas all other units become slaves. With the current Bluetooth Specification, up to seven slaves can actively communicate with one master. However, there can be almost an unlimited number of units virtually attached to a master, enabling communication to start instantly.

Security

Because radio signals can be easily intercepted, Bluetooth devices have built-in security to prevent eavesdropping or falsifying the origin of messages (spoofing). The main security features include the following:

- **Challenge-response routine** For authentication, which prevents spoofing and unwanted access to critical data and functions.
- **Stream cipher** For encryption, which prevents eavesdropping and maintains link privacy.
- **Session key generation** Session keys can be changed at any time during a connection.

Three entities are used in the security algorithms:

- **Bluetooth device address (BD_ADDR) (48 bits)** A public entity that is unique for each device. The address can be obtained through the inquiry procedure.
- **Private user key (128 bits)** A secret entity. The private key is derived during initialization and is never disclosed.
- **Random number (128 bits)** Different for each new transaction. The random number is derived from a pseudorandom process in the Bluetooth unit.

In addition to these link-level functions, frequency hopping and the limited transmission range also help to prevent eavesdropping.

Hardware Architecture

The Bluetooth hardware consists of an analog radio part and a digital part —the Host Controller. The Host Controller has a hardware digital signal processing part called the Link Controller, a central processing unit (CPU) core, and interfaces to the host environment.

The Link Controller consists of hardware that performs baseband processing and PHY protocols such as the Automatic Repeat Request (ARQ) protocol and FEC coding. The function of the Link Controller includes asynchronous transfers, synchronous transfers, audio coding, and encryption.

The CPU core enables the Bluetooth module to handle inquiries and filter page requests without involving the host device. The Host Controller can be programmed to answer certain page messages and authenticate remote links.

The Link Manager software runs on the CPU core. The Link Manager discovers other Link Managers and communicates with them via the Link Manager Protocol (LMP) to perform its service provider role and to use the services of the underlying Link Controller.

Software Architecture

In order to make different hardware implementations compatible, Bluetooth devices use the Host Controller Interface (HCI) as a common interface between the Bluetooth host (for example, a portable PC) and the Bluetooth core.

Higher-level protocols like the Service Discovery Protocol (SDP), Radio Frequency Communications (RFCOMM)[10] (emulating a serial port connection), and the Telephony Control Protocol are interfaced to baseband services via the Logical Link Control and Adaptation Protocol (L2CAP). L2CAP takes care of segmentation and reassembly (SAR) to enable larger data packets to be carried over a Bluetooth baseband connection.

The SDP enables applications to find out about available services and their characteristics when devices are moved or switched off.

Competing Technologies

There is no single WPAN competitor covering the entire concept of the Bluetooth wireless technology, but in certain market segments, other technologies exist. For cable replacement, the IR standard IrDA has been around for some years and is quite well known and widespread. IrDA is faster than the Bluetooth wireless technology, but is limited to point-to-point connections and above all it requires a clear LOS. In the past, IrDA has had problems with incompatible standard implementations—a lesson that the Bluetooth SIG has had to learn.

Licensing Technologies

Ericsson initially developed Bluetooth wireless technology. Now they license the Bluetooth chip solutions through packages of intellectual property, which comprises hardware and software in the form of the Bluetooth Core Product, the Bluetooth Host Stack, and the Bluetooth Radio. Thus, Ericsson is able to deliver total Bluetooth solutions to promote faster product time to market. Bluetooth intellectual property from Ericsson that can be used in products is aimed at three prime customer groups:

- High-volume silicon manufacturers/vendors
- Other equipment manufacturers (OEMs) that are able to develop and implement their own solutions for Bluetooth devices

- OEMs that require total Bluetooth solutions to match their product demands

Samples of baseband application-specific integrated circuits (ASICs) are currently available from Philips Semiconductors. Samples of RF modules are currently available from Ericsson Microelectronics. Volume production was planned for 2002.

WWAN Approaches

This section provides some additional details on WWAN technologies. Up to now, these architectures have focused on voice services or at most low-speed (for example, 9.6 Kbps) circuit-mode data. The plans for the future are to add higher-speed data services. Hotspot networks continue to be best served by WLANs and WPANs for the next two to three years rather than by WWANs.

Current Baseline

As already noted, the major cellular architectures now in place encompass the following:

- Time Division Multiple Access (TDMA) (IS-136)
- cdmaOne (see Figure 3-11)
- Global System for Mobile Communications (GSM) with General Packet Radio Services (GPRS) (see Figures 3-12 and 3-13)

These architectures have rather limited support for data services. Evolving architectures include

- **CDMA2000** Evolving third-generation (3G) in the United States (see Figures 3-14 and 3-15)
- **Wideband CDMA (W-CDMA)** Evolving 3G in the rest of the world

Table 3-5 summarizes some key capabilities of first-generation (1G), second-generation (2G), and 3G WWAN systems.

Time Division Multiple Access (TDMA)

In the early 1990s, the Telecommunications Industry Association/Electronics Industry Association (TIA/EIA) developed a TDMA standard that

Figure 3-11 Circuit-mode data over cellular phone

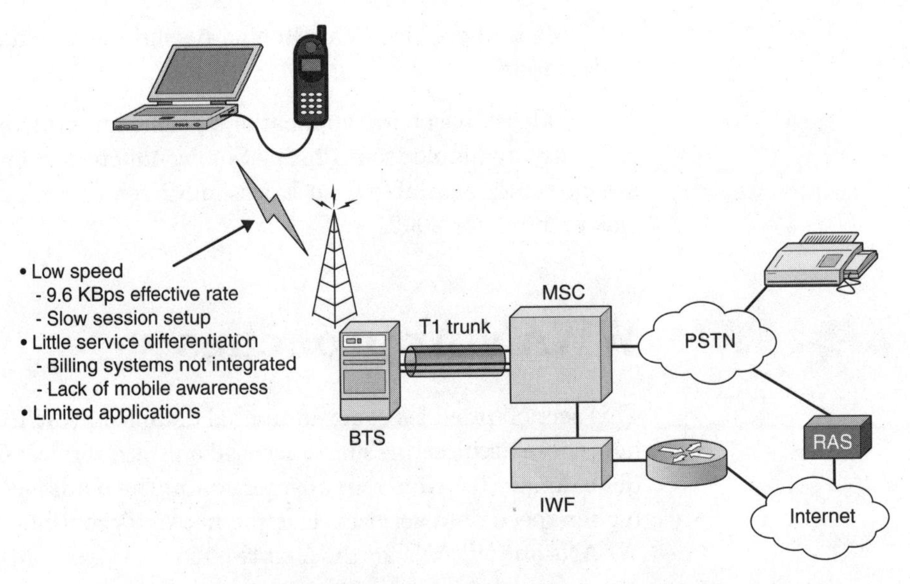

Present-day CDMA systems in the United States are known as cdmaOne and are based on IS-95 standards. Spectral allocation is 50 MHz for cellular and 120 MHz for PCS.

became known as IS-54 (1992). Systems based on this technology were introduced in the United States in 1993. This standard was eventually superceded by a newer version called TIA/EIA IS-136.1 and IS-136.2 in 1994. With the standard, a TDMA frame is 40 milliseconds long and consists of six time slots (each, therefore, are 6.67 milliseconds). Three callers can be accommodated in a 30 kHz band. For IS-136, the spectrum allocation is 824 to 849 MHz in the inbound link and 869 to 894 MHz in the outbound link. The approach, therefore, is to use TDMA over each FDMA 30 kHz channel.

cdmaOne

The feasibility of a commercial CDMA-based system based on code division multiplexing techniques that had been used in military communications was demonstrated in 1998. cdmaOne is based on TIA/EIA specifications IS-95A and IS-95B; cellular applications operate at 850 MHz (50 MHz bandwidth) and personal communications service (PCS) applications operate at 1,800 MHz (120 MHz bandwidth). Each caller is assigned a unique

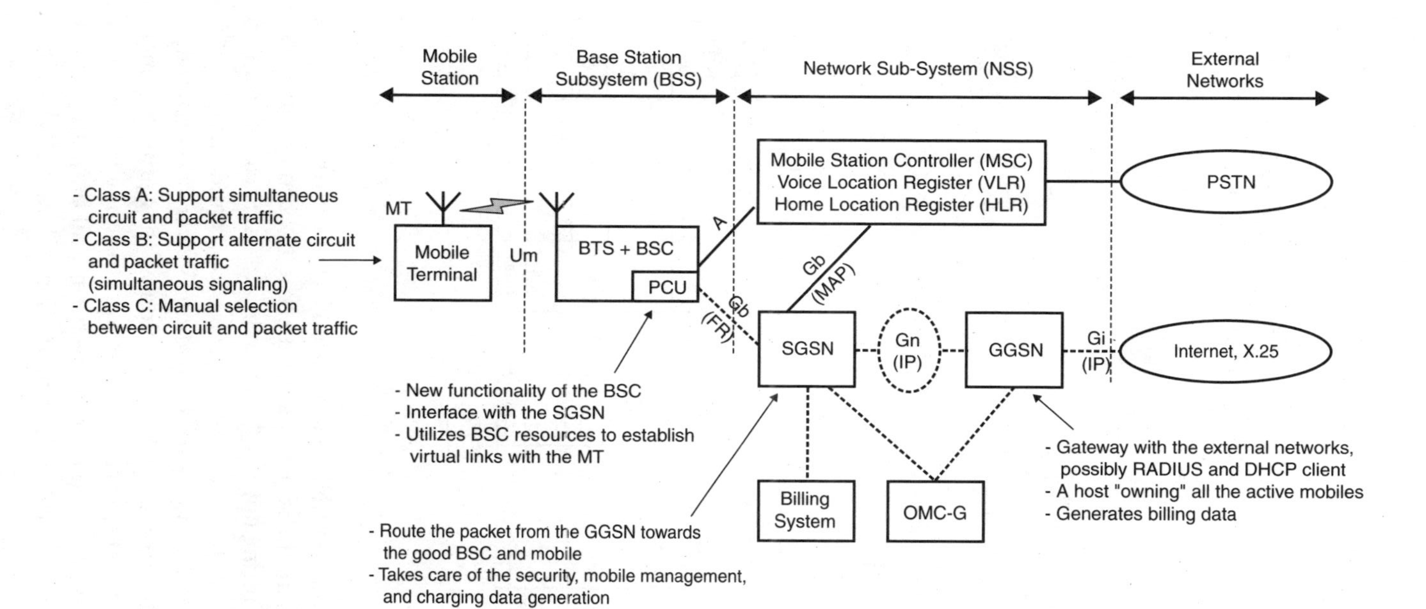

Figure 3-12 *GPRS architecture*

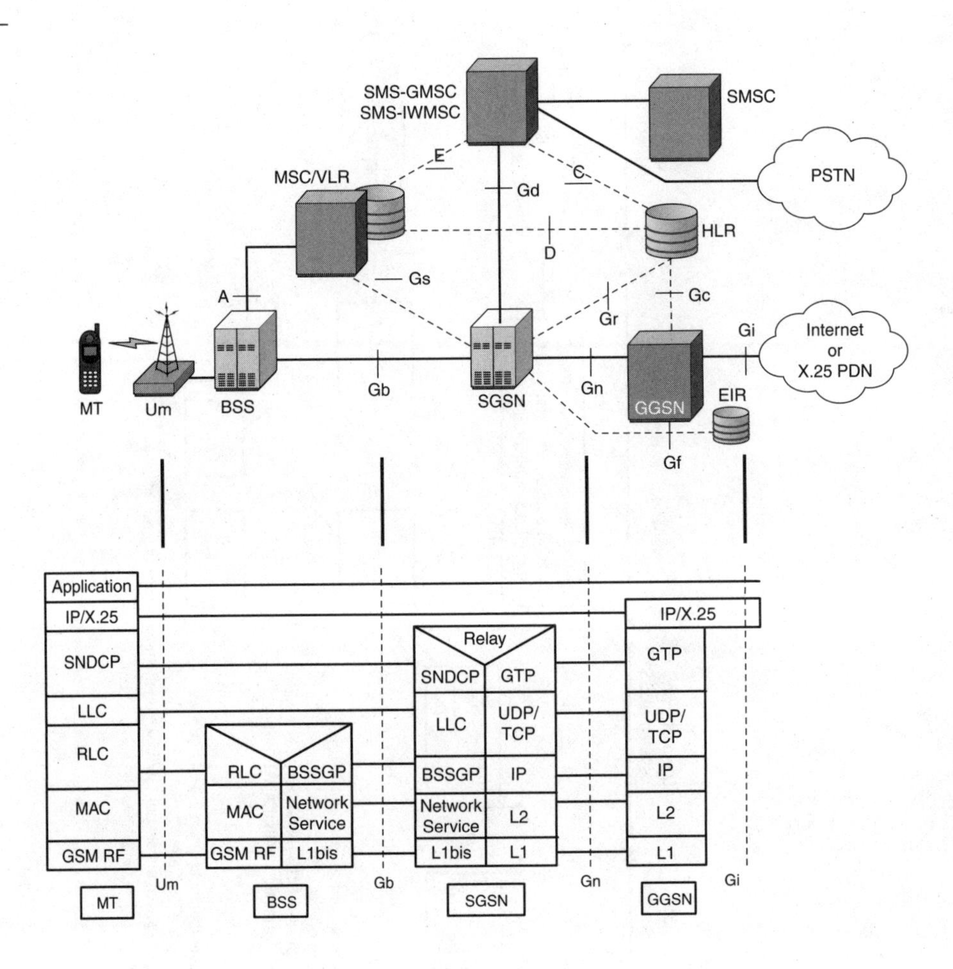

Figure 3-13
GPRS protocol stack

pseudonoise (PN) code whose clock rate (that is, the chip rate) is much higher than the user data rate. The PN code modulates the user data and the resulting output phase modulates a carrier. The spectrum is partitioned into a number of channels, each with a bandwidth of 1.25 MHz. CDMA systems have a number of advantages over other systems. They can provide much larger bandwidth per channel, they provide a more robust communication of data services employing efficient channel coding, they make more efficient utilization of the spectrum, and they support better speech quality using low-bit-rate linear predictive coders.

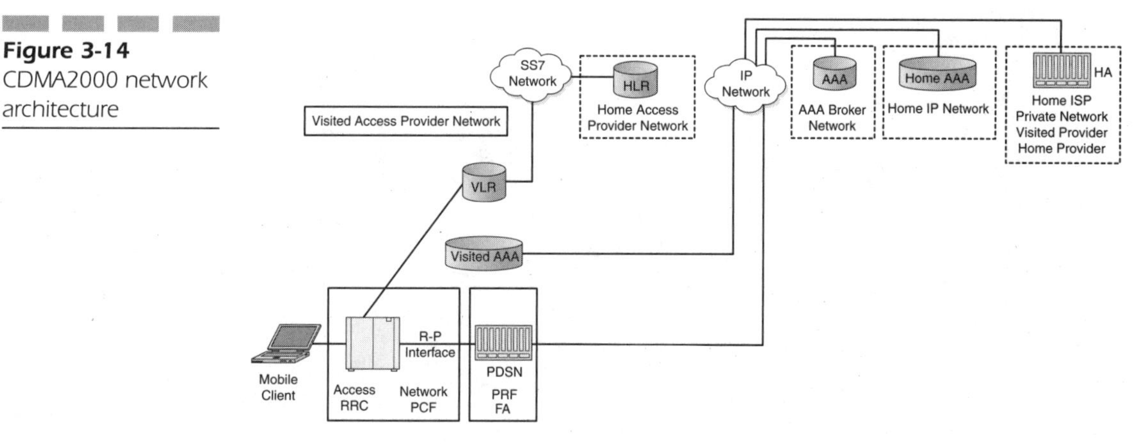

Figure 3-14 CDMA2000 network architecture

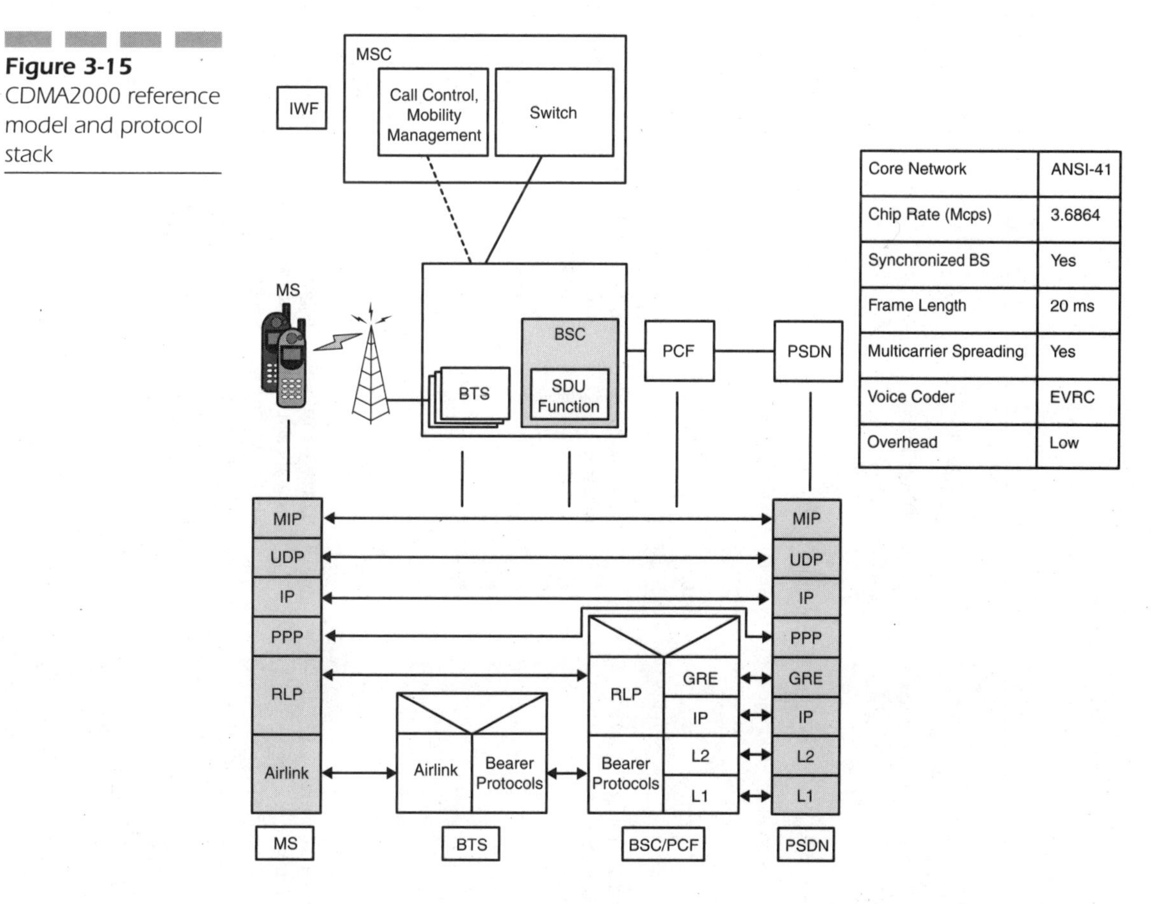

Core Network	ANSI-41
Chip Rate (Mcps)	3.6864
Synchronized BS	Yes
Frame Length	20 ms
Multicarrier Spreading	Yes
Voice Coder	EVRC
Overhead	Low

Figure 3-15 CDMA2000 reference model and protocol stack

Table 3-5
Key capabilities of 1G, 2G, and 3G systems

System	Original Analog Cellular System	IS-136 (Originally IS-54)	GSM	cdmaONE	W-CDMA	CDMA2000
Generation	1G	1G	2G	2G	3G	3G
Multiple access scheme	Frequency Division Multiple Access (FDMA) with frequency division duplex (FDD)	TDMA	TDMA	CDMA and FDD	FDD and time division duplex (TDD)	FDD
Spectrum range and channel bandwidth	30 kHz	824–849 MHz inbound 869–894 MHz outbound 30 kHz	890–915 MHz inbound 935–960 MHz outbound 200 kHz	Cellular CDMA: 824–849 MHz inbound 869–894 MHz outbound PCS CDMA: 1,850–1,910 MHz inbound 1,930–1,990 MHz outbound 1.23 MHz	FDD mode: 1,920–1,980 MHz inbound 2,110–2,170 MHz outbound TDD mode: 1,900–1,920 MHz inbound 2,010–2,025 MHz outbound 5 MHz	1,850–1,910 MHz inbound 1,930–1,990 MHz outbound 1.25×N MHz. In Phase 1, the value of N may be 1, 2, or 3. For later phases, the value of N could be 6, 9, or 12.
Number of users per channel	1	Three for full-rate speech and six for half-rate speech. There are six time slots per frame.	Eight for full-rate speech	About 16		
RF channel modulation data rate	10 Kbps (only for control)	48.6 Kbps	270.8333 Kbps	1.2288 Mbps		

Table 3-5 cont.
Key capabilities of 1G, 2G, and 3G systems

System	Original Analog Cellular System	IS-136 (Originally IS-54)	GSM	cdmaONE	W-CDMA	CDMA2000
Speech handling	Analog (300–3,000 Hz)	Vector sum-excited linear predictive (VSELP) coder at 7.95 Kbps with 159 bits per 20 ms frame.	Regular pulse excitation with long-term predictor (RPE-LTP) at 13 Kbps for full-rate coding	Code-excited linear predictive (CELP) coder—1.2, 2.4, 4.8, and 9.6 Kbps for cellular IS-95 and CELP. PCS-based IS-95 uses 14.4 Kbps.	Adaptive multirate rate (AMR) coding	AMR coding
Modulation	Frequency modulation (FM) for speech and frequency-shift keying (FSK) for data	$\pi/4$-Shifted differential quadrature phase-shift keying (DQPSK)	GMSK	Quadrature phase-shift keying (QPSK) and offset quadrature phase-shift keying (OQPSK)	QPSK	QPSK
User data capabilities	None	Basic capabilities, such as short messages on a dedicated control channel (DCCH)	Packet data 300–9.6 Kbps (with GPRS), Short Messaging Service (SMS), and circuit-switched data	Packet data at 9.6 and 14.4 Kbps. IS-95B supports higher data rates in increments of 8 Kbps.	Circuit mode—up to 144 Kbps, 384 Kbps, and 2.048 Mbps Packet-mode—144 Kbps, 384 Kbps, and 2.048 Mbps	144 Kbps, 384 Kbps, and 2.048 Mbps

Global System for Mobile Communications (GSM)

GSM was developed in 1990s for European digital cellular communications. GSM-based systems were first deployed in 18 European countries in 1991. By the end of 1993, it was adopted in 9 additional countries of Europe, as well as South America, Australia, and much of Asia. Some deployment also exists in the United States. For example, AT&T Wireless has been converting its network to GSM and the conversion was 60 percent complete by the end of 2002 (the entire network will then run on GSM, though AT&T will keep its older network operational so that customers are not forced to replace their cell phones.) Cingular and VoiceStream also operate on GSM technology. GSM technology is seeing rapid deployment in many parts of the world and is evolving to provide advanced features and data communications. GSM combines FDMA and TDMA access schemes and uses two frequency bands in the 900 MHz area. The reverse link operates at 890 to 915 MHz and the forward link operates at 935 to 960 MHz. Each channel has a bandwidth of 200 kHz and is comprised of eight time slots, each of which is assigned to an individual user.[11] Speech is digitally encoded at 13 Kbps using linear predictive coding (LPC); information is transmitted in frames, which are each 4.615 milliseconds long and divided into eight equal time slots. Table 3-6 identifies the key capabilities of GSM. It supports voice, circuit-switched data, and SMS. MTs can transmit or receive short messages during both idle and active call states. A message can contain up to 160

Table 3-6

Capabilities of GSM

Service	Features
Teleservices	Mobile telephony with interworking with Public Switched Telephone Networks (PSTNs), emergency calling, and voice messaging.
Bearer services	Data services and short messaging services. Data services are either circuit mode or packet mode. In circuit-mode data, the Mobile Switching Center (MSC) may be connected to a circuit-switched PSTN via a modem and have speeds up to 9.6 Kbps. In packet-mode data, the MSC is connected to the Internet or a public data network (via a packet assembler/diassembler device [PAD]) and connectivity is supported at 2.4, 4.8, and 9.6 Kbps. Each physical channel is shared by multiple users. Users can request a QoS-based service from the network; however, only a limited number of QoS profiles are supported.
Supplementary services	Call forwarding, call hold, call waiting, call transfer, calling number identification, three-party conference calls, and so on.

alphanumeric characters; shorter messages of 93 characters can be broadcast repetitively by a base station to all mobiles in a serving area.

GSM supports circuit-switched data services at rates up to 9.6 Kbps (it can support a maximum of 76.8 Kbps data rate by inverse-multiplexing eight channels). The development of standards for providing a packet-mode data service in GSM started in 1994 and was completed in 1997. The new system specified by these standards is called GPRS. GPRS supports a new set of bearer services for GSM. It provides packet-mode transmission within the Public Land Mobile Network (PLMN) environment and allows interworking with Transmission Control Protocol/Internet Protocol (TCP/IP) networks (such as the Internet) and with X.25 networks. GPRS

Table 3-7

Key GPRS protocols

Interface	Protocol	Notes
Base station subsystem/mobile terminal (BSS/MT)	Radio Layer Control (RLC)	Defines procedures for the retransmission of unsuccessfully delivered data units
	MAC	Defines procedures for multiple MTs to share a common transmission link (several MTs on one transmission or one MT to several transmissions)
Gateway GPRS Support Node/ Serving GPRS Service Node (GGSN/SGSN) (Gn)	IP/X.25	Interworks with external data network
	GPRS Tunneling Protocol (GTP)	Tunnels from/to entry GGSN to/from SGSN supporting the MT
	User Datagram Protocol (UDP)/TCP	Transports data and signaling information in the GPRS environment
SGSN/BSS/MT	Subnetwork Data Convergence Protocol (SNDCP)	Supports non-GSM layer 2 services, the compression of headers, and SAR
	LLC	Supports non-GSM layer 1 services and logical link connections between SGSN and MT
	Base Station System GPRS Protocol (BSSGP)	Provisioning at the radio level
	Network service	Frame relay

also supports QoS as defined by metrics such as priority (precedence), delay, throughput, and reliability/availability. Table 3-7 depicts the key GPRS protocols and interfaces.

GPRS aims at making efficient use of valuable air interface resources. In a radio environment, it makes perfect sense to be efficient in transmission; many times designers go out of their way to be super efficient on terabit-per-second optical fiber systems, where it makes no sense to trade bandwidth for complexity—in wireless systems, however, this design criterion is legitimate. GPRS uses excess voice capacity for data transmissions. Unlike standard TDMA systems, with GPRS, radio resources are utilized only when the user is actually sending or receiving data. This approach clearly lowers the cost of data transmission as measured in cost per bit. This also enables the operators to charge fees based on the packets transmitted rather than on air time, if they so choose. This also provides a simpler environment for application development and deployment.

Basic applications include access to the Internet or an intranet. Customers can be given cost-effective access to WAP-based applications. The goal of carriers is to increase the average revenue per user (ARPU). (Vodafone reportedly increased ARPU by 25 percent over five years by offering wireless Internet applications.) Services will be introduced in two phases. Phase 1 will support point-to-point services, interworking with TCP/IP, and X.25 services between the MT and the GGSN (using X.28 between the MT and the GGSN) and between the GGSN and external X.25 networks. Phase 2 is expected to support point-to-multipoint multicast/broadcast services.

Table 3-8 identifies some key elements of today's data WWAN networks. Figure 3-16 shows a typical 2G network, specifically of GSM. The air interface is known as reference point Um; it exists between a mobile station and a base transceiver station (BTS). Each cell is served by a BTS that consists of one or more radio receivers and transmitters. Base station controllers (BSCs) perform radio control functions; this includes power control as received from the mobile station and potential handoff to another cell. Each BSC can connect to one or more BTSs over the A-bis interface. There may be one or more BSCs in a serving area. BTSs and the associated BSCs for a given area are known as a base station subsystem (BSS). (Note that BSS does not stand for basic service set in this context.)

The MSC is responsible for call controls, call routing to/from PSTNs, call switching, and call handover. The MSC connects to a BSS over the A interface. The MSC interconnects with a number of other systems, namely with the Visitor Location Register (VLR), the Home Location Register (HLR), the Equipment Identity Register (EIR), and the Operations and Maintenance Center (OMC). The MSC also makes use of the Authentication Center (AuC), which is associated with the HLR. The HLR is a database system of

Table 3-8

Key elements of today's data WWAN networks

AuC	Authentication Center—Retains information for the authentication of mobile users and for encrypting the data (and voice) session.
BSC	Base station controller—It makes up the guts of a base station. It controls the radio equipment in the BSS.
BTS	Base transceiver station—It is the radio portion of the BSS.
GMSC	Gateway Mobile Switching Center—Supports switching between the wireline and the wireless networks.
HLR	Home Location Register—A system providing subscriber profiles and authorization information in the wireless network.
MS	Mobile subscriber
MSC	Mobile Switching Center—The (classical) (voicecentric) circuit switch that connects the mobile radio access to the PSTN.
SCP	Service control point—The CCSS7 system that contains the information required to complete some kinds of calls (for example, 800/toll free or Advanced Intelligent Network [AIN]).
SMS-C	Short Message Service Center—A system to support SMS, which is a globally accepted wireless capability that enables the transmission of messages between MSs and external systems (such as v-mail, e-mail, paging, and so on).
VLR	Visitor Location Register—A database that contains temporary information about MSs that is needed by the MSC in order to support visiting subscribers.
BSS	Base station subsystem—It includes the BTS and BSC.
PSTN	Public Switched Telephone Network—The existing wireline telephone network.

all mobile subscribers who are registered in a PLMN. (The HLR may be implemented in a distributed fashion.) The VLR contains the database of all mobile subscribers who are visiting a specified serving area.

Whenever a mobile station roams into a foreign serving area, the VLR[12] of the visited system requests the database of that mobile station from its HLR and saves this information in its memory so that it can serve that mobile as long as it is in that area. At the same time, the MSC of this foreign serving area instructs the HLR of the home system with regards to the location of this mobile station so that the home area of this subscriber can route calls to it when necessary. The AuC authenticates the user at call inception time; the AuC also stores appropriate authentication parameters.

The EIR contains the International Mobile station Equipment Identity (IMEI) numbers of all MSs that are registered. Each mobile station is assigned a unique IMEI number that can be used to determine whether the equipment is legitimate.

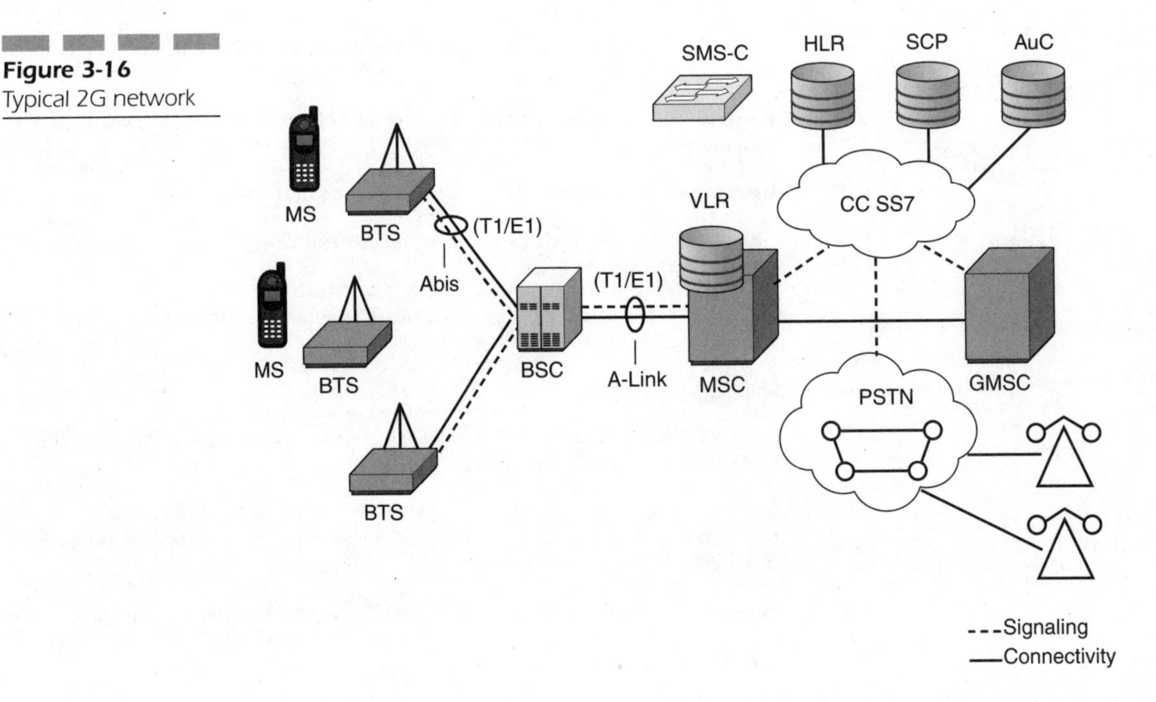

Figure 3-16 Typical 2G network

The OMC is a centralized network management system that provides the capability of remote system administration and maintenance. The Mobile Application Part (MAP) protocol of Common Channel Signaling System 7 (CCSS7) provides signaling between an MSC, the VLR, the HLR, and the AuC.

GPRS requires two new network elements: an SGSN and GGSN. The SGSN provides GPRS services to a mobile station in the serving area of its associated MSC (when there are multiple SGSNs, they are connected together over an IP-based Gn interface). An SGSN node locates MSs that subscribe to GPRS services and adds this information to the HLR. An SGSN connects to a GGSN via a Gn interface and to its BSS over a Gb interface (this uses a frame relay protocol at layer 2). The GGSN provides an interface between a GPRS network and any external network. The GGSN contains the routing information of all of the mobile stations attached to it and forwards an incoming packet appropriately en route to its destination. In the 2.5G version of the system, GPRS is supported by adding packet-handling capabilities to the BSC. This is done by means of an interface called the packet control unit (PCU). In a 3G system, the interface to a GPRS network is expected to be integrated into the BSS.

Third Generation (3G)

The brief snapshot of 3G technology in this section is based on an exposition of the topic by Zahariadis, Vaxevanakis, Tsantilas, Zervos, from which the following text has been reprinted.[13]

Next-generation mobile/wireless networks are expected to provide a substantially wider and enhanced range of services compared to current WWAN networks. Global convergence, interoperability, and mobility are some of the differentiating factors. Moreover, the inherent IP support will encourage new personalized interactive multimedia services and broadband applications, such as video telephony, videoconferencing, and mobile Internet. Deployment of a global all-IP wireless/mobile network, however, is expected through evolutionary rather than revolutionary steps.

Vendors promote the new profitable IP services that a network will allow, whereas post, telephone, and telegraph (PTT) companies look to maximize the profit and return on investment (ROI) based on existing equipment. As a result, the wireless network infrastructure may be organized in a cell hierarchy based on technology that is either already deployed or still under development, as depicted in Figure 3-17. Starting from the home cell, coverage in private buildings (such as a house or office) or in hotspot locations (such as an airport, train station, or conference center) may be provided by APs. IEEE 802.11, high-performance radio local area network (HIPERLAN), Bluetooth, and HomeRF are alternative technologies that may be deployed. The APs may also provide connectivity in picocells, whereas a combination with pico-GSM or Digital Enhanced Cordless Telecommunications (DECT) can also be considered. Moreover, fixed wireless access via central stations (CSs) and remote stations (RSs) may provide wireless access up to macrocells in suburban areas. Horizontal mobility to MTs that move with different speeds in micro- or macrocells may be provided by utilizing 2G and 2G+ (also known as 2.5G) networks (such as GSM, high-speed circuit-switched data [HSCSD], GPRS, Enhanced Data rates for Global Evolution [EDGE], Cellular Digital Packet Data [CDPD], IS-95, and CDMA). Connectivity and mobility in satellites cells are provided via Geostationary Earth Orbit (GEO), Medium Earth Orbit (MEO), or Low Earth Orbit (LEO), and Fixed Earth Stations (FESs) or mobile satellite terminals (STs).

In these environments, roaming is critical. In order to support both horizontal and vertical roaming in such complex environments, the first step is to gain connectivity at the physical layer. In this respect, either multimode or adaptive terminals are considered. For example, terminals equipped with

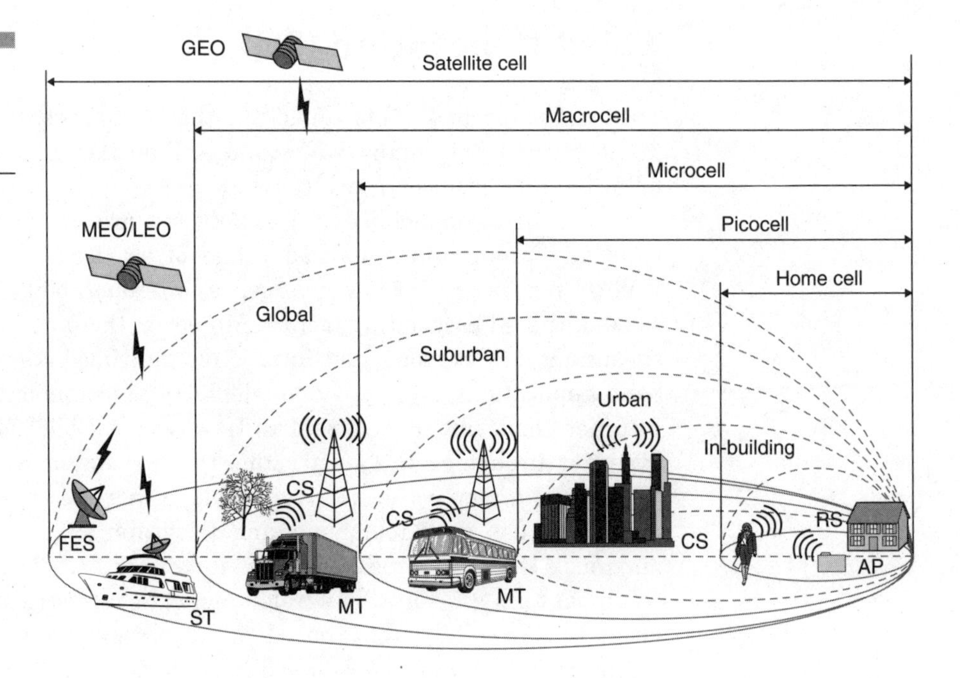

Figure 3-17 Cell hierarchy of a next-generation network

a commercial WLAN (such as IEEE 802.11b), cellular (such as GSM/GPRS), or satellite network interface cards (NICs) may be introduced. Soft radio techniques have also been proposed. Global roaming, however, requires the integration and interoperation of the mobility management processes of each independent network. IP is the most widely accepted protocol; thus, mobility based on IP will be facilitated by the use of an established technology.

The detailed architecture of an all-IP wireless/mobile network architecture is shown in Figure 3-18. WLAN, 2G and 3G cellular, and satellite networks are selected as alternative radio access networks. Due to different physical and protocol characteristics, each radio access network consists of different base stations and radio control nodes, which are connected to the common core network via a service support node (SSN). This may be an MSC+ for the cellular networks, an IP L1/L2 switch for the WLAN, or an FES. The SSN also provides the VLR or foreign agent (FA) functionality, respectively, in cooperation with an extended HLR+ or home subscriber server (HSS). HSS maintains user profiles and may integrate or cooperate with a Remote Authentication Dial-In User Server (RADIUS) and/or authentication, authorization, and accounting (AAA) server for user authentication and authorization.

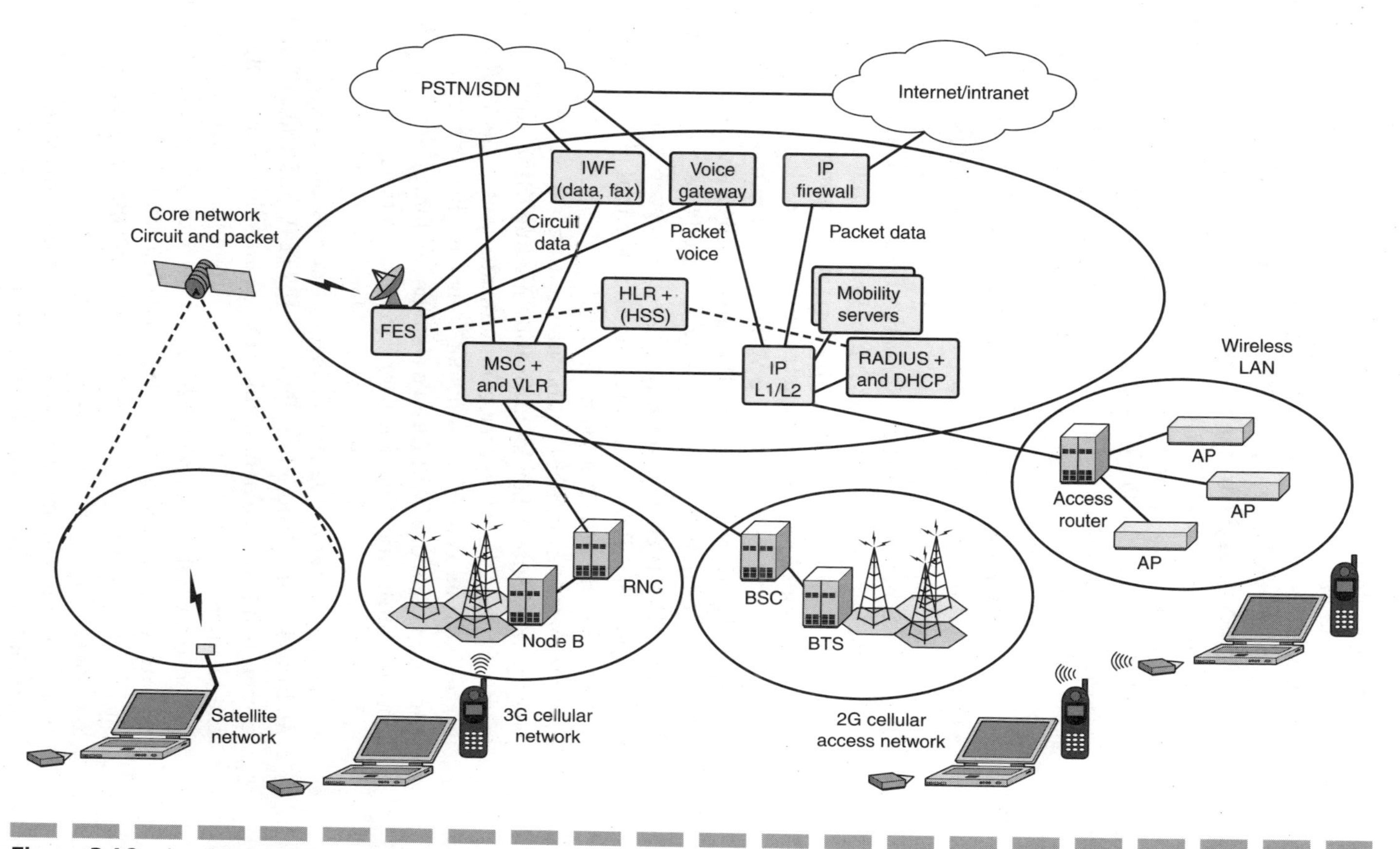

Figure 3-18 *An all-IP wireless/mobile network*

Interoperability between circuit-switched and packet-based networks is also mandatory; thus, the common core network of the proposed all-IP architecture supports both circuit-based connections and packet-based transmission. Access to both PSTN/Integrated Services Digital Network (PSTN/ISDN) and the Internet is provided via interworking function (IWF) units, voice gateways, firewalls, or generally Gateway Support Nodes (GSNs). Additional servers (such as the Dynamic Host Configuration Protocol [DHCP] and the Domain Name Server [DNS]) provide complementary services in the IP domain, whereas mobility servers provide mapping between Universal Personal Telecommunications (UPT) numbers and dynamic IP addresses.

The architecture is based on enhancements of existing equipment so assume that horizontal roaming will be handled by specific network roaming mechanisms. For example, a direct extension toward roaming of IP traffic in GPRS networks proposed by the GSM Association in the form of a GPRS Roaming Exchange (GRX) architecture that carries traffic between mobile operators' networks. In the vertical roaming scenario, however, the terminal should have a more active role and initiate the specific roaming mechanisms. Starting from the WLAN cell, whenever a terminal is activated, it has to obtain a valid IP address. This may be a preconfigured IP address or most likely a dynamically allocated one via a local or distributed DHCP/DNS server. Moreover, the mobility server responsible for the specific hotspot or AP may authenticate the terminal via a centralized or distributed RADIUS/AAA server.

User/terminal authentication and authorization based on MAC/password pair or Subscriber Identification Module (SIM) card could be considered. The RADIUS+ server communicates with the VLR and/or HLR+ servers and associates the hotspot user with his or her cellular database entry; thus, the user will receive a single bill for all services. After registration and authentication, the user can roam in the hotspot while communicating via the WLAN connection. Moreover, a virtual private network (VPN) can be set up via higher-layer protocols (such as IP security [IPSec] and the Layer 2 Tunneling Protocol [L2TP]), and the MT can access the corporate intranet. When the user moves outside the hotspot coverage area, the terminal has to initiate a vertical roaming mechanism and soft handover to a GSM; a Universal Mobile Telecommunications Service (UMTS) or satellite network can be activated. The mobile network intermediate nodes (node B, RNC in UMTS, BSC, BTS in GSM, and FES in satellite) are considered transparent to mobile Internet traffic because they control radio resources and handover decisions only at the physical level. The terminal communicates with the mobility servers located in the IP part of the network, and a new authentication/authorization process is initialized. After the connection at the physical layer has been established, various Mobile IP extensions may be applied to keep connections uninterrupted.

This topic is revisited in additional detail in Chapter 9, "Migrating to 3G WWANs."

Antenna Basics for Hotspot Services

This section introduces antenna technology and serves as an overview of antenna considerations for hotspots. Additional information can be found in Chapter 8, "Designing Nomadic and Hotspot Networks."

The following are three baseline concepts for antennas:

- **Directionality** Where the signal is concentrated.
  - Omnidirectional (360-degree coverage)
  - Directional (limited angle of coverage)
- **Gain** The amount of increase in energy that an antenna appears to add to an RF signal. If the gain of an antenna goes up, the coverage area or angle goes down. Coverage areas or radiation patterns are measured in degrees.

  Gain is measured in dBi and dBd (0dBd = 2.14dBi). There are different methods for measuring gain, depending on the reference point selected. A common measure is dBi, which is gain using a theoretical isotropic antenna as a reference point. Some antennas are rated in dBd, which uses a dipole-type antenna in place of an isotropic antenna as the reference point. To convert from dBd to dBi, add 2.14 to the dBd figure of merit.
- **Polarization** The physical orientation of the element on the antenna that actually emits the RF energy. For example, an omnidirectional antenna is usually a vertical polarized antenna.

Coverage angles are referred to as *beamwidth* and have a horizontal and vertical measurement. A theoretical isotropic antenna has a perfect 360-degree vertical and horizontal beamwidth. Antennas are measured compared to what is known as an isotropic antenna, which is a theoretical antenna. An isotropic antenna's coverage can be thought of as a balloon, extending the signal in all directions equally.

In point-to-multipoint systems, the FCC has limited the maximum effective radiated power (EIRP) in the 900 MHz, 2.4 GHz, and 5.7 GHz spread spectrum bands to 36 dBm. EIRP is the sum of the transmit power and antenna gain. At 2.4 GHz, a typical transmitter power is 20 dBm; hence, the largest antenna gain allowable is 16 dBi. Under the Professional Installer clause, a trained installer may attach bigger antennas, if it is verified that the system is operating within the FCC rules and guidelines, and only if

$$\text{Antenna gain} + \text{transmitter power} + \text{cable losses} \leq +36\text{dBm}.$$

In point-to-point system directional antenna applications, the rules are as follows. For 900 MHz, the EIRP limit is the same as the multipoint system; for 2.4 and 5.7 GHz, the EIRP limit has a 3:1 ratio for additional antenna gain (over 6 dBi) compared to transmit power reduction. At 30 dBm transmit power, a 6 dBi antenna is still the maximum. At 27 dBM transmit power (–3dB), the antenna can be 9 dB above the initial 6 dB, namely 15 dBi. At 24 dBm transmit power, a 24 dBi antenna can be used.

Dipoles

To obtain omnidirectional gain from an isotropic antenna, the energy lobes are pushed in from the top and bottom and forced out in a doughnut-type pattern. The higher the gain, the smaller the vertical beamwidth and the larger the horizontal lobe area. The gain of a dipole is 2.14 dBi (0 dBd). When we design an omnidirectional antenna to have gain, we lose coverage in certain areas. Visualize the radiation pattern of an isotropic antenna as a balloon, which extends from the antenna equally in all directions. Then imagine pressing in the top and bottom of the balloon; this causes the balloon to expand in an outward direction, covering more area in the horizontal dimension, but reducing the coverage area above and below the antenna. Focusing the energy yields a higher gain as the antenna appears to extend to a larger coverage area. The higher the gain, the smaller the vertical beamwidth.

High-Gain Omnidirectional Antennas

High-gain omnidirectional antennas create more coverage area horizontally away from the antenna, but the energy level directly below the antenna becomes lower. If you continue to push in on the ends of the isotropic balloon, you get a pancake effect with very narrow vertical beamwidth, but very large horizontal coverage. This type of antenna design can support relatively long distances, but has the drawback of poor coverage below the antenna. This problem can be partially addressed by designing the antenna in a downtilt. An antenna that uses downtilt is designed to radiate at a slight angle rather that at 90 degrees from the vertical element. This helps local coverage, but reduces the horizontal length of coverage.

Directional Antennas

In directional antennas, lobes are pushed in a certain direction, causing the energy to be concentrated there. Specifically, very little energy is found in the back of a directional antenna.

Conclusion

This chapter examined a gamut of available and evolving technologies for public hotspot services. The opinion of the author is that the IEEE 802.11-family of services will be the most important technology for hotspots through 2006-7, as illustrated in Figure 3-19.

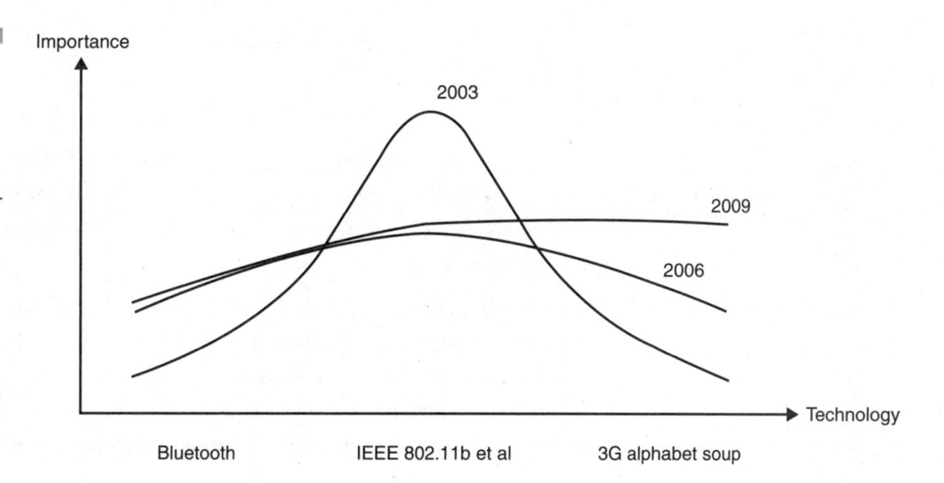

Figure 3-19 Relative importance/penetration of various technologies for hotspot services

End Notes

1. Glenn Fleishman, "New Wireless Standards Challenge 802.11b," www.oreillynet.com/pub/a/wireless/2001/05/08/standards.html, June 8, 2001.
2. Princy C. Mehta, "Wired Equivalent Privacy Vulnerability," http://rr.sans.org/wireless/equiv.php, April 4, 2001.
3. Material for this subsection has been reprinted from promotional information from Wireless LAN Association (www.wlana.com/learn/educate1.htm). The Wireless LAN Association is a nonprofit educational trade association comprised of the thought leaders and technology innovators in the local area wireless technology industry. Through the vast knowledge and experience of sponsor and affiliate members, WLANA provides a clearinghouse of information about wireless local area applications, issues, and trends, and serves as a resource to customers and prospects of wireless local area products and wireless personal area products and to industry press and analysts.
4. Jim Zyren and Al Petrick, "IEEE 802.11 Tutorial," www.wirelessethernet.org/downloads/IEEE_80211_Primer.pdf.
5. L. Munoz, M. Garcia, J. Choque, R. Aguero, and P. Mahonen, "Optimizing Internet Flows over IEEE 802.11b Wireless Local Area Networks: A Performance-Enhancing Proxy Based on Forward Error Correction," *IEEE Communications Magazine* (December 2001): 60 ff.
6. ANSI/IEEE 802.11 Standard, 1999 Edition [ISO/IEC 8802-11: 1999].
7. M. G. Arranz et al., "Behavior of UDP-Based Applications over IEEE 802.11 Wireless Networks," 12th IEEE International Symposium on Personal Indoor and Mobile Radio Communication, San Diego, September 2001.
8. A list of qualified products can be found at http://qualweb.opengroup.org/Template.cfm?LinkQualified=QualifiedProducts.
9. Information in this section is taken from promotional materials promulgated by the Bluetooth SIG (www.bluetooth.com) and Ericsson (www.ericsson.com).
10. RFCOMM is a serial line emulation protocol to emulate RS-232 control and data signals over Bluetooth™ link. Legacy applications can seamlessly communicate wirelessly over RFCOMM layer. RFCOMM is a simple transport protocol that provides emulation of RS232 serial ports over the L2CAP protocol. The protocol is based on the ETSI

standard transmission 07.10. Only a subset of the transmission 07.10 standard is used and an RFCOMM—specific extension is added in the form of a mandatory credit based flow control scheme. The RFCOMM protocol supports up to 60 simultaneous connections between two base transceiver devices. The number of connections that can be used simultaneously in a base transceiver device is implementation specific. For the purposes of RFCOMM, a complete communication path involves two applications running on different devices (the communication endpoints) with a communication segment between them.

11. GSM operates in the FDD mode, using one band for inbound links and a separate one for outbound links. The 25-MHz spectrum in either direction is partitioned into 125 physical channels, each with a bandwidth of 200 kHz.

12. There is a VLR for each serving area controlled by an MSC.

13. T. B. Zahariadis, K. G. Vaxevanakis, C. P. Tsantilas, and N. A., Zervos, "Global Roaming in Next-Generation Networks," *IEEE Communications Magazine* (February 2002): 145 ff.

CHAPTER 4

Security Considerations for Hotspot Services

This chapter focuses on the key issue of security. Security is an ongoing concern. In 1999, 70 percent of companies reported cyber-attacks, with more than a third experiencing six or more incidents. The American Society for Industrial Security (ASIS) and PriceWaterhouseCoopers reported that Fortune 1000 companies lost more than $45 billion to theft of proprietary information. In an article on wireless local area networks (WLANs), Kim Getgen reinforces this point: "We began to hear [in 2001] about wireless network drive-by hacking incidents. From the Highway 101 corridor that connects San Francisco to the Silicon Valley, to the financial and technology districts of New York, Boston, and London, similar reports have been published by a number of different independent researchers. Sitting in the parking lots of reputable companies, or even driving down city streets, reporters, researchers, and ethical hackers were able to retrieve data . . . ."[1]

In a recent experiment, techie web site ExtremeTech set up a wireless laptop on a roof in Manhattan and discovered it could access 61 wireless networks, of which 48 were completely unsecured.[2] In Silicon Valley, ExtremeTech found 100 networks, many accessible from within a car zooming down Highway 101, of which 66 lacked any kind of security.

This chapter aims at providing the designers of hotspot services with approaches, alternatives, tools, and methods to incorporate adequate security in their systems.

Problems in Security Protocols

This section highlights some of the issues that impact security in wireless local area networks and WLAN-based hotspot networks used as wireless wide area networks (WWANs).

Wired Equivalent Privacy (WEP) Issues

To provide security features in an intrinsically insecure environment, Institute of Electrical and Electronics Engineers (IEEE) 802.11 specified the (optional) WEP protocol mechanism to support confidentiality and integrity of the traffic in the WLAN.[3] However, researchers from Intel, the University of California at Berkeley, the University of Maryland, and Zero-Knowledge Systems soon independently exposed flaws with WEP.[4,5] Fluhrer, Mantin, and Shamir outlined a passive attack that Stubblefield, Ioanndis, and Rubin at AT&T Labs and Rice University implemented by capturing a hidden WEP key based on the attacks proposed in the Shamir et al. paper; this attack took just hours to implement. Furthermore, making things even worse, it is even reported that "most enterprises do not turn WEP on . . . the preliminary reports from the independent surveys taking place in London, New York, and the Silicon Valley suggest that the majority of WLANs deployed do not use WEP at all."[1]

As discussed earlier in the book, WEP depends on a secret key shared between the communicating parties (mobile station and access point [AP]) to protect the payload of a transmitted frame in each direction. Ron's code 4 (RC4) Pseudorandom Number Generator (PRNG) algorithm used by WEP includes an integrity check vector (ICV) to check the integrity of each packet. This process is summarized here from the earlier discussion and from Mehta.[10]

WEP computes the ICV by performing a 32-bit cyclical redundancy check (CRC-32) of the frame and appends the vector to the original frame, resulting in the plaintext. Then the message plus the ICV is encrypted via the RC4 PRNG algorithm (which is symmetric) using a long sequence key stream, a long sequence of pseudorandom bits. This key stream is a function of the 40-bit secret key (which is shared between all authorized stations in the WLAN) and a 24-bit initialization vector (IV). Consequently, an exclusive OR (XOR) operation is made between the plaintext and the key stream to produce the ciphertext. Finally, it is the ciphertext that is sent over the radio link. Theoretically, the ciphertext provides data integrity because of the ICV and confidentiality due to encryption. The receiver performs the same procedure described here, but in reverse to retrieve the original message frame. Specifically, the ciphertext is decrypted using a duplicated key stream to recover the plaintext. The recipient then validates the checksum on this plaintext by computing the ICV and comparing it to the last 32 bits of the plaintext, thus ensuring that only frames with a valid checksum will be accepted by the receiver.

Borisov, Goldberg, and Wagner demonstrated a number of security flaws in WEP.[6] WEP fails to specify how IVs for RC4 are specified. Several PC cards reset IVs to zero every time they are initialized, and then the cards increment them by one for every use. This results in a high likelihood that key streams will be reused, leading to simple cryptanalytic attacks against the cipher and the decryption of message traffic. Furthermore, the space from which IVs are chosen is too small, virtually guaranteeing reuse, and leading to the same cryptanalytic attacks just described.[7]

As covered earlier in the book, a stream cipher operates by expanding a short key into an infinite, pseudorandom key stream. The sender uses an XOR on the key stream with the plaintext to produce ciphertext. The receiver has a copy of the same key and uses it to generate an identical key stream. XORing the key stream with the ciphertext yields the original plaintext. This mode of operation makes stream ciphers vulnerable to several attacks. If an attacker flips a bit in the ciphertext upon decryption, the corresponding bit in the plaintext will be flipped. Also, if an eavesdropper intercepts two ciphertexts encrypted with the same key stream, it is possible to obtain the XOR of the two plaintexts. Knowledge of this XOR can enable statistical attacks to recover the plaintexts. The statistical attacks become increasingly more practical as more ciphertexts that use the same key stream are known. Once one of the plaintexts becomes known, it is trivial to recover all the others.[8]

Experts attribute the WEP-related problems to the fact that the WEP algorithm did not go through formal testing; this left imperfections in WEP's cryptography approaches.[9] Another misjudgment that can now compromise integrity was in the selection of the 32-bit cyclic redundancy (CRC-32) checksums to perform integrity checks. Borisov, Goldberg, and Wagner documented the importance of using a cryptographically secure message authentication code (MAC), such as SHA1-HMAC, to protect the integrity of transmissions: the use of CRC is inappropriate for this purpose since an attacker simply needs to flip selected bits of a message to yield the same integrity check vector as the original one.[6] A secure MAC is especially critical when considering a composition of protocols because the lack of message integrity in one portion of a system can breach secrecy in the entire system.[10]

WEP can be implemented with the classic 40-bit key and 24-bit IV or a vendor-dependent (hence, proprietary) extended version that affords a larger key. The shorter key length can be relatively easy to compromise via a brute-force attack, even with modest computing resources; however, a larger key such as a 128-bit key would render brute-force attacks more daunting, even for sophisticated computing systems. Nevertheless, alternative attacks are possible that do not require a brute-force strategy, thereby diminishing the strength of the key length. The subsequent four sections describe each of the potential attacks, as described by Mehta, on which we base this discussion.[11]

Passive Attack to Decrypt Traffic The IV is a 24-bit field intended to randomize part of the key (since all stations on a WLAN share the 40-bit key). This means that an 802.11b AP transmitting at 11 Mbps can exhaust all IV combinations within 5 hours, as shown in Equation 4-1.

Although 802.11b APs generate a theoretical maximum of 11 Mbps, its observed rates are usually much less due to overhead and packet collisions, which can increase the amount of time before an IV is reused. However,

Equation 4-1
Time to exhaust 24-bit IV

$$\frac{1500\ bytes}{packet} \times \frac{8\ bits}{1\ byte} \times \frac{1\ \text{sec}}{11\ Mb} \times \frac{1\ Mb}{10^6\ bits} \times 2^{24}\ packets \approx 18{,}300\ \text{sec} \approx 5\ hrs$$

packets are usually not at the Ethernet maximum of 1500 bytes, which reduces the IV reuse. In any event, this small space of IVs *guarantees* that a key stream will be reused in less than half of one day.

The importance of this relatively small time is that an attacker can collect two ciphertext packets that are encrypted with the same key stream. Then the attacker can perform statistical attacks to recover the plaintext, and once he has positively matched a ciphertext message with its plaintext counterpart, then an XOR operation will reveal the key, enabling him to comprehend all other ciphertext messages. Hence, it is fruitful for the passive eavesdropper to intercept all wireless traffic until an IV collision is observed.

Even if the threat cannot understand all the contents, the attacker can infer them by exploiting the predictable nature and redundancy of IP traffic. Further educated guesses about the contents of a plaintext message also allow the threat to statistically diminish the space of possible messages to get a clear idea of the exact contents. As the attacker ascertains more collisions using the same IV, the attacker can get an even better understanding of the concealed messages, facilitating the success rate of statistical analysis.

A variation of this attack is to use a host to send traffic from outside the WLAN to the AP. Because the attacker knows the plaintext (which he fabricated), he has everything necessary to recover the crucial key stream once he taps its ciphertext from the air.

Active Attack to Insert Traffic The problems from the attack mentioned previously can be worsened. If an attacker knows the precise plaintext and ciphertext pair, he would manifestly be able to generate the key stream. With this knowledge, the threat can build correctly encrypted packets by constructing a message, calculating its CRC-32, and executing an XOR operation with the newly discovered key stream. This ciphertext packet can be sent to the AP or mobile station to deceive it into thinking that it is a valid packet.

A minor modification to this attack can make it much more pernicious. Even if the attacker has not attained complete knowledge of the packet, he can alter selected bits of the message and successfully adjust the encrypted

ICV to obtain a correct encrypted version of the modified packet. This is a lethal attack of the packet's integrity, since all the attacker requires is partial knowledge of the packet's contents to perform selective modification.

Active Attack from Both Ends The previous active attack can be extended even further to decrypt arbitrary traffic. Suppose the attacker speculates about the frame's header only, not necessarily its actual contents. For example, he may be able to predict the destination IP address with a high degree of confidence. Equipped with just this information, the attacker can modify appropriate bits to transform the destination IP address to send the packet to a machine in his control via a rogue mobile station. Then the packet could be successfully decrypted by the AP and forwarded as plaintext to the attacker's machine.

Even if the AP is positioned behind a firewall, once the attacker can deduce the Transmission Control Protocol (TCP) headers of the packet, he can change it to port 80 (generally indicating World Wide Web service), thereby allowing it to be forwarded after decryption through most firewalls and to his machine somewhere on the Internet.

Table-based Attack The small space of potential IVs can allow an attacker to construct a decryption table. Once the plaintext of a packet is realized, an attacker can compute the key stream produced by the IV utilized. Accordingly, this key stream can be used to decrypt all other packets employing the same IV. Eventually, the threat can generate a table of IVs and corresponding key streams. Then the black hat can decrypt any and *all* packets sent over the wireless link, regardless of their IVs.

Because the table can contain up to 2^{24} (over 16 million) values, and each entry is a maximum of 1500 bytes, the table will be no larger than 24 GB. Thus, it is conceivable that a committed attacker can accumulate enough data to build a full decryption "dictionary." Although this would be an arduous undertaking, his motivation is that once the table is formed, it is possible to immediately decrypt every subsequent ciphertext with little effort. Worse yet, this table can be distributed to other black hats. The flaw here is in the 24-bit IV; if the IV is expanded in a future version of WEP, constructing such a dictionary would be exponentially more laborious.

Workarounds A number of workarounds are being tested to make WLANs secure (see the following box).

Summary of WEP Issues

Adam Stubblefield, John Ioannidis, and Aviel D. Rubin implemented an attack against the WEP, the link-layer security protocol for 802.11 networks. Fluhrer, Mantin, and Shamir also described the attack in a paper of their own. They were able to recover the 128-bit secret key used in a production network with a passive attack. The WEP standard uses RC4 IVs improperly, and the attack exploits this design failure. They conclude that 802.11 WEP is totally insecure, and we provide some recommendations.[7]

Stubblefield et al. were able to implement the attack described by Fluhrer et al. in several hours. According to the authors, it took a few days to figure out which tools to use and what equipment to buy to successfully read keys off of 802.11 wireless networks. The attack used off-the-shelf hardware and software, and the only piece the authors provided was the implementation of the RC4 attack, along with some optimizations. Stubblefield et al. have demonstrated the ultimate break of WEP, which is the recovery of the secret key by the observation of traffic. Given this attack, 802.11 networks should be viewed as insecure.

Stubblefield et al. recommend the following for people using such wireless networks:

- Assume that the link layer offers no security.
- Use higher-level security mechanisms such as IP Security (IPSec)[12] and Secure Shell (SSH)[13] for security, instead of relying on WEP.
- Treat all systems that are connected via 802.11 as external. Place all APs outside the firewall.
- Assume that anyone within the physical range can communicate on the network as a valid user. Keep in mind that an adversary may utilize a sophisticated antenna with a much longer range than found on a typical 802.11 PC card.

Wireless systems that rely on a 64-bit key (used in many homes and in the original WLAN hardware) can be broken in 60 seconds or less. The original choice of encryption by the IEEE 802.11 group may also be related to U.S. export law restrictions.[14] In the United States, vendors are advancing proprietary 128-bit solutions. Although this may provide stronger encryption security, it impacts interoperability with equipment from different vendors. Also, poorly chosen passwords can still be cracked with an old technique known as a dictionary attack: using a list of common passwords and a dictionary of words, the potential intruder can try various combinations until the password is broken.[15]

Some stakeholders are working on upgrading WEP to a WEP2 version that separates encryption and authentication functions in such a manner that the same static key does not need to be shared within a WLAN.[16] Recognizing that WEP2 will not be the WLAN's final security solution, the IEEE 802.11 group has approved a draft to establish a stronger authentication and 128-bit key management system, tentatively called the Enhanced Security Network (ESN). ESN's encryption will augment the RC4 PRNG algorithm with the Advanced Encryption Standard (AES).[17] ESN is expected to be finalized in 2002. Security work is going on both in IEEE 802.11e and 802.11i.

In the meantime, security experts recommend the deployment of multiple security measures. For example, the recommendation is to place the WLAN outside the firewall and to run a virtual private network (VPN) inside the firewall. The Remote Access Dial-In User Service (RADIUS) protocol can be implemented to provide another level of security designed to authenticate remote clients to a centralized server.[18]

Summary of the Issue and Planned Fixes

Summarizing the previous discussion, the vulnerabilities exposed in WEP can be traced back to two problems in the standard:[19] (1) the limitations of the IV combined with (2) the use of static WEP keys where the odds of collisions are very high. IV collisions produce so-called "weak" WEP keys when the same IV is used with the same WEP key on more than one data frame. When a number of these weak keys can be analyzed, WEP can be attacked to expose the shared secret. Some early reports inferred that the stream cipher used for WEP encryption, RC4, was the weakness. But this turned out not the case: the vulnerabilities exposed in WEP can be traced back to the way the initialization vector and the WEP key are combined to get a per-packet RC4 key.

Some IVs produce weak RC4 keys that leak information on the WEP key. Free tools (or scripts) like AirSnort and WEPCrack have appeared on the Internet that anyone can use to attack WEP. AirSnort authors claim their code can capture WEP keys after gathering information from just 2,000 packets with weak keys. It is estimated that out of 16 million keys generated using 128-bit WEP encryption, 3,000 are weak. (Keep in mind that 802.11b actually calls for the use of 40-bit WEP encryption, which is even more vulnerable. Many vendors are going one step ahead of the spec and providing 128-bit WEP encryption in their products today, but even this tighter security is vulnerable to the new tools.) Network sniffers like AirSnort analyze the weak keys to discover the shared secret between wireless clients and APs. Once the shared secret is discovered, a malicious attacker would have access to the WLAN network and be able to go back and decrypt data packets being passed on the exposed network.

In a public statement responding to the weakness discovered in WEP, Ron Rivest, inventor of the RC4 algorithm, recommends that "users consider strengthening the key scheduling algorithm by preprocessing the base key and any counter or initialization vector by passing them through a hash function such as MD5. Alternatively, weaknesses in the key scheduling algorithm can be prevented by discarding the first 256 output bytes of the pseudo-random generator before beginning encryption. Either or both of these techniques suffice to defeat the Fluhrer, Mantin, and Shamir attacks on WEP and WEP2."

Network security is only as strong as the authentication system it is based on. Proper authentication techniques thwart popular attacks like the man-in-the-middle and denial of service attacks. In the current 802.11b standard, turning WEP on allows client authentication to take place via a password-based system, but leaves the system vulnerable to the attacks described earlier. What's worse is that no mechanism for AP authentication is identified.

The Protected Extensible Authentication Protocol (EAP) in the new 802.1X standard improves the authentication mechanisms in 802.11 standards like 802.11b or 802.11a (the standard was introduced by RSA Security, Cisco, and Microsoft). The standard solves three key problems:

- It protects the network from rogue APs being introduced.
- It outlines a way users can strongly authenticate themselves to the AP using a variety of popularly used authentication methods.
- It allows users to strongly authenticate themselves while roaming between the separate AP that make up a network's WLAN.

The Protected EAP proposal calls for EAP to be used in combination with the Transport Layer Security (TLS) protocol. Both EAP and TLS are popularly used Internet Engineering Task Force (IETF) standards on the Internet. The combination of the two protocols results in client and server authentication that protects the WLAN network against passive eavesdroppers.

Protected EAP works in two phases. A TLS phase authenticates APs using an encrypted tunnel to protect authentication information being exchanged, even when users are roaming between different APs. Next, an EAP phase authenticates the users of wireless clients.

Phase One A TLS handshake (see Figure 4-1) is used to authenticate the AP to the wireless client. First, the wireless client sends a message to a backend server announcing that it is connected to the wireless AP. The message tells the server that a new connection should be initiated, and it also tells which cryptographic algorithms the client understands so that secure messages sent between the two can be understood.

After receiving this message, the backend server responds with a new session ID, a list of algorithms that will be used to correspond, and a public key certificate that enables the client to trust the AP it has used to establish the connection to the network. The wireless client verifies the signature

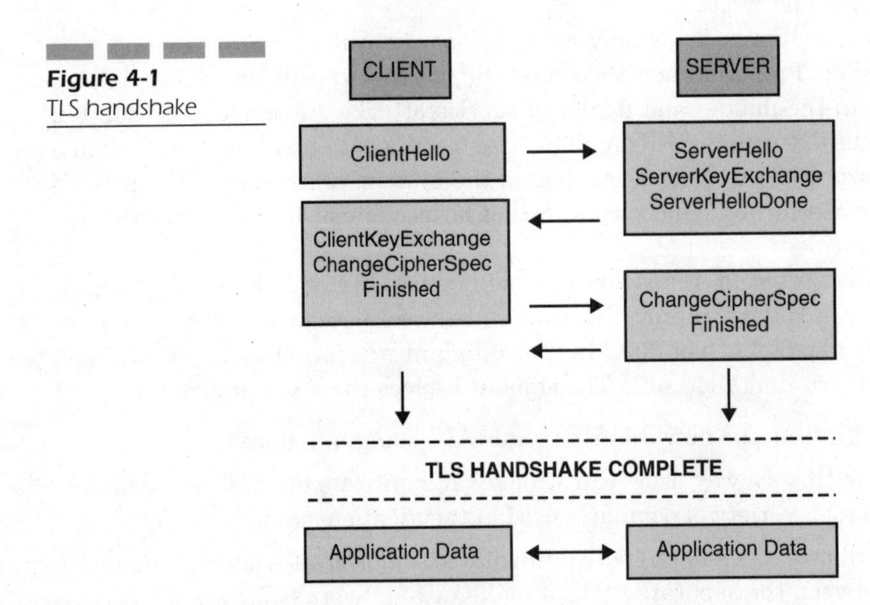

Figure 4-1 TLS handshake

and validity of the server certificate and then responds by generating a secret key and encrypting it with the public key obtained from the server certificate. This protected information is sent back to the server and, if the server is able to decrypt this information, it is authenticated: only the server's private key would be able to decrypt messages encrypted with its public key. After this last exchange, authentication of the AP is complete and a secure TLS session is established to protect the user authentication credentials that will be passed in phase two.

Phase Two An added layer of protection is provided in the second phase of Protected EAP through the use of EAP, which is able to strongly authenticate the end user of the wireless client by challenging the user with a suitable EAP mechanism (see Figure 4-2). Suitable EAP mechanisms include the use of passwords, smart cards, digital certificates, or time-synchronous tokens like the RSA SecurID token, which produces one-time passcode challenges. The EAP challenges are passed to backend authentication servers connected to the wireless APs sitting on a company's WLAN. The versatility here is really the key, because EAP enables enterprises to continue to use any appropriate EAP method already deployed to their employees and extend its use to the WLAN.

Figure 4-2 Protected EAP authentication phase

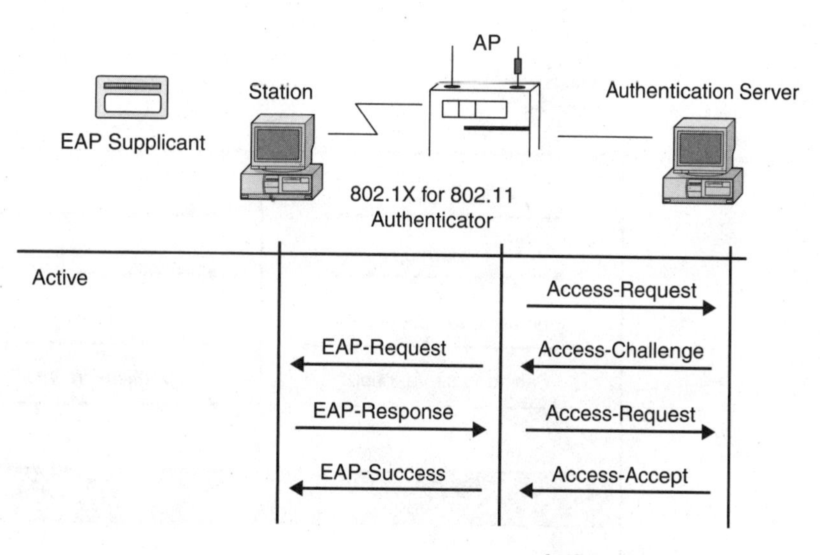

The real benefit will be felt by users who are roaming within a corporation's WLAN and require a seamless connection. Users want mobility and convenience. If they are asked to authenticate themselves each time they pass from one conference room to another, they will want to give up security in favor of convenience, and that is the reason for choosing TLS for Protected EAP in the 802.1X standard (see "Moving Forward"). Using the connection reestablishment mechanism provided in the TLS handshake enables users to have one seamless connection while roaming between different APs connected to the same backend server. If the session ID is still valid, the wireless client and server can share old secrets to negotiate a new handshake and keep the connection alive and secure (see Figure 4-3).

Moving Forward Moving forward, one can expect 802.1X to be adopted into a variety of 802.11 standards such as 802.11a, which is good news from a security perspective, because products being built today can use 802.1X. These products will offer advantages to enterprises by enabling them to

Figure 4-3 Roaming in protected EAP

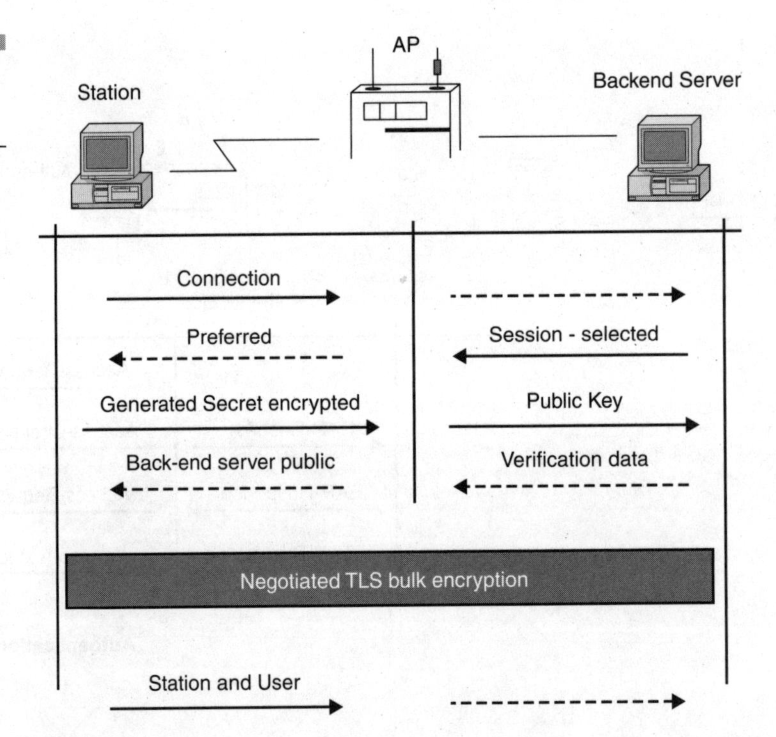

take the same strong authentication solutions currently deployed in enterprises and use them with WLAN products. In addition, implementing 802.1X in WLAN products will offer the advantage of enabling users to roam more conveniently and securely. It is a long-term fix to solve user and AP authentication in a roaming environment.

On the encryption side, Task Group I is working out a solution to fix WEP for the long term. Vendors are following up with announcements about how they plan to incorporate the merits of the Task Group's work into a quick fix for WEP to clean up the issue in all of the 802.11b hardware currently deployed.

For the longer-term solution, Task Group I work may roll into an 802.11i standard that will eventually replace 802.11b. It seems logical to draw the conclusion that Task Group I's work will also be incorporated into the 802.11a standard to secure WLAN products that will be capable of offering higher bandwidth to users sometime in 2003. Vendors working with 802.11a should also start evaluating the Protected EAP proposal in the 802.1X standard right away.

802.1X: Port-based Network Access Control

Authentication as defined in the IEEE 802.11 standard is focused more on WLAN connectivity than on verifying station identity. For WLAN systems to scale to a large number of users, the current method of authentication must be replaced with an authentication framework that supports centralized user authentication. As just noted, Task Group I of the IEEE 802.11 committee recently developed 802.1X, an IEEE standard that provides an authentication framework for 802-based LANs.

IEEE 802.1X is a standard drafted by the IEEE that is designed to provide enhanced security for users of 802.11b WLANs. 802.1X is a supplement to ISO/IEC 15802-3:1998 (IEEE standard 802.1D-1998) that defines the changes necessary to the operation of a MAC bridge in order to provide port-based network access control capability. The activity was launched in early 2000 and by January 2002 Draft 11 incorporated comments received from the industry. 802.1aa-802.1X Maintenance is an amendment to IEEE standard 802.1X-2001 and is intended to document maintenance items identified in the text of IEEE standard 802.1X-2001.

The standard provides port-level authentication for any wired or wireless Ethernet client system. IEEE 802.1X was originally designed as a standard for wired Ethernet, but is applicable to WLANs. It leverages many of the security features used with dial-up networking.[20] When 802.1X is applied to 802.11 networks, it has the potential to simplify security management for WLAN environments; however, it is only one element of the security layers needed in WLANs. When it is coupled with an authentication algorithm and data frame encryption, it can support scalable, manageable, and mobile network services.

IEEE 802.1X is a *port-based authentication protocol*, but it can be used in any scenario where one can define any abstract notion of a port. It requires entities to play three roles in the authentication process: a supplicant, an authenticator, and an authentication server. Figure 4-4 recapitulates the basic scenario.[21] A port access entity (PAE) is an entity that has access or is capable of gaining or controlling access to some port that offers some services.

As noted in the previous section, WLANs based on the 802.11 standards have come under intense scrutiny as their security based on WEP has been found to be unreliable. The new 802.1X has a key management protocol built into its specification that provides keys automatically. Keys can also be changed rapidly at set intervals.[20] 802.1X takes advantage of EAP.[22] 802.1X takes EAP, which is written around PPP, and ties it to the physical medium, be it Ethernet or WLAN. EAP messages are encapsulated in 802.1X messages and referred to as EAP over LAN (EAPOL).[23] It should be noted that 802.1X alone lacks the components that 802.11-based LANs

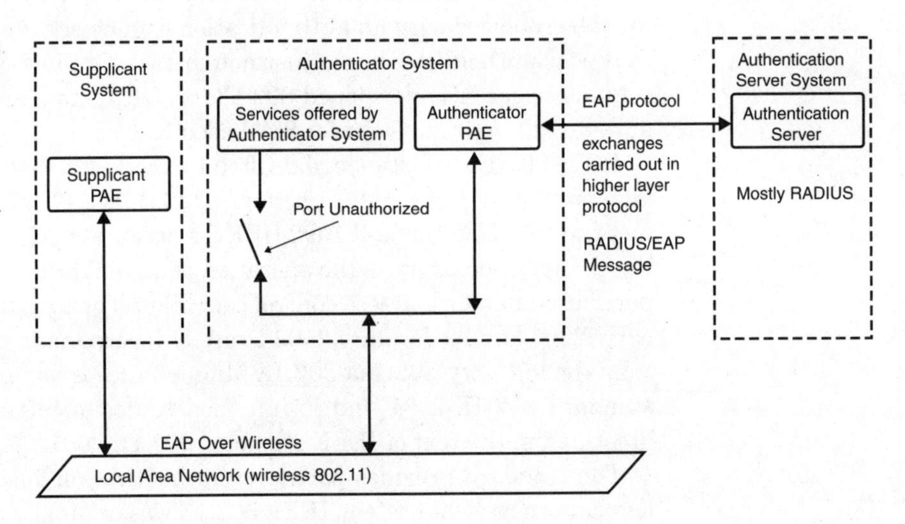

Figure 4-4 The setup using the IEEE 802.1X port-based authentication mechanism

need for user-based authentication. Task Group I was in the process of drafting amendments to the 802.11 specifications at press time to incorporate 802.1X services. Currently,[24] 802.1X specifies that it operates on physical ports only in systems that support link aggregation; ports cannot aggregate until they have been authorized.[24]

As previously mentioned, 802.1X authentication for WLANs has three components (see Figure 4-5[25]): the supplicant (the client software), the authenticator (the AP), and the authentication server (a RADIUS server, although RADIUS is not specifically required by 802.1X). The client attempts to connect to the AP, which detects the client and enables the client's port. At this juncture, it forces the port into an unauthorized state, meaning that only 802.1X traffic is forwarded; traffic such as the Dynamic Host Configuration Protocol (DHCP), Hypertext Transfer Protocol (HTTP), File Transfer Protocol (FTP), Simple Mail Transfer Protocol (SMTP), and Post Office Protocol 3 is blocked.

The client then sends an EAP-start message (see Figure 4-6). The AP replies with an EAP-request identity message to obtain the client's identity. The client's EAP-response packet containing the client's identity is forwarded to the authentication server. The authentication server is configured to authenticate clients with a specific authentication algorithm. The result is an *accept* or *reject* packet from the authentication server to the AP. Upon receiving the accept packet, the AP transitions the client's port to an

Figure 4-5
General topology of IEEE 802.1X

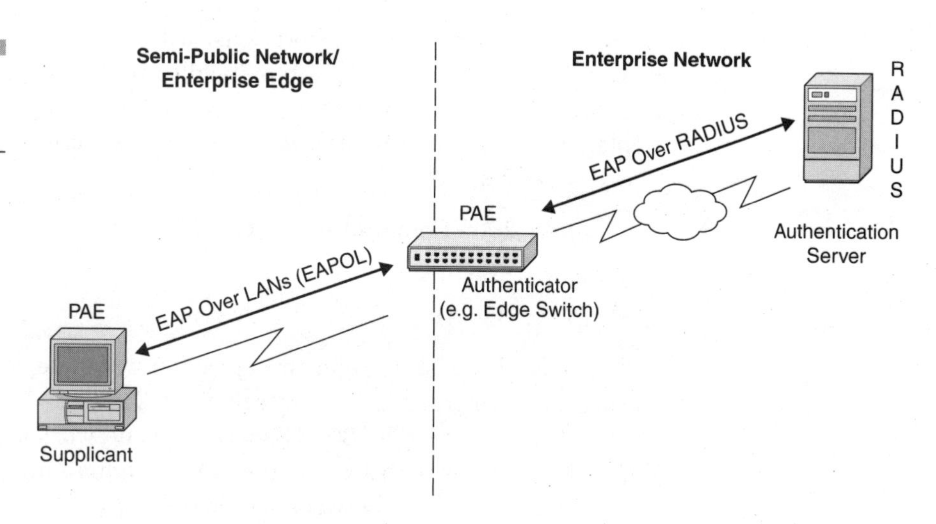

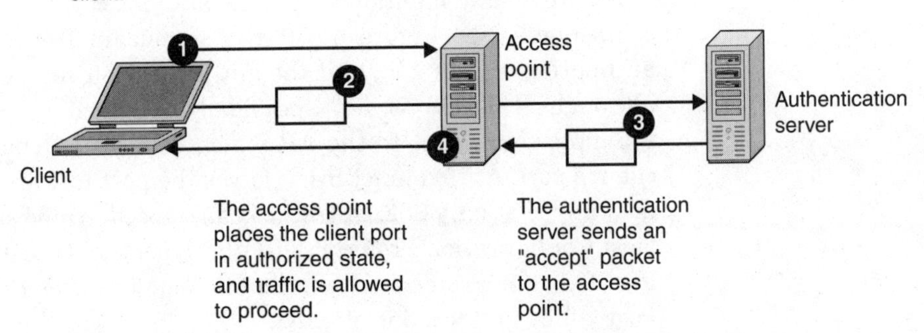

Figure 4-6 IEEE 802.1X

authorized state, and traffic will be forwarded. At logoff, the client will send an EAP-logoff message. This forces the AP to transition the client port to an unauthorized state.[23]

Proponents claim that Windows XP and 802.1X provide for what is called *zero configuration support*, enabling laptops with a wireless adapter card to automatically detect and connect to wireless APs within range. The combination allows link layer authentication, enabling seamless user authentication. As an example, corporations will be able to use their active directories and databases to automatically authenticate employees.[20] Cisco and Microsoft have partnered in creating an early commercial 802.1X implementation. The Microsoft/Cisco implementation requires three components (see Figure 4-7[26]):

- Windows 2000 Domain Controller
- Windows XP Client
- Cisco 340/350 AP

You may also wish to refer to Figure 2-5 which depicts another view of the actual commercial implementation of IEEE 802.1X by Cisco Systems.

The client side of IEEE 802.1X, port-based network access control, solves the problem of key distribution in WLANs by enabling public key authentication and encryption between wireless APs and roaming stations. It also enables network managers to control 802.1X user profiles from a centralized RADIUS server.[27]

Figure 4-7 Initial Cisco implementation of IEEE 802.1X

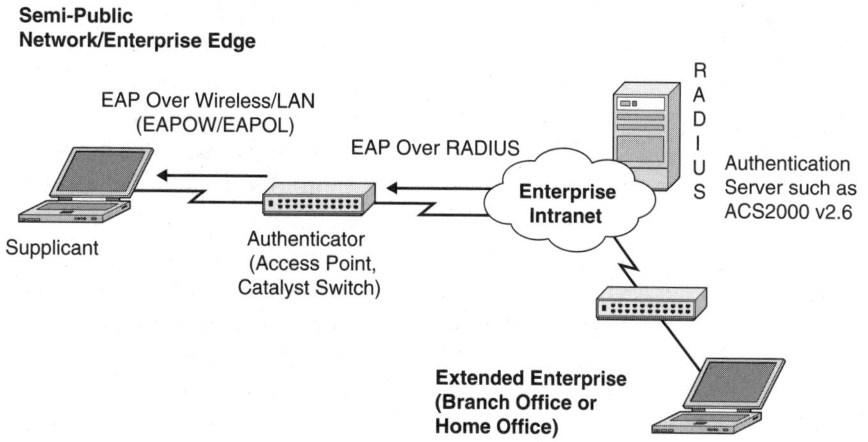

Figure 4-8 An infrastructure mode network

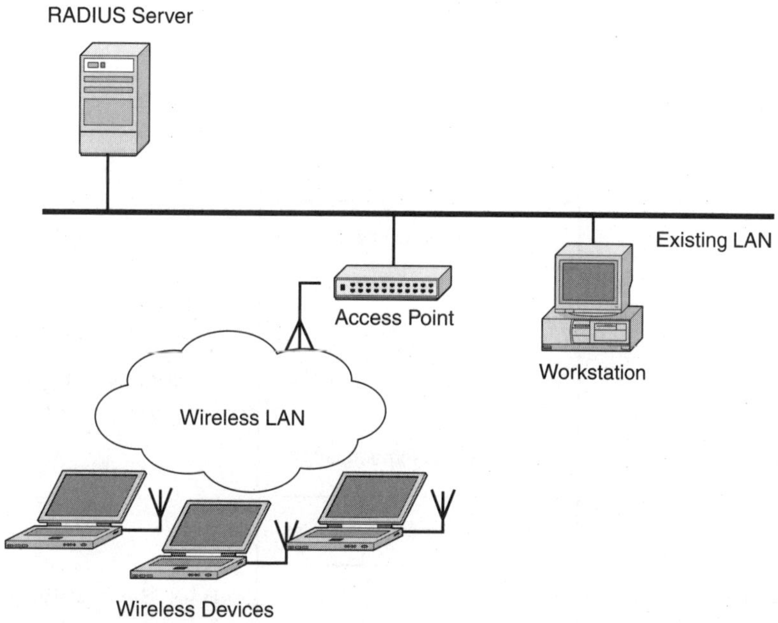

Figure 4-8 shows the network elements involved in a typical WLAN. When 802.1X is running, a wireless device must authenticate itself with the AP in order to get access to the existing LAN. With respect to the terms used in the 802.1X standard, wireless APs function as *authenticators* and wireless devices function as *supplicants*. The authenticator keeps a control port status for each supplicant it is serving. If the supplicant has been authenticated, then the control port status is said to be *authorized*, and the supplicant can send application data to the LAN through the wireless APs. Otherwise, the control port status is said to be *unauthorized*, and application data cannot traverse the wireless APs.

Figure 4-9 is a typical message exchange when the device and the wireless AP support 802.1X (note that this figure amplifies Figure 4-2). When a wireless AP acting as an authenticator detects a wireless station on the LAN, it sends an EAP-request for the user's identity to the device. (The EAP is an authentication protocol that runs before network layer protocols transmit data over the link; it will be examined in the next subsection.) In turn, the device responds with its identity, and the wireless AP relays this identity to an authentication server, which is typically an external RADIUS server, as depicted in Figure 4-8. This allows the RADIUS server to act as the central repository of user profile information, which allows the user to

Figure 4-9 Typical message exchange

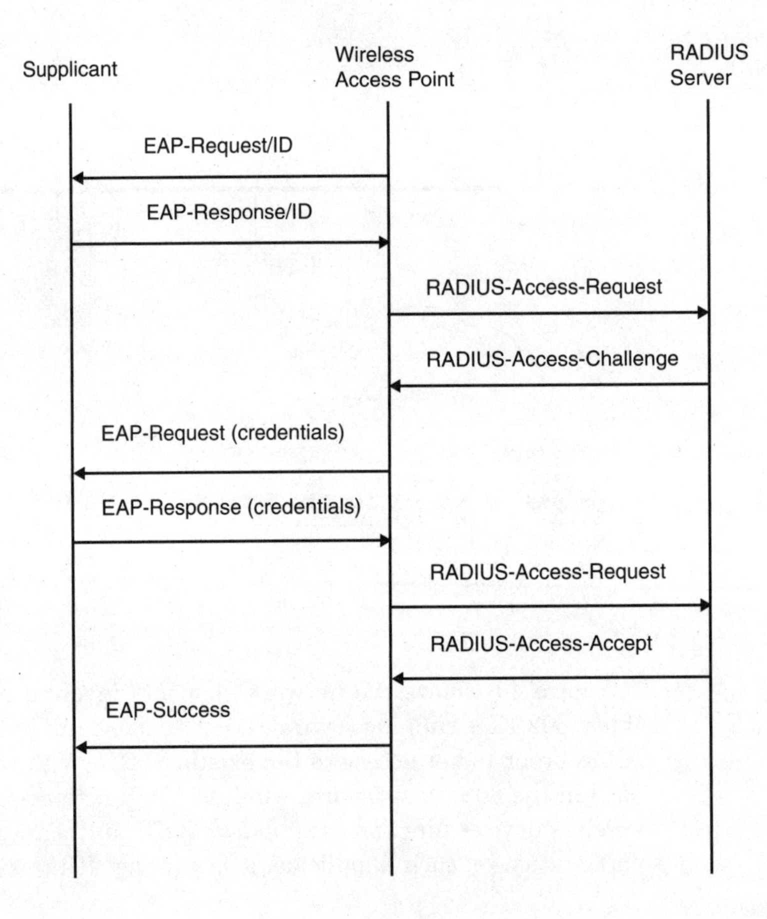

access WLANs at many different points, but still be authenticated against the same server.

In response to the access-request, the RADIUS server sends an access-challenge to the wireless AP, which is then relayed in the form of an EAP-request to the device. The device sends it credentials to the wireless AP, which in turn relays them to the RADIUS server. The RADIUS server determines whether or not access to the network is accepted or denied based on the supplicant's credentials.

An example of a product in this space is SecureSupplicant™ by Meetinghouse Data Communications.[28] SecureSupplicant enables network administrators to continue to use RADIUS as their centralized authentication server. In 802.11b, authentication took place between the AP and the station; there was no concept of passing credentials from the AP to an authentication server. For traditional LANs, this was fine, but as users began to use their devices in remote locations, this became inadequate. 802.1X solves this problem by enabling APs to pass through client credentials to the appropriate authentication server.

For example, Figure 4-10 represents the authentication flow for a mobile user who wishes to create a VPN in his home office. By using the SecureSupplicant, the user can associate with a wireless network provided by a third party, in this case the Internet service provider (ISP). We assume that the company and the ISP have established a service relationship beforehand. When the ISP receives the user's credentials, it proxies these credentials to the company's RADIUS server, which either accepts or denies the user. This response is then propagated to the remote user.

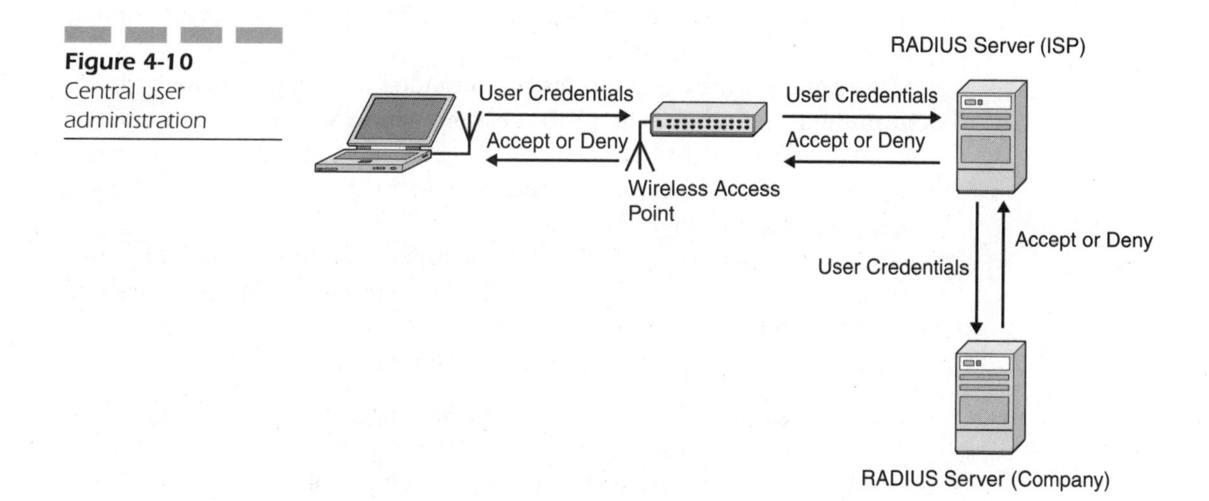

Figure 4-10 Central user administration

One problem with WEP is that the shared key used by the station and the AP is inherently static. That is, this shared key will only change if it is manually reconfigured on both devices. Software such as the SecureSupplicant remedies this by supporting the TLS protocol. TLS ensures that a new shared key is generated each time a station associates itself with an AP. TLS has proven itself to be an excellent authentication and encryption protocol in commercial environments.

PPP Extensible Authentication Protocol (EAP)

We now turn to EAP, mentioned several times in previous sections, but will now be discussed in some detail.

The Point-to-Point Protocol (PPP)[29] provides a standard method for transporting multiprotocol datagrams over point-to-point links. PPP also defines an extensible Link Control Protocol (LCP) that enables the negotiation of a protocol for authenticating its peer before letting Network Layer protocols transmit over the link. RFC 2284[30] defines the PPP EAP.

In order to establish communication over a point-to-point link, each end of the PPP link must first send LCP packets to configure the data link during the link establishment phase. After the link has been established, PPP provides for an optional authentication phase before proceeding to the Network Layer Protocol phase. By default, authentication is not mandatory. If authentication of the link is desired, an implementation must specify the authentication-protocol configuration option during the link establishment phase.

These authentication protocols are intended for use primarily by hosts and routers that connect to a PPP network server via switched circuits or dial-up lines, but might be applied to dedicated links as well. The server can use the identification of the connecting host or router in the selection of options for Network layer negotiations.

This section defines the PPP EAP. The link establishment and authentication phases, as well as the authentication-protocol configuration option, are defined in PPP.[29,31]

The RFC frequently uses the following terms:

- **Authenticator** The end of the link requiring the authentication. The authenticator specifies the authentication protocol to be used in the configure-request during the link establishment phase.

- **Peer** The other end of the point-to-point link—the end that is being authenticated by the authenticator.
- **Silently discard** This means the implementation discards the packet without further processing. The implementation should provide the capability of logging the error, including the contents of the silently discarded packet, and should record the event in a statistics counter.
- **Displayable message** This is interpreted to be a human-readable string of characters and must not affect the operation of the protocol. The message encoding must follow the UTF-8 transformation format.[32]

The PPP EAP is a general protocol for PPP authentication that supports multiple authentication mechanisms. EAP does not select a specific authentication mechanism at the link control phase, but rather postpones this until the authentication phase. This allows the authenticator to request more information before determining the specific authentication mechanism. This also permits the use of a backend server that actually implements the various mechanisms while the PPP authenticator merely passes through the authentication exchange. The authentication consists of the following steps:

1. After the link establishment phase is complete, the authenticator sends one or more requests to authenticate the peer. The request has a type field to indicate what is being requested. Examples of request types include identity, MD5-challenge, one-time passwords, and generic token card. The MD5-challenge type corresponds closely to the Challenge Handshake Authentication Protocol (CHAP). Typically, the authenticator will send an initial identity request followed by one or more requests for authentication information. However, an initial identity request is not required and may be bypassed in cases where the identity is presumed (leased lines, dedicated dial-ups, and so on).
2. The peer sends a response packet in reply to each request. As with the request packet, the response packet contains a type field that corresponds to the type field of the request.
3. The authenticator ends the authentication phase with a success or failure packet.

Advantages

EAP can support multiple authentication mechanisms without having to prenegotiate a particular one during the LCP phase. Certain devices (such as a network access server [NAS]) do not necessarily have to understand each request type and may be able to simply act as a passthrough agent for a backend server on a host. The device only needs to look for the success/failure code to terminate the authentication phase.

Disadvantages

EAP does require the addition of a new authentication type to LCP and thus PPP implementations will need to be modified to use it. It also strays from the previous PPP authentication model of negotiating a specific authentication mechanism during LCP.

Configuration Option Format

A summary of the authentication-protocol configuration option format for negotiating EAP appears in Figure 4-11. Fields are transmitted from left to right.

Type 3
Length 4
Authentication Protocol C227 (Hex) for PPP EAP

Packet Format

Exactly one PPP EAP packet is encapsulated in the Information field of a PPP Data Link layer frame where the protocol field indicates type hex C227 (PPP EAP). A summary of the EAP packet format is shown in Figure 4-12. The fields are transmitted from left to right.

- **Code** The Code field is one octet and identifies the type of EAP packet. EAP codes are assigned as follows:

Figure 4-11 Authentication-protocol configuration option format

```
 0                   1                   2                   3
 0 1 2 3 4 5 6 7 8 9 0 1 2 3 4 5 6 7 8 9 0 1 2 3 4 5 6 7 8 9 0 1
+-+-+-+-+-+-+-+-+-+-+-+-+-+-+-+-+-+-+-+-+-+-+-+-+-+-+-+-+-+-+-+-+
|     Type      |    Length     |    Authentication-Protocol    |
+-+-+-+-+-+-+-+-+-+-+-+-+-+-+-+-+-+-+-+-+-+-+-+-+-+-+-+-+-+-+-+-+
```

Figure 4-12
EAP packet format

```
 0                   1                   2                   3
 0 1 2 3 4 5 6 7 8 9 0 1 2 3 4 5 6 7 8 9 0 1 2 3 4 5 6 7 8 9 0 1
+-+-+-+-+-+-+-+-+-+-+-+-+-+-+-+-+-+-+-+-+-+-+-+-+-+-+-+-+-+-+-+-+
|     Code      |  Identifier   |            Length             |
+-+-+-+-+-+-+-+-+-+-+-+-+-+-+-+-+-+-+-+-+-+-+-+-+-+-+-+-+-+-+-+-+
|    Data ...
+-+-+-+-+
```

1 Request

2 Response

3 Success

4 Failure

- **Identifier** The Identifier field is one octet and aids in matching responses with requests.
- **Length** The Length field is two octets and indicates the length of the EAP packet including the Code, Identifier, Length, and Data fields. Octets outside the range of the Length field should be treated as Data Link layer padding and should be ignored on reception.
- **Data** The Data field is zero or more octets. The Code field determines the format of the Data field.

Request and Response

- **Description** The request packet is sent by the authenticator to the peer. Each request has a Type field that serves to indicate what is being requested. The authenticator must transmit an EAP packet with the Code field set to 1 (request). Additional request packets must be sent until a valid response packet is received, or an optional retry counter expires. Retransmitted requests must be sent with the same identifier value in order to distinguish them from new requests. The contents of the Data field depend on the request type. The peer must send a response packet in reply to a request packet. Responses must only be sent in reply to a received request and never be retransmitted on a timer. The Identifier field of the response must match that of the request.
- **Implementation note** Because the authentication process will often involve user input, some care must be taken when deciding upon retransmission strategies and authentication timeouts. It is suggested

that a retransmission timer of 6 seconds with a maximum of 10 retransmissions be used as default. One may wish to make these timeouts longer in certain cases (when token cards are involved). Additionally, the peer must be prepared to silently discard received retransmissions while waiting for user input.

See a summary of the request and response packet format in Figure 4-13. Fields are transmitted from left to right.

- **Code** 1 for request; 2 for response.
- **Identifier** The Identifier field is one octet. The Identifier field must be the same if a request packet is retransmitted due to a timeout while waiting for a response. Any new (nonretransmission) requests must modify the Identifier field. If a peer receives a duplicate request for which it has already sent a response, it must resend its response. If a peer receives a duplicate request before it has sent a response to the initial request (it is waiting for user input), it must silently discard the duplicate request.
- **Length** The Length field is two octets and indicates the length of the EAP packet including the Code, Identifier, Length, Type, and Type-Data fields. Octets outside the range of the Length field should be treated as Data Link layer padding and should be ignored on reception.
- **Type** The Type field is one octet. This field indicates the type of request or response. Only one type must be specified per EAP request or response. Normally, the Type field of the response will be the same as the type of the request. However, there is also a Nak response type for indicating that a request type is unacceptable to the peer. When sending a Nak in response to a request, the peer may indicate an alternative desired authentication type that it supports.
- **Type-Data** The Type-Data field varies with the type of request and the associated response.

Figure 4-13 Request and response packet format

```
 0                   1                   2                   3
 0 1 2 3 4 5 6 7 8 9 0 1 2 3 4 5 6 7 8 9 0 1 2 3 4 5 6 7 8 9 0 1
+-+-+-+-+-+-+-+-+-+-+-+-+-+-+-+-+-+-+-+-+-+-+-+-+-+-+-+-+-+-+-+-+
|     Code      |  Identifier   |            Length             |
+-+-+-+-+-+-+-+-+-+-+-+-+-+-+-+-+-+-+-+-+-+-+-+-+-+-+-+-+-+-+-+-+
|     Type      |  Type-Data ...
+-+-+-+-+-+-+-+-+-+-+-+-+-+-+-+-+-+-+-+-+-+-+-+-+
```

Success and Failure

- **Description** The success packet is sent by the authenticator to the peer to acknowledge successful authentication. The authenticator must transmit an EAP packet with the Code field set to 3 (success).

  If the authenticator cannot authenticate the peer (unacceptable responses to one or more requests), then the implementation must transmit an EAP packet with the Code field set to 4 (failure). An authenticator may wish to issue multiple requests before sending a failure response in order to allow for human typing mistakes.
- **Implementation note** Because the success and failure packets are not acknowledged, they may be potentially lost. A peer must allow for this circumstance. The peer can use a Network Protocol packet as an alternative indication of success. Likewise, the receipt of a LCP Terminate-Request can be taken as a failure.

A summary of the success and failure packet format is shown in Figure 4-14. The fields are transmitted from left to right.

- **Code** 3 for success; 4 for failure.
- **Identifier** The Identifier field is one octet and aids in matching replies to responses. The Identifier field must match the Identifier field of the response packet that it is sent in response to.
- **Length** 4.

Initial EAP Request/Response Types

Let's define an initial set of EAP types used in request/response exchanges. Additional types may be defined at a future point in follow-on documents issued by the IETF. The Type field is one octet and identifies the structure of an EAP request or response packet. The first three types are considered special case types. The remaining types define authentication exchanges.

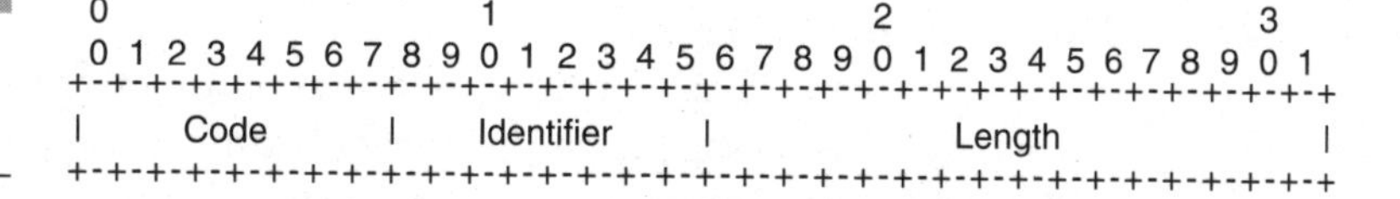

Figure 4-14 Success and failure packet format

The Nak type is valid only for response packets; it must not be sent in a request. The Nak type must only be sent in response to a request that uses an authentication type code. All EAP implementations must support types 1 through 4. These types, as well as types 5 and 6, are defined:

1 Identity

2 Notification

3 Nak (response only)

4 MD5-challenge

5 One-time password (OTP) (RFC 1938)

6 Generic token card

Identity

- **Description** The Identity type is used to query the identity of the peer. Generally, the authenticator will issue this as the initial request. An optional, displayable message may be included to prompt the peer in the case of interaction with a user. A response must be sent to this request with a type of 1 (identity).
- **Implementation note** The peer may obtain the identity via user input. It is suggested that the authenticator retry the identity request in the case of an invalid identity or authentication failure to allow for potential typos on the part of the user. It is suggested that the identity request be retried a minimum of three times before terminating the authentication phase with a failure reply. The notification request may be used to indicate an invalid authentication attempt prior to transmitting a new identity request (optionally, the failure may be indicated within the message of the new identity request itself).
- **Type** 1
- **Type-Data** This field may contain a displayable message in the request. The response uses this field to return the identity. If the identity is unknown, this field should be zero bytes in length. The field must not be null terminated. The length of this field is derived from the Length field of the request/response packet and hence a null is not required.

Notification

- **Description** The Notification type is optionally used to convey a displayable message from the authenticator to the peer. The peer should display this message to the user or log it if it cannot be displayed. It is intended to provide an acknowledged notification of some imperative nature. Examples include a password with an expiration time that is about to expire, an OTP sequence integer that is nearing zero, an authentication failure warning, and so on. In most circumstances, notification should not be required.

- **Type** 2
- **Type-Data** The Type-Data field in the request contains a displayable message greater than zero octets in length. The length of the message is determined by the Length field of the request packet. The message must not be null terminated. A response must be sent in reply to the request with a Type field of 2 (notification). The Type-Data field of the response is zero octets in length. The response should be sent immediately (independent of how the message is displayed or logged).

Nak

- **Description** The Nak type is valid only in response messages. It is sent in reply to a request when the desired authentication type is unacceptable. Authentication types are numbered 4 and above. The response contains the authentication type desired by the peer.
- **Type** 3
- **Type-Data** This field must contain a single octet indicating the desired authentication type.

MD5-Challenge

- **Description** The MD5-challenge type is analogous to the PPP CHAP protocol[33] (with MD5 as the specified algorithm). The PPP CHAP RFC[33] should be referred to for further implementation specifics. The request contains a "challenge" message to the peer. A response must be sent in reply to the request. The response may be either of Type 4 (MD5-challenge) or Type 3 (Nak). The Nak reply indicates the peer's desired authentication mechanism type. All EAP implementations must support the MD5-challenge mechanism.
- **Type** 4
- **Type-Data** The contents of the Type-Data field are summarized in Figure 4-15. For reference on the use of this field, see the PPP CHAP.[33]

One-Time Password (OTP)

- **Description** The OTP system is defined in RFC 1938, "A One-Time Password System."[34] The request contains a displayable message containing an OTP challenge. A response must be sent in reply to the

Figure 4-15 Contents of the Type-Data field

```
 0                   1                   2                   3
 0 1 2 3 4 5 6 7 8 9 0 1 2 3 4 5 6 7 8 9 0 1 2 3 4 5 6 7 8 9 0 1
+-+-+-+-+-+-+-+-+-+-+-+-+-+-+-+-+-+-+-+-+-+-+-+-+-+-+-+-+-+-+-+-+
|  Value-Size   |  Value ...                                    |
+-+-+-+-+-+-+-+-+-+-+-+-+-+-+-+-+-+-+-+-+-+-+-+-+-+-+-+-+-+-+-+-+
|  Name ...
+-+-+-+-+-+-+-+-+-+-+-+-+-+-+-+-+
```

request. The response must be of Type 5 (OTP) or Type 3 (Nak). The Nak reply indicates the peer's desired authentication mechanism Type.

- **Type** 5
- **Type-Data** The Type-Data field contains the OTP challenge as a displayable message in the request. In the response, this field is used for the six words from the OTP dictionary.[34] The messages must not be null terminated. The length of the field is derived from the Length field of the request/reply packet.

Generic Token Card

- **Description** The generic token card type is defined for use with various token card implementations requiring user input. The request contains an ASCII text message and the reply contains the token card information necessary for authentication. Typically, this would be information read by a user from the token card device and entered as ASCII text.
- **Type** 6
- **Type-Data** The Type-Data field in the request contains a displayable message greater than zero octets in length. The length of the message is determined by the Length field of the request packet. The message must not be null terminated. A response must be sent in reply to the request with a Type field of six (generic token card). The response contains data from the token card required for authentication. The length of the data is determined by the Length field of the response packet.

Security Considerations for PPP

Security issues are the primary topic of RFC 2284. The interactions of the authentication protocols within PPP are highly implementation dependent. For example, upon failure of authentication, some implementations do not terminate the link. Instead, the implementation limits the kind of traffic in the Network Layer protocols to a filtered subset, which in turn gives the user the opportunity to update secrets or send mail to the network administrator, indicating a problem.

There is no provision for retries of failed authentication. However, the LCP state machine can renegotiate the authentication protocol at any time, thus allowing a new attempt. It is recommended that any counters used for authentication failure not be reset until after successful authentication, or subsequent termination of the failed link.

There is no requirement that authentication be full duplex or that the same protocol be used in both directions. It is perfectly acceptable for different protocols to be used in each direction. This will, of course, depend on the specific protocols negotiated.

In practice, within or associated with each PPP server, it is not anticipated that a particular named user would be authenticated by multiple methods. This would make the user vulnerable to attacks that negotiate the least secure method from among a set (such as the Password Authentication Protocol [PAP] rather than EAP). Instead, for each named user, there should be an indication of exactly one method used to authenticate that user name. If a user needs to make use of different authentication methods under different circumstances, then distinct identities should be employed, each of which identifies exactly one authentication method.[35]

Authentication, Authorization, and Accounting (AAA)

The Mobile IP and Authentication, Authorization, and Accounting (AAA) working groups in IETF have been looking at defining the requirements for AAA. RFC 2977[36] contains the requirements that would have to be supported by an AAA service to aid in providing Mobile IP services.

Introduction

Clients[37] obtain Internet services by negotiating a point of attachment to a "home domain," generally from an ISP or another organization from which service requests are made and fulfilled. With the increasing popularity of mobile devices, a need has been generated to allow users to attach to any domain convenient to their current location. In this way, a client needs access to resources being provided by an administrative domain different than their home domain (called a foreign domain). The need for service from a foreign domain requires, in many models, authorization, which leads directly to authentication, and accounting (hence, AAA). There is some discussion as to which of these leads to the others, or is derived from the others, but there is common agreement that the three AAA functions are closely interdependent.

An agent in a foreign domain, being called on to provide access to a resource by a mobile user, is likely to request or require the client to provide credentials that can be authenticated before access to resources is permitted. The resource may be as simple as a conduit to the Internet or may be as complex as access to specific private resources within the foreign domain. Credentials can be exchanged in many different ways, all of which are beyond the scope of this chapter. Once authenticated, the mobile user may be authorized to access services within the foreign domain. An accounting of the actual resources may then be assembled.

Mobile IP is a technology that enables a network node (a mobile node) to migrate from its home network to other networks, either within the same administrative domain or to other administrative domains. The possibility of movement between domains that require AAA services has created an immediate demand to design and specify AAA protocols. Once available, the AAA protocols and infrastructure will provide the economic incentive for a wide-ranging deployment of Mobile IP. This document will identify, describe, and discuss the functional and performance requirements that Mobile IP places on AAA protocols. The formal description of Mobile IP can be found in three articles by Perkins and one by Solomon and Glass.[38,39,40,41]

In RFC 2977, an attempt is made to exhibit the requirements in a progressive fashion. After showing the basic AAA model for Mobile IP, it lists the model's requirements as follows:

- It must be based on the general model.
- It must be based on providing IP service for mobile nodes.
- It must be derived from specific Mobile IP protocol needs.

Terminology

RFC 2977 frequently uses the following terms in addition to those defined in RFC 2002:[39]

- **Accounting** The act of collecting information on resource usage for the purpose of trend analysis, auditing, billing, or cost allocation.
- **Administrative domain** An intranet or a collection of networks, computers, and databases under a common administration. Computer entities operating in a common administration may be assumed to share administratively created security associations.
- **Attendant** A node designed to provide the service interface between a client and the local domain.
- **Authentication** The act of verifying a claimed identity in the form of a preexisting label from a mutually known name space, as the originator of a message (message authentication) or as the end point of a channel (entity authentication).
- **Authorization** The act of determining if a particular right, such as access to some resource, can be granted to the presenter of a particular credential.
- **Billing** The act of preparing an invoice.
- **Broker** An intermediary agent, trusted by two other AAA servers, that is able to obtain and provide security services from those AAA servers. For instance, a broker may obtain and provide authorizations or assurances that credentials are valid.
- **Client** A node wishing to obtain service from an attendant within an administrative domain.
- **Foreign domain** An administrative domain, visited by a Mobile IP client, and containing the AAA infrastructure needed to carry out the necessary operations enabling Mobile IP registrations. From the point of view of the foreign agent, the foreign domain is the local domain.
- **Interdomain accounting** Interdomain accounting is the collection of information on the resource usage of an entity within an administrative domain, for use within another administrative domain. In interdomain accounting, accounting packets and session records will typically cross administrative boundaries.
- **Intradomain accounting** Intradomain accounting is the collection of information on resources within an administrative domain, for use within that domain. In intradomain accounting, accounting packets and session records typically do not cross administrative boundaries.
- **Local domain** An administrative domain containing the AAA infrastructure of immediate interest to a Mobile IP client when it is away from home.

- **Real-time accounting** Real-time accounting involves the processing of information on resource usage within a defined time window. Time constraints are typically imposed in order to limit financial risk.
- **Session record** This represents a summary of the resource consumption of a user over an entire session. Accounting gateways creating the session record may do so by processing interim accounting events.

Basic Model

This section covers the main features of a basic model for the operation of AAA servers that have good support within the Mobile IP working group. On the Internet, a client belonging to one administrative domain (called the home domain) often needs to use resources provided by another administrative domain (called the foreign domain). An agent in the foreign domain that attends to the client's request (the agent is also known as the attendant) is likely to require that the client provide some credentials that can be authenticated before access to the resources is permitted. These credentials may be something the foreign domain understands, but in most cases they are assigned by, and understood only by, the home domain and may be used for setting up secure channels with the mobile node (see Figure 4-16).

The attendant often does not have direct access to the data needed to complete the transaction. Instead, the attendant is expected to consult an authority (typically in the same foreign domain) in order to request proof that the client has acceptable credentials. Since the attendant and the local authority are part of the same administrative domain, they are expected to

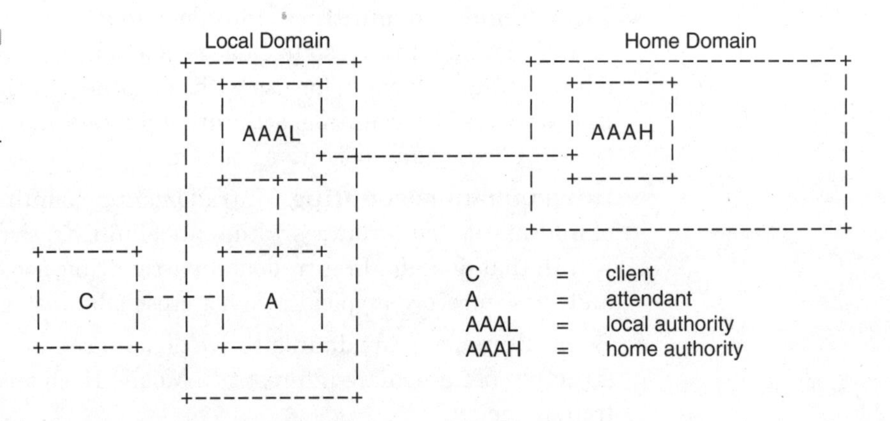

Figure 4-16 AAA servers in home and local domains

have established, or be able to establish for the necessary lifetime, a secure channel for the purposes of exchanging sensitive (access) information and keep it private from (at least) the visiting mobile node.

The local authority (AAA Local [AAAL]) itself may not have enough information stored locally to carry out the verification of the client's credentials. In contrast to the attendant, the AAAL is to be configured with enough information to negotiate the verification of client credentials with external authorities. The local and the external authorities should be configured with sufficient security relationships and access controls so that they, possibly without the need for any other AAA agents, can negotiate the authorization that may enable the client to have access to any/all requested resources. In many typical cases, the authorization depends only upon secure authentication of the client's credentials.

Once the local authority has obtained authorization and notified the attendant about the successful negotiation, the attendant can provide the requested resources to the client. There may be many attendants for each AAAL in the picture, and there might be many clients from many different home domains. Each home domain provides an AAAH that can check credentials originating from clients administered by that home domain.

An implicit security model is shown in Figure 4-16, and identifying the specific security associations it assumes is quite important. First, it is natural to assume that the client has a security association with the AAAH, since that is roughly what it means for the client to belong to the home domain. Second, as Figure 4-16 suggests, AAAL and AAAH have to share a security association, otherwise they could not rely on the authentication results, authorizations, or even the accounting data that might be transacted between them. Requiring such bilateral security relationships is, however, not scalable in the end; the AAA framework must provide for better scalability.

Finally, the figure clearly shows that the attendant can naturally share a security association with the AAAL. This is necessary in order for the model to work because the attendant has to know that it is permissible to allocate the local resources to the client.

As an example from today's Internet, one can cite the deployment of RADIUS[42] as an instance in which mobile computer clients have access to the Internet by way of a local ISP. The ISP wants to make sure that the mobile client can pay for the connection. Once the client has provided his or her credentials (identification, unique data, and an unforgettable signature), the ISP checks with the client's home authority to verify the signature and to obtain assurance that the client will pay for the connection.

Here the attendant function can be carried out by the NAS, and the local and home authorities can use RADIUS servers. Credentials allowing authorization at one attendant should be unusable in any future negotiations at the same or any other attendant.

From the previous description and example, several requirements fall out:

- Each local attendant has to have a security relationship with the local AAA server (AAAL). The local authority has to share, or dynamically establish, security relationships with external authorities that are able to check client credentials.
- The attendant has to keep state (retain information) for pending client requests while the local authority contacts the appropriate external authority. Since the mobile node may not necessarily initiate network connectivity from within its home domain, it must be able to provide complete yet unforgeable credentials without ever having been in touch with its home domain. Since the mobile node's credentials have to remain unforgeable, intervening nodes (neither the attendant nor the local authority [AAAL] or any other intermediate nodes) must not be able to learn any (secret) information that may enable them to reconstruct and reuse the credentials.

It is easy to see the reasons for the natural requirement that the client has to share, or dynamically establish, a security relationship with the external authority in the home domain. Otherwise, it is technically infeasible (given the implied network topology) for the client to produce unforgeable signatures that can be checked by the AAAH. Figure 4-17 illustrates

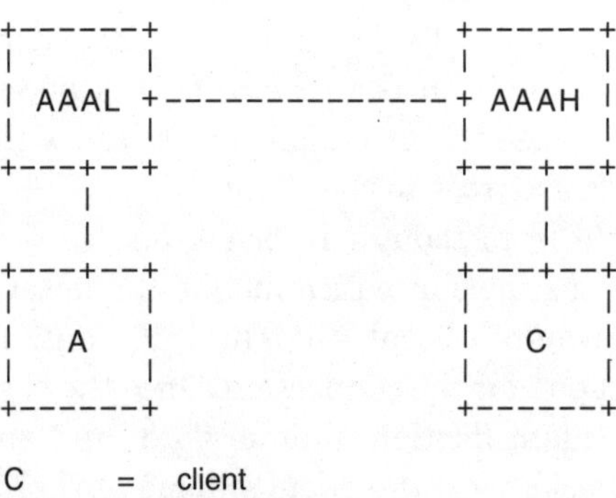

Figure 4-17 Security associations

the natural security associations one understands from the proposed model. Note that there may, by mutual agreement between AAAL and AAAH, be a third party inserted between AAAL and AAAH to help them arbitrate secure transactions in a more scalable fashion.

In addition to the requirements listed, one specifies the following requirements that derive from operational experience with today's roaming protocols:

- An attendant will have to manage requests for many clients at the same time.
- The attendant must protect against replay attacks.
- The attendant equipment should be as inexpensive as possible, since it will be replicated as many times as possible to handle as many clients as possible in the foreign domain.
- Attendants should be configured to obtain authorization from a trusted local AAA server (AAAL) for QoS requirements placed by the client.

Nodes in two separate administrative domains (for instance, AAAH and AAAL) often must take additional steps to verify the identity of their communication partners, or alternatively to guarantee the privacy of the data making up the communication. Although these considerations lead to important security requirements, as mentioned previously in the context of security between servers, the exact choice of security associations between the AAA servers is beyond the scope of this chapter. The choices are unlikely even to depend upon any specific features of the general model illustrated in Figure 4-16. On the other hand, the security associations needed between Mobile IP entities will be of central importance in the design of a suitable AAA infrastructure for Mobile IP. The general model is generally compatible with the needs of Mobile IP. However, some basic changes are needed in the security model of Mobile IP.

Lastly, recent discussion on the Mobile IP working group indicates that the attendant must be able to terminate service to the client based on policy determination by either the AAAH or AAAL server.

AAA Protocol Roaming Requirements

This section details additional requirements based on issues discovered through operational experience with existing roaming RADIUS networks. The AAA protocol must satisfy these requirements in order for providers to offer a robust service. These requirements have been identified by TR45.6 as part of their involvement with the Mobile IP working group:

- A reliable AAA transport mechanism must be supported:
  - There must be an effective hop-by-hop retransmission and failover mechanism so that reliability does not solely depend on end-to-end retransmission.
  - This transport mechanism must be able to indicate to an AAA application that a message was delivered to the next peer AAA application or that a timeout occurred.
  - Retransmission is controlled by the reliable AAA transport mechanism, and not by lower-layer protocols such as TCP.
  - Even if the AAA message is to be forwarded, or the message's options or semantics do not conform to the AAA protocol, the transport mechanism will acknowledge that the peer received the AAA message.
  - Acknowledgements may be piggybacked in AAA messages.
  - AAA responses have to be delivered in a timely fashion so that Mobile IP does not timeout and retransmit.
- A digital certificate in an AAA message must be transported in order to minimize the number of round trips associated with AAA transactions. Note that this requirement applies to AAA applications and not mobile stations. Certificates could be used by foreign and home agents to establish an IPSec association empowered to protect the mobile node's tunneled data. In this case, the AAA infrastructure could assist by obtaining the revocation status of such a certificate (either by performing online checks or otherwise validating the certificate) so that home and foreign agents could avoid a costly online certificate status check.
- Message integrity and identity authentication on a hop-by-hop (AAA node) basis must be provided.
- Replay protection and optional nonrepudiation capabilities for all authorization and accounting messages must be supported. The AAA protocol must provide the capability for accounting messages to be matched with prior authorization messages.
- Accounting must be supported via both bilateral arrangements and broker AAA servers providing accounting clearinghouse services and reconciliation between serving and home networks. There is an explicit agreement that if the private network or home ISP authenticates the mobile station requesting service, then the private network or home ISP network also agrees to reconcile charges with the home service provider or broker. Real-time accounting must be supported. Timestamps must be included in all accounting packets.

Requirements for Basic IP Connectivity

The requirements listed in the previous section pertain to the relationships between the functional units and do not depend on the underlying network addressing. On the other hand, many nodes (mobile or merely portable) are programmed to receive some IP-specific resources during the initialization phase of their attempt to connect to the Internet.

The RFC places the following additional requirements on the AAA services in order to satisfy such clients:

- Any AAA server must be able to obtain, or to coordinate the allocation of, a suitable IP address for the customer, upon request by the customer.
- AAA servers must be able to identify the client by some means other than its IP address.

Policy in the home domain may dictate that the home agent instead of the AAAH manages the allocation of an IP address for the mobile node. AAA servers must be able to coordinate the allocation of an IP address for the mobile node at least in this way.

AAA servers today identify clients by using the Network Access Identifier (NAI).[43] A mobile node can identify itself by including the NAI along with the Mobile IP registration request.[44] The NAI is of the form "user@realm." It is unique and well suited for use in the AAA model illustrated in Figure 4-16. Using a NAI such as user@realm allows AAAL to easily determine the home domain ("realm") for the client. Both the AAAL and the AAAH can use the NAI to keep records indexed by the client's specific identity.

AAA for Mobile IP

Clients using Mobile IP require specific features from the AAA services in addition to the requirements already mentioned for basic AAA functionality and IP connectivity. To understand the application of the general model for Mobile IP, consider the mobile node to be the client in Figure 4-16, and the attendant to be the foreign agent. If a situation arises that no foreign agent is present (as in the case of an IPv4 mobile node with a co-located care-of address or an IPv6 mobile node), the equivalent attendant functionality is to be provided by the address allocation entity, such as a DHCP server, for instance. Such attendant functionality is outside the scope of this chapter.

The home agent, while important to Mobile IP, is allowed to play a role during the initial registration that is subordinate to the role played by the AAAH. For an application used with Mobile IP, one modifies the general model (as illustrated in Figure 4-18). After the initial registration, the mobile node is authorized to continue using Mobile IP at the foreign domain without requiring further involvement by the AAA servers. Thus, the initial registration will probably take longer than subsequent Mobile IP registrations.

In order to reduce this extra time overhead as much as possible, it is important to reduce the time taken for communications between the AAA servers. A major component of this communication latency is the time taken to traverse the wide area Internet that is likely to separate the AAAL and the AAAH. This leads to a further strong motivation for the integration of the AAA functions themselves as well as the integration of AAA functions with the initial Mobile IP registration. In order to reduce the number of messages that traverse the network for the initial registration of a mobile node, the AAA functions in the visited network (AAAL) and the home network (AAAH) need to interface with the foreign agent and the home agent to handle the registration message. Latency would be reduced as a result of the initial registration being handled in conjunction with AAA and the Mobile IP mobility agents. Subsequent registrations, however, would be handled according to RFC 2002.[39] Another way to reduce latency as to accounting would be the exchange of small records.

As attendants may provide many different types of subservices to mobile clients, there must be extensible accounting formats. In this way, the specific services being provided can be identified, including accounting support, should more services be identified in the future.

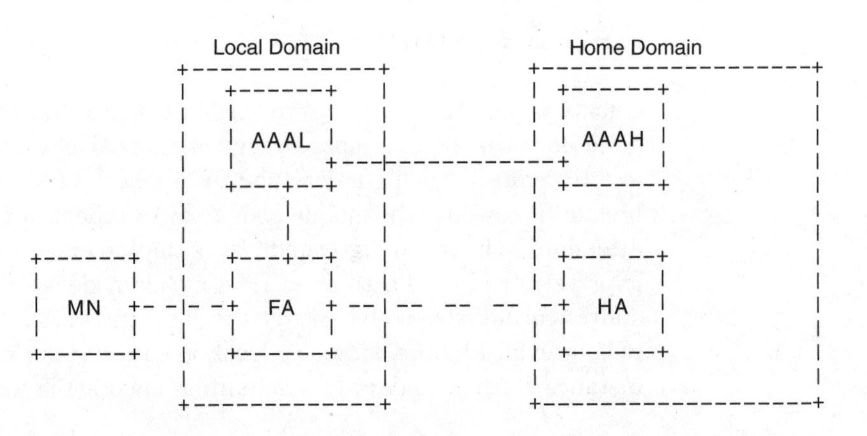

Figure 4-18 AAA Servers with Mobile IP Agents

The AAA home domain and the home agent home domain of the mobile node need not be part of the same administrative domain. Such a situation can occur if the home address of the mobile node is provided by one domain, such as an ISP that the mobile user uses while at home, and the authorization and accounting by another (specialized) domain, such as a credit card company. The foreign agent sends only the authentication information of the mobile node to the AAAL, which interfaces with the AAAH.

After a successful authorization of the mobile node, the foreign agent is able to continue with the mobile IP registration procedure. Such a scheme introduces more delay if the access to the AAA functionality and the mobile IP protocol is sequentialized. Subsequent registrations would be handled according to RFC 2002[39] without further interaction with the AAA.

Whether to combine or separate the Mobile IP protocol data with or from the AAA messages is ultimately a policy decision. A separation of the Mobile IP protocol data and the AAA messages can be successfully accomplished only if the IP address of the mobile node's home agent is provided to the foreign agent performing the attendant function.

All of the needed AAA and Mobile IP functions should be processed during a single Internet traversal. This must be done without requiring AAA servers to process protocol messages sent to Mobile IP agents. The AAA servers must identify the Mobile IP agents and security associations necessary to process the Mobile IP registration, pass the necessary registration data to those Mobile IP agents, and remain uninvolved in the routing and authentication processing steps particular to Mobile IP registration.

For Mobile IP, the AAAL and the AAAH servers have the following additional general tasks:

- Enable (re)authentication for Mobile IP registration.
- Authorize the mobile node (once its identity has been established) to use at least the set of resources for minimal Mobile IP functionality, plus potentially other services requested by the mobile node.
- Initiate accounting for service utilization.
- Use AAA protocol extensions specifically for including Mobile IP registration messages as part of the initial registration sequence to be handled by the AAA servers.

These tasks, and the resulting more specific tasks to be listed later in this section, are beneficially handled and expedited by the AAA servers shown in Figure 4-16 because the tasks often happen together, and task processing needs access to the same data at the same time.

In the model in Figure 4-16, the initial AAA transactions are handled without needing the home agent, but Mobile IP requires every registration to be handled between the home agent and the foreign agent, as shown by the sparse dashed (lower) line between them in the figure. This means that during the initial registration, something has to happen that enables the home agent and foreign agent to perform subsequent Mobile IP registrations. After the initial registration, the AAAH and AAAL in Figure 4-18 would not be needed, and subsequent Mobile IP registrations would only follow the lower control path between the foreign agent and the home agent.

Any Mobile IP data that is sent by the foreign agent through the AAAL to AAAH must be considered opaque to the AAA servers. Authorization data needed by the AAA servers then must be delivered to them by the foreign agent from the data supplied by the mobile node. The foreign agent becomes a translation agent between the Mobile IP registration protocol and AAA.

As we have discussed, nodes in two separate administrative domains often must take additional steps to guarantee their security and privacy, as well as the security and privacy of the data they are exchanging. In today's Internet, such security measures can be provided via several different algorithms. Some algorithms rely on the existence of a public-key infrastructure (Housley, Ford, Polk, and Solo); others rely on the distribution of symmetric keys to the communicating nodes (Kohl and Neuman). AAA servers should be able to verify credentials using either style in their interactions with Mobile IP entities.

In order to enable subsequent registrations, the AAA servers must be able to perform some key distribution during the initial Mobile IP registration process from any particular administrative domain in order to provide the following security functions:

- Identify or create a security association between the mobile node and the home agent; this is required for the mobile node to produce the (re)authentication data for the mobile node/home agent authentication extension, which is mandatory on Mobile IP registrations.
- Identify or create a security association between the mobile node and foreign agent for use with subsequent registrations at the same foreign agent, so that the foreign agent can continue to obtain assurance that the same mobile node has requested the continued authorization for Mobile IP services.
- Identify or create a security association between the home agent and foreign agent for use with subsequent registrations at the same foreign

agent, so that the foreign agent can continue to obtain assurance that the same home agent has continued the authorization for Mobile IP services for the mobile node.

- Participate in the distribution of the security association (and Security Parameter Index, or SPI) to the Mobile IP entities.
- Validate certificates provided by the mobile node and provide reliable indication to the foreign agent.
- Accept on the part of AAAL an indication from the foreign agent about the acceptable lifetime for its security associations with the mobile node and/or the mobile node's home agent. The lifetime for those security associations should be an integer multiple of the registration lifetime offered by the foreign agent to the mobile node. This may allow for Mobile IP reauthentication to take place without the need for reauthentication to take place on the AAA level, thereby shortening the time required for mobile node reregistration.
- Conditionally accept a Mobile IP registration authorization according to whether the registration requires broadcast or multicast service to the mobile node tunneled through the foreign agent.
- In addition, reverse tunneling may also be a necessary requirement for mobile node connectivity. Therefore, AAA servers should also be able to condition their acceptance of Mobile IP registration authorization depending upon whether the registration requires reverse tunneling support to the home domain through the foreign agent.

The lifetimes of any security associations distributed by the AAA server for use with Mobile IP should be great enough to avoid a too frequent initiation of the AAA key distribution, since each invocation of this process is likely to cause lengthy delays between (re)registrations.[45] Registration delays in Mobile IP cause dropped packets and noticeable disruptions in service. Note that any key distributed by AAAH to the foreign agent and home agent may be used to initiate Internet Key Exchange (IKE).[46]

Note further that the mobile node and home agent may well have a security association established that does not depend upon any action by the AAAH.

Mobile IP with Dynamic IP Addresses Many people would like their mobile nodes to be identified by their NAI and to obtain a dynamically allocated home address for use in the foreign domain. These people may often be unconcerned with details about how their computers implement Mobile IP and indeed may not have any knowledge of their home agent or any security association except for the one between themselves and the AAAH (see Figure 4-17). In this case, the Mobile IP registration data has to be carried along with the AAA messages. The AAA home domain and the HA home domain have to be part of the same administrative domain.

Mobile IP requires the home address assigned to the mobile node belong to the same subnet as the home agent providing service to the mobile node. For the effective use of IP home addresses, the home AAA (AAAH) should be able to select a home agent for use with the newly allocated home address. In many cases, the mobile node will already know the address of its home agent, even if the mobile node does not already have an existing home address. Therefore, the home AAA (AAAH) must be able to coordinate the allocation of a home address with a home agent that might be designated by the mobile node.

Allocating a home address and a home agent for the mobile further simplifies the configuration needs of a client's mobile node. Currently in the Proposed Standard Mobile IP specification, a mobile node has to be configured with a home address and the address of a home agent, as well as with a security association with that home agent. In contrast, the proposed AAA features would only require the mobile node to be configured with its NAI and a secure shared secret for use by the AAAH. The mobile node's home address, the address of its home agent, the security association between the mobile node and the home agent, and even the identity (Domain Name Server [DNS] name or IP address) of the AAAH can all be dynamically determined as part of Mobile IP initial registration with the mobility agent in the foreign domain (a foreign agent with AAA interface features). Nevertheless, the mobile node may choose to include the mobile node/home agent security extension as well as AAA credentials, and the proposed Mobile IP and AAA server model must work when both are present.

The reason for all this simplification is that the NAI encodes the client's identity as well as the name of the client's home domain; this follows the existing industry practice for the way NAIs are used today. The home domain name is then available for use by the local AAA (AAAL) to locate the home AAA serving the client's home domain. In the general model, the AAAL would also have to identify the appropriate security association for use with that AAAH. The section entitled "Broker Model" suggests a way to reduce the number of security associations that have to be maintained between pairs of AAA servers.

Firewalls and AAA Mobile IP has encountered some deployment difficulties related to firewall traversal; see Montenegro and Gupta,[47] for instance. Since the firewall and AAA server can be part of the same administrative domain, the RFC proposes that the AAA server should be able to issue control messages and keys to the firewall at the boundary of its administrative domain that will configure the firewall to be permeable to Mobile IP registration and data traffic from the mobile node.

Mobile IP with Local Home Agents In some Mobile IP models, mobile nodes boot on subnets that are technically foreign subnets, but the services they need are local, and hence communication with the home subnet as if they were residing on the home is not necessary. As long as the mobile node can get an address routable from within the current domain, it can use Mobile IP to roam around that domain, calling the subnet on which it booted its temporary home. This address is likely to be dynamically allocated upon request by the mobile node.

In such situations, when the client is willing to use a dynamically allocated IP address and does not have any preference for the location of the home network either geographical or topological, the local AAA server may be able to offer this additional allocation service to the client. Then the home agent will be located in the local domain, which is likely to reduce delays for new Mobile IP registrations.

In Figure 4-19, AAAL has received a request from the mobile node to allocate a home agent in the local domain. The new home agent receives keys from AAAL to enable future Mobile IP registrations. From the figure, it is evident that such a configuration avoids problems with firewall protection at the domain boundaries. On the other hand, this configuration makes it difficult for the mobile node to receive data from any communications partners in the mobile node's home administrative domain. Note that, in this model, the mobile node's home address is affiliated with the foreign domain for routing purposes. Thus, any dynamic update to DNS, to associate the mobile node's home Fully Qualified Domain Name (FQDN)[48] with

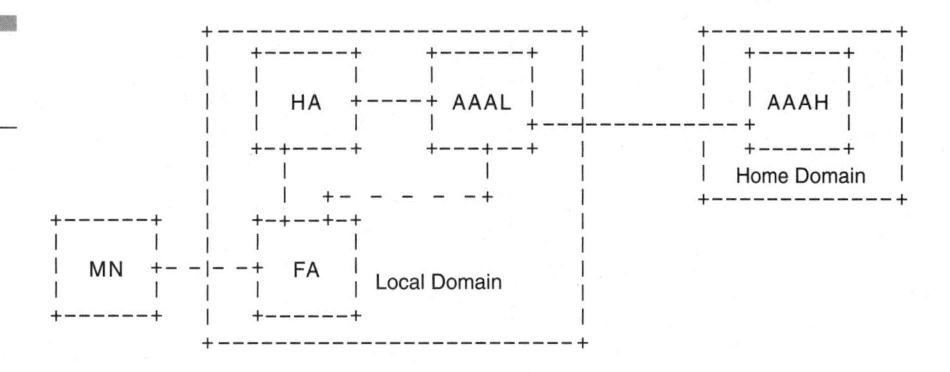

Figure 4-19 Home agent allocated by AAAL

its new IP address, will require the insertion of a foreign IP address into the home DNS server database.

Mobile IP with Local Payments Since the AAAL will be able to allocate a local home agent upon demand, one can make a further simplification. In cases where the AAAL can manage any necessary authorization function locally (if the client pays with cash or a credit card), then there is no need for an AAA protocol or infrastructure to interact with the AAAH. The resulting simple configuration is illustrated in Figure 4-20.

In this simplified model, one may consider that the role of the AAAH is taken over either by a national government (in the case of a cash payment), by a card authorization service if payment is by credit card, or some authority acceptable to all parties. Then the AAAL expects those external authorities to guarantee the value represented by the client's payment credentials (cash or credit). There are likely to be other cases where clients are granted access to local resources, or access to the Internet, without any charges at all. Such configurations may be found in airports and other common areas where business clients are likely to spend time. The service provider may find sufficient rewards in the goodwill of the clients or from advertisements displayed on Internet portals that are to be used by the clients. In such situations, the AAAL should still allocate a home agent, appropriate keys, and the mobile node's home address.

Fast Handover Since movement from coverage area to coverage area may be frequent in Mobile IP networks, it is imperative that the latency involved in the handoff process be minimized. See, for instance, Perkins and Johnson's Route Optimization article[49] for one way to do this using binding updates. When the mobile node enters a new visited subnet, it would be desirable for the mobile node to provide the previous foreign agent's NAI.

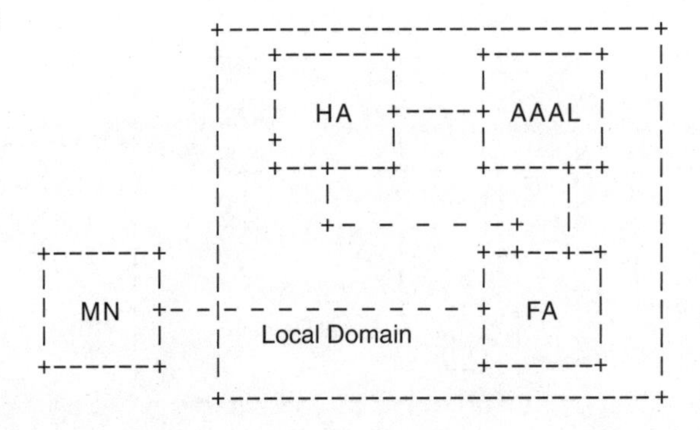

Figure 4-20 Local payment for local Mobile IP services

The new foreign agent can use this information to either contact the previous foreign agent to retrieve the Key Distribution Center (KDC) session key information, or it can attempt to retrieve the keys from the AAAL. If the AAAL cannot provide the necessary keying information, the request will have to be sent to the mobile node's AAAH to retrieve the new keying information. After initial authorization, further authorizations should be done locally within the local domain.

When a mobile node moves into a new foreign subnet as a result of a handover and is now served by a different foreign agent, the AAAL in this domain may contact the AAAL in the domain that the mobile node has just been handed off from to verify the authenticity of the mobile node and/or to obtain the session keys. The new serving AAAL may determine the address of the AAAL in the previously visited domain from the previous foreign agent NAI information supplied by the mobile node.

Broker Model

Figure 4-1 displays a configuration in which the local and the home authority have to share a trust. Depending on the security model used, this configuration can cause a quadratic growth in the number of trust relationships, as the number of AAA authorities (AAAL and AAAH) increases. This has been identified as a problem by the IETF roaming working group,[50] and any AAA proposal must solve this problem. Using brokers solves many of the scalability problems associated with requiring direct business/roaming relationships between every two administrative domains. In order to provide scalable networks in highly diverse service provider networks in which there are many domains (many service providers and large numbers of private networks), multiple layers of brokers must be supported for both of the broker models described.

The integrity or privacy of information between the home and serving domains may be achieved by either hop-by-hop security associations or end-to-end security associations established with the help of the broker infrastructure. A broker may play the role of a proxy between two administrative domains that have security associations with the broker and that relay AAA messages back and forth securely.

Alternatively, a broker may also enable the two domains with which it has associations, but the domains themselves do not have a direct association when establishing a security association, thereby bypassing the broker for carrying the messages between the domains. This may be established by virtue of having the broker relay a shared secret key to both the domains that are trying to establish secure communication and then having the domains use the keys supplied by the broker in setting up a security association.

Assuming that an AAA broker (AAAB) accepts responsibility for payment to the serving domain on behalf of the home domain, the serving domain is assured of receiving payments for the services offered. However, the redirection broker will usually require a copy of authorization messages from the home domain and accounting messages from the serving domain in order for the broker to determine if it is willing to accept responsibility for the services being authorized and utilized. If the broker does not accept such responsibility for any reason, then it must be able to terminate service to a mobile node in the serving network. In the event that multiple brokers are involved, in most situations all brokers must be so copied. This may represent an additional burden on foreign agents and AAALs.

Though this mechanism may reduce latency in the transit of messages between the domains after the broker has completed its involvement, there may be many more messages involved as a result of additional copies of authorization and accounting messages to the brokers involved. There may also be additional latency for initial access to the network, especially when a new security association needs to be created between AAAL and AAAH (for example, from the use of the Internet Security Association and Key Management Protocol [ISAKMP]). These delays may become important factors for latency-critical applications.

The AAAB in Figure 4-21 is the broker's authority server. The broker acts as a settlement agent, providing security and a central point of contact for many service providers and enterprises.

The AAAB enables the local and home domains to cooperate without requiring each of the networks to have a direct business or security rela-

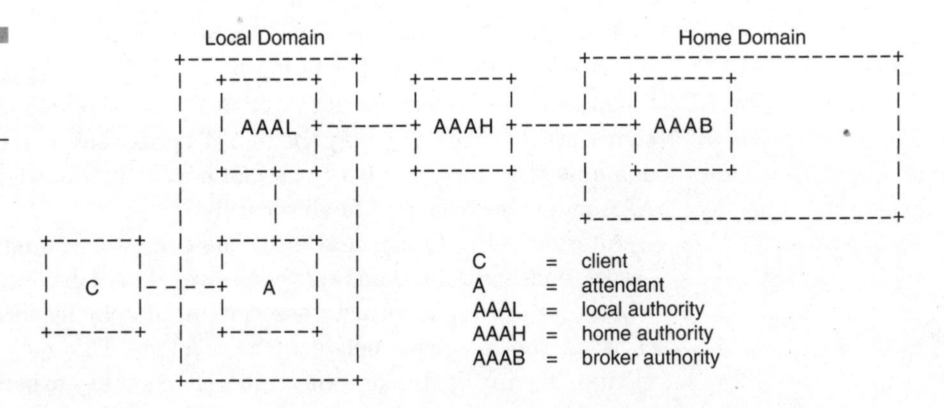

Figure 4-21
AAA servers using a broker

tionship with all the other networks. Thus, brokers offer the needed scalability for managing trust relationships between otherwise independent network domains. Use of the broker does not preclude managing separate trust relationships between domains, but it does offer an alternative to doing so. Just as with the AAAH and AAAL, data specific to Mobile IP control messages must not be processed by the AAAB. Any credentials or accounting data to be processed by the AAAB must be present in AAA message units, not extracted from Mobile IP protocol extensions.

The following requirements come mostly from Aboba and Vollbrecht[51], who discuss the use of brokers in the particular case of authorization for roaming dial-up users:

- The management of trust with external domains by way of brokered AAA.
- Accounting reliability—accounting data that traverses the Internet may suffer substantial packet loss. Since accounting packets may traverse one or more intermediate authorization points (brokers), retransmission is needed from intermediate points to avoid long end-to-end delays.
- End-to-end security—the local domain and home domain must be able to verify signatures within the message, even though the message is passed through an intermediate authority server.
- Since the AAAH in the home domain may be sending sensitive information, such as registration keys, the broker must be able to pass encrypted data between the AAA servers.

The need for end-to-end security results from the following attacks that were identified when brokered operations use RADIUS (see Figure 4-17 for more information on the individual attacks):

- Message editing
- Attribute editing
- Theft of shared secrets
- Theft and modification of accounting data
- Replay attacks
- Connection hijacking
- Fraudulent accounting

These are serious problems that no acceptable AAA protocol or infrastructure can permit to persist.

Security Considerations for AAA

Because AAA is security driven, most of this document addresses the security considerations AAA must make on behalf of Mobile IP. As with any security proposal, adding more entities that interact using security protocols creates new administrative requirements for maintaining the appropriate security associations between the entities. In the case of the AAA services proposed, however, these administrative requirements are natural and already well understood on today's Internet because of experience with dial-up network access.

IPv6 Considerations

The main difference between Mobile IP for IPv4 and Mobile IPv6 is that in IPv6 there is no foreign agent. The attendant function therefore has to be located elsewhere. Logical repositories for that function are either at the local router, for stateless address autoconfiguration, or at the nearest DHCPv6 server, for stateful address autoconfiguration. In the latter case, it is possible that there would be a close relationship between the DHCPv6 server and the AAALv6, but the RFC believes that the protocol functions should still be maintained separately. The MN-NAI would be equally useful for identifying the mobile node to the AAALv6, as is described in earlier sections of the RFC.[52]

Remote Authentication Dial-In User Service (RADIUS)

Managing a large number of dispersed users on WLANs, wireless personal area networks (WPANs), and wireless wide area networks (WWANs) in general, and on serial line and modem pools in particular, can create the need for significant administrative support. Since WLANs, WPANs, WWANs, and modem pools are by definition a link to the outside world, they require careful attention to security, authorization, and accounting. This can be best achieved by managing a single database of users, which allows for authentication as well as configuration information for the type of service to deliver to the user (such as the Serial Line Interface Protocol [SLIP], PPP, telnet, or rlogin). What follows is a brief summary of RADIUS functions, based on RFC 2138.[55] Figure 4-22 applies RADIUS to the hotspot service environment.

Key features of RADIUS are as follows:

- **Client/server model** A network access server (NAS) operates as a client of RADIUS. The client is responsible for passing user information to designated RADIUS servers and then acting on the response that is returned. RADIUS servers are responsible for receiving user connection requests, authenticating the user, and then

Figure 4-22 RADIUS application

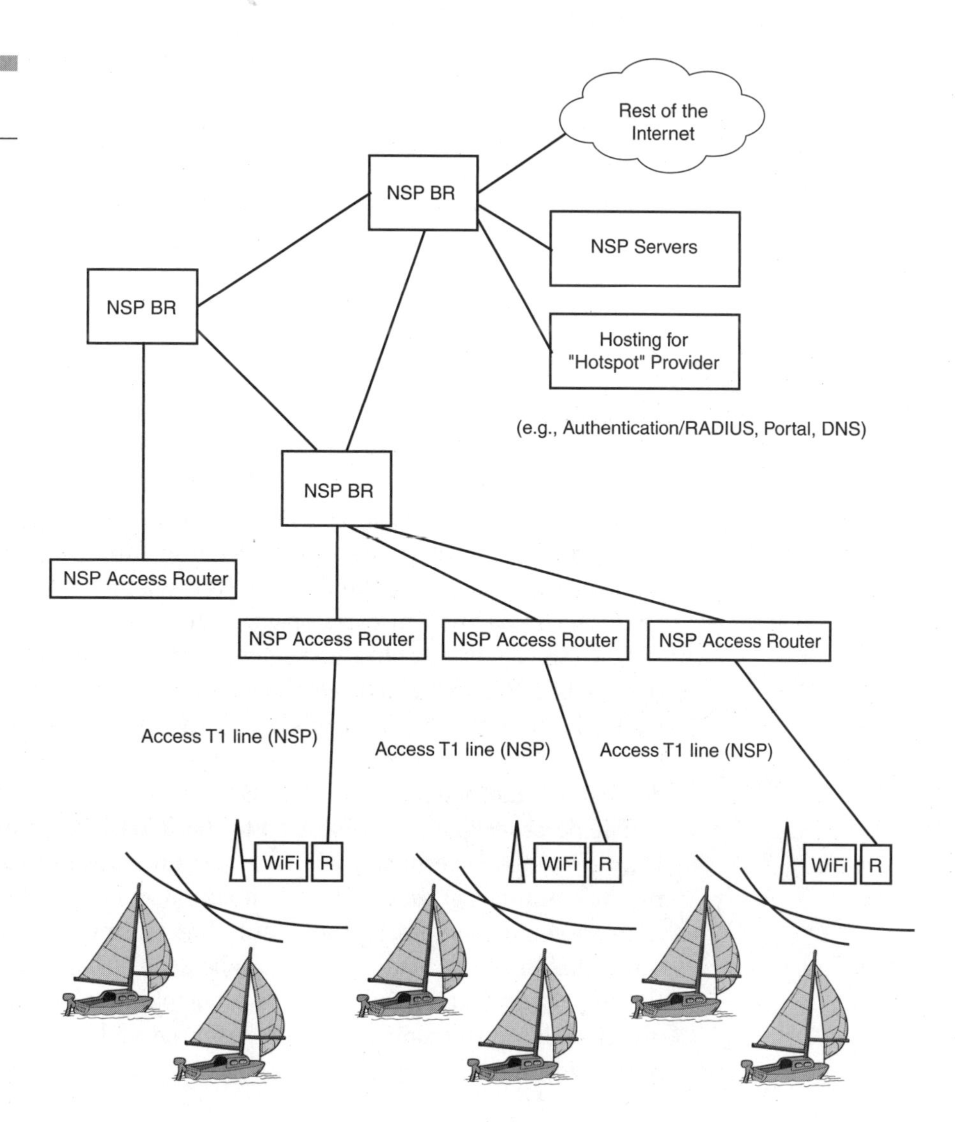

returning all configuration information necessary for the client to deliver service to the user. A RADIUS server can act as a proxy client to other RADIUS servers or other kinds of authentication servers.

- **Network security** Transactions between the client and RADIUS server are authenticated through the use of a shared secret, which is never sent over the network. In addition, any user passwords are sent encrypted between the client and RADIUS server to eliminate the possibility that someone snooping on an unsecured network could determine a user's password.
- **Flexible authentication mechanisms** The RADIUS server can support a variety of methods for authenticating a user. When it is provided with the user name and original password given by the user, it can support PPP PAP or CHAP, UNIX login, and other authentication mechanisms.
- **Extensible protocol** All transactions are comprised of variable-length, attribute-length-value three-tuples. New attribute values can be added without disturbing the existing implementations of the protocol.

Operation

When a client is configured to use RADIUS, any user of the client presents authentication information to the client. This might be with a customizable login prompt, where the user is expected to enter his or her username and password. Alternatively, the user might use a link-framing protocol such as PPP, which has authentication packets to carry this information.

Once the client has obtained such information, it may choose to authenticate using RADIUS. To do so, the client creates an *access-request* containing such attributes as the user's name, the user's password, the client's ID, and the port ID the user is accessing. When a password is present, it is hidden using a method based on the RSA Message Digest algorithm MD-5.[53]

The access-request is submitted to the RADIUS server via the network. If no response is returned within a certain length of time, the request is resent a number of times. The client can also forward requests to an alternate server or servers in the event that the primary server is down or unreachable. An alternate server can be used either after a number of tries to the primary server fail or in a round-robin fashion. Retry and fallback algorithms are the topic of current research and are not specified in detail in this document.

Once the RADIUS server receives the request, it validates the sending client. A request from a client for which the RADIUS server does not have a shared secret should be silently discarded.[54] If the client is valid, the RADIUS server consults a database of users to find the user whose name matches the request. The user entry in the database contains a list of requirements to be met before the user is granted access. It always includes verification of the password, but can also specify the client(s) or port(s) to which the user is allowed access. The RADIUS server may make requests of other servers in order to satisfy the request, in which case it acts as a client.

If any condition is not met, the RADIUS server sends an *access-reject* response indicating that this user request is invalid. If desired, the server may include a text message in the access-reject that may be displayed by the client to the user. No other attributes are permitted in an access-reject.

If all conditions are met and the RADIUS server wishes to issue a challenge to which the user must respond, the RADIUS server sends an access-challenge response. It may include a text message to be displayed by the client to the user prompting for a response to the challenge and it may include a state attribute. If the client receives an access-challenge and supports a challenge/response, it may display the text message, if any, to the user and then prompt the user for a response. The client then resubmits its original access-request with a new request ID, with the user-password attribute replaced by the response (encrypted), and including the state attribute from the access-challenge, if any. Only zero or one instance of the state attributes should be present in a request. The server can respond to this new access-request with an access-accept, an access-reject, or another access-challenge.

If all conditions are met, the list of configuration values for the user is placed into an access-accept response. These values include the type of service (such as SLIP, PPP, and login user) and all necessary values to deliver the desired service. For SLIP and PPP, this may include values such as the IP address, subnet mask, Maximum Transmission Unit (MTU), desired compression, and desired packet filter identifiers. For character mode users, this may include values such as the desired protocol and host.

Challenge/Response In challenge/response authentication, the user is given an unpredictable number and challenged to encrypt it and give back the result. Authorized users are equipped with special devices such as smart cards or software that facilitates calculation of the correct response with ease. Unauthorized users, lacking the appropriate device or software as well as knowledge of the secret key necessary to emulate such a device or software, can only guess at the response.

The access-challenge packet typically contains a reply message, including a challenge to be displayed to the user, such as a numeric value unlikely ever to be repeated. Typically, this is obtained from an external server that knows what type of authenticator should be in the possession of the authorized user and can therefore choose a random or nonrepeating pseudorandom number of an appropriate radix and length.

The user then enters the challenge into his or her device (or software) and it calculates a response, which the user enters into the client that forwards it to the RADIUS server via a second access-request. If the response matches the expected response, the RADIUS server replies with an access-accept, or otherwise with an access-reject.

Let's look at an example. The NAS sends an access-request packet to the RADIUS server with NAS-identifier, NAS-port, user-name, and user-password (which may just be a fixed string like "challenge" or ignored). The server sends back an access-challenge packet with state and a reply message along the lines of "Challenge 12345678, enter your response at the prompt," which the NAS displays. The NAS prompts for the response and sends a new access-request to the server (with a new ID) with NAS-identifier, NAS-port, user-name, user-password (the response just entered by the user, encrypted), and the same state attribute that came with the access-challenge. The server then sends back either an access-accept or access-reject based on whether the response matches what it should be, or it can even send another access-challenge.

Interoperation with PAP and CHAP For PAP, the NAS takes the PAP ID and password and sends them in an access-request packet as the user-name and user-password. The NAS may include the attributes service-type = framed-user and framed-protocol = PPP as a hint to the RADIUS server that PPP service is expected.

For CHAP, the NAS generates a random challenge (preferably 16 octets) and sends it to the user, who returns a CHAP response along with a CHAP ID and CHAP username. The NAS then sends an access-request packet to the RADIUS server with the CHAP username as the user-name and with the CHAP ID and CHAP response as the CHAP-password (attribute 3). The random challenge can either be included in the CHAP-challenge attribute or, if it is 16 octets long, it can be placed in the Request Authenticator field of the access-request packet. The NAS may include the attributes service-type = framed-user and framed-protocol = PPP as a hint to the RADIUS server that PPP service is expected.

The RADIUS server looks up a password based on the user-name, encrypts the challenge using MD-5 on the CHAP ID octet, the password, and the CHAP challenge (from the CHAP-challenge attribute if present, otherwise from the request authenticator), and compares that result to the CHAP-password. If they match, the server sends back an access-accept; otherwise, it sends back an access-reject.

If the RADIUS server is unable to perform the requested authentication, it should return an access-reject. For example, CHAP requires that the user's password be available in cleartext to the server so that it can encrypt the CHAP challenge and compare that to the CHAP response. If the password is not available in cleartext to the RADIUS server, then the server must send an access-reject to the client.

Use of User Datagram Protocol (UDP) A frequently asked question is why RADIUS uses the User Datagram Protocol (UDP) instead of TCP as a transport protocol. UDP was chosen for strictly technical reasons.

A number of issues must be understood. RADIUS is a transaction-based protocol that has several interesting characteristics:

- If the request to a primary authentication server fails, a secondary server must be queried. To meet this requirement, a copy of the request must be kept above the transport layer to allow for alternate transmission. This means that retransmission timers are still required.
- The timing requirements of this particular protocol are significantly different than TCP provides. At one extreme, RADIUS does not require a "responsive" detection of lost data. The user is willing to wait several seconds for the authentication to complete. The generally aggressive TCP retransmission (based on the average round trip time) is not required, nor is the acknowledgement overhead of TCP.

  At the other extreme, the user is not willing to wait several minutes for authentication. Therefore, the reliable delivery of TCP data two minutes later is not useful. The faster use of an alternate server allows the user to gain access before giving up.
- The stateless nature of this protocol simplifies the use of UDP. Clients and servers come and go. Systems are rebooted or are power cycled independently. Generally, this does not cause a problem and with creative timeouts and the detection of lost TCP connections, code can be written to handle anomalous events. UDP, however, completely eliminates any of this special handling. Each client and server can open their UDP transport just once and leave it open through all types of failure events on the network.

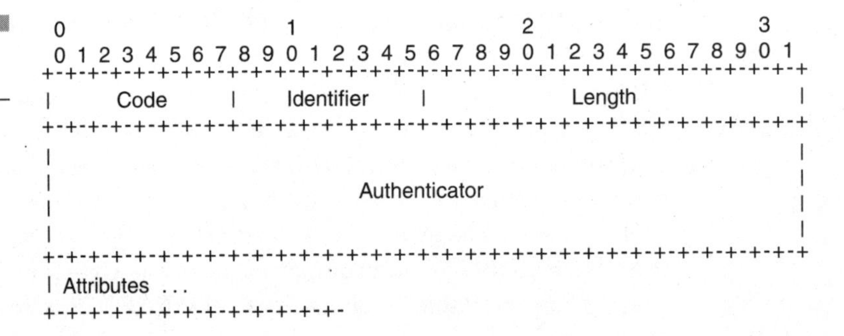

Figure 4-23
RADIUS data format

- UDP simplifies the server implementation. In the earliest implementations of RADIUS, the server was single threaded. This means that a single request was received, processed, and returned. This was found to be unmanageable in environments where the backend security mechanism took real time (one or more seconds). The server request queue would fill and in environments where hundreds of people were being authenticated every minute, the request turn-around time increased to longer than users were willing to wait (this was especially severe when a specific lookup in a database or over DNS took 30 or more seconds). The obvious solution was to make the server multithreaded. Achieving this was simple with UDP. Separate processes were spawned to serve each request and these processes could respond directly to the client NAS with a simple UDP packet to the original transport of the client.

It is not all a panacea. As noted, using UDP requires one thing that is built into TCP: with UDP we must artificially manage retransmission timers to the same server, although they don't require the same attention to timing provided by TCP. This one penalty is a small price to pay for the advantages of UDP in this protocol.

Without TCP we would still probably be using tin cans connected by string. But for this particular protocol, UDP is a better choice.

Packet Format

Exactly one RADIUS packet is encapsulated in the UDP Data field,[55] where the UDP Destination Port field indicates 1812 (decimal). When a reply is generated, the source and destination ports are reversed.

A summary of the RADIUS data format is shown in Figure 4-23. Fields are transmitted from left to right.

The Code field is one octet and identifies the type of RADIUS packet.

Table 4-1

WLAN security choices

Technique	Quality	Description
EAP	Good	EAP is an extension of PPP defined in RFC 2284. EAP is a general authentication protocol that supports multiple authentication methods, including traditional passwords, token cards, Kerberos, digital certificates, and public-key authentication.
802.11i	Good	IEEE 802.11 Task Group I is responsible for enhancing the existing 802.11 MAC standard to provide improved security. This eventually will include strong encryption and standards-based authentication.
802.1X	Good	This IEEE standard can be used as a basis for authentication on all 802 networks, including Ethernet, Token Ring, and WLANs. IEEE 802.1X specifies how EAP information should be encapsulated in frames. To be useful in enabling WLAN security, 802.1X must be supported by WLAN infrastructure equipment as well as mobile-device operating systems.
MAC ACLs	Medium	These ACLs are implemented based on the MAC address of a device, which is normally set in read-only memory (ROM) by the manufacturer. Many 802.11 product manufacturers provide capabilities for restricting access to the WLAN based on a table of MAC addresses stored on the AP. Some vendors provide management utilities that let these MAC ACLs be distributed to multiple APs within an organization.
SSID	Low	SSID is a unique identifier that wireless APs and wireless nodes use to communicate with each other. The SSID is contained within the header of all packets exchanged within a defined WLAN basic service set (BSS). A device cannot be permitted to join the BSS unless it can provide the unique SSID. However, because most APs broadcast their SSIDs and the SSID is contained in plain text in all packets (even if WEP encryption is used), there is no effective way to secure SSIDs.

Table 4-1 cont.

WLAN security choices

Technique	Quality	Description
VPN	Good to excellent	A VPN provides access to secure information over insecure networks using one of a variety of tunneling protocols, including the Point-to-Point Tunneling Protocol (PPTP), the Layer Two Tunneling Protocol (L2TP), and IPsec. A VPN gateway is a device that acts as the interface between secure and insecure networks. To gain access to resources, network devices must support the appropriate tunneling and authentication protocols. VPNs have traditionally been deployed to securely interconnect sites using the public Internet instead of leased lines or Frame Relay and to provide secure dial-up access to secure systems over the public Internet.
WEP	Medium to low	WEP is an optional encryption standard defined by the IEEE 802.11 committee and is implemented in most WLAN products. To gain Wireless Fidelity (WiFi) certification by the Wireless Ethernet Compatibility Alliance (WECA), products must support 40-bit WEP. Most vendors also support 128-bit WEP. WEP was designed to provide the security equivalent to a wired LAN and was not originally envisioned as a bulletproof security architecture. WEP's architecture has been shown to be flawed, and tools are available that can effectively break WEP encryption through passive hacking.

Source: D. Molta, School of Information Studies at Syracuse University.

When a packet is received with an invalid Code field, it is silently discarded. RADIUS codes (decimal) are assigned as follows:

1 Access-request
2 Access-accept
3 Access-reject
4 Accounting-request
5 Accounting-response
11 Access-challenge
12 Status-server (experimental)
13 Status-client (experimental)
255 Reserved

The reader is referred to the RFC for additional protocol details.

Practical Security Aspects

Some WLAN managers view the 802.11 service set ID (SSID) as an element of their security implementations.[56] Each 802.11 WLAN AP must be assigned an SSID, and WLAN clients use the SSID when they associate with the AP. This mechanism does not provide much security, however, because most APs broadcast their SSIDs. Hence, any WLAN client that can be configured to scan for SSIDs will recognize the availability of APs and often present the available systems to the network interface card (NIC) looking to associate in a pick list. Some APs can be configured to suppress the broadcast of SSIDs. On systems configured in this manner, only clients that know the SSID will be able to associate with that AP. (Table 4-1 provides a summary of the various techniques discussed herewith.)

Some network managers take advantage of access-control lists (ACLs) based on MAC addresses, a feature supported in most APs. This is an effective solution for small networks, but it has several problems. First, hackers can spoof MAC addresses, thereby overcoming the access-control restrictions. Second, the number of MAC address entries that any given AP can support usually has limits, a potential problem in environments with thousands of wireless nodes. Finally, in multiple access-point environments, you need to have a system in place to automatically distribute all MAC address entries to all APs.

WEP provides a base level of data encryption. Although WEP's encryption system has been shown to be vulnerable, some vendors are now shipping enhanced versions that inhibit the use of "easily guessable" WEP keys. Even if one is comfortable depending on WEP for privacy, key distribution can be a challenge. If the network includes more than 100 clients, maintaining WEP keys on APs and clients will likely be a significant administrative burden.

The lack of advanced, standards-based security solutions coming from the IEEE, coupled with the acknowledged weaknesses of the existing WEP encryption mechanism, has led some vendors to recommend physical or logical separation of wired and wireless nodes. This will be done through the use of either a dedicated wireless backbone or a virtual LAN (VLAN) running on an existing wired network infrastructure. Wireless nodes have Internet access as well as access to low-security intranet applications, but a VPN gateway controls access to secure applications and data. Like a remote-access VPN, a WLAN implementation requires that VPN client software is installed on all WLAN clients and used to gain authenticated access to secure resources.

Since VLANs and VPNs are standards based, no proprietary elements are included in this solution, enabling a planner to implement a multivendor WLAN environment and still provide standards-based secure access. This solution provides access control, privacy based on strong encryption, and, in some cases, device- and subnet-based access control. This is a reasonable solution that provides the option of selecting best-of-breed solutions both for wireless infrastructure as well as for VPN-based security. VPN implementations, however, are not simply a plug-and-play solution, and they are not inexpensive.

If the planner wants to avoid maintaining yet another authentication database, the planner will need to find an appropriate method of interfacing the VPN system with an external directory server. VPN software must be installed and configured on all the client devices. For some operating systems, the VPN client is included, so this is a relatively simple configuration and support challenge, but for other devices, such as PDAs, one may need to rely on third-party providers for VPN client software. Finally, if the planner chooses to deploy a single VLAN that provides enterprise coverage, the planner must make sure that the Ethernet switching infrastructure and VLAN implementation are secure. Also, because the VLAN is a single IP subnet with APs acting as MAC-layer bridges, one needs to monitor levels of broadcast traffic.

Several of the leading WLAN vendors[57] have developed security framework solutions for their WLAN implementations. These products are based on standard protocols; however, each is proprietary to the extent that it relies on client software available only for that vendor's wireless NICs.

By way of illustration, Cisco's wireless security solution provides mutual authentication of wireless clients and APs using proprietary extensions to the Internet-standard EAP (RFC 2284), which Cisco calls Lightweight EAP (LEAP). LEAP requires the use of either the Cisco Secure Access Control Server (ACS) or a compatible RADIUS server. Cisco Secure ACS has hooks into external directory services, including Microsoft ADS and Novell NDS. With 802.1X and EAP, wireless clients and a remote authentication dial-in user server on the wired LAN perform mutual authentication through APs using one of several supported authentication methods. Once authentication is complete, the RADIUS server sends a unique per-session WEP key for data-stream encryption. Third-party solutions have also emerged to address the issue; these include both software and hardware systems. According to observers, there is cause for optimism that more mature security systems based on interoperable standards should be available in 2002.

End Notes

1. Kim Getgen, "Securing the air: Don't let your wireless LAN be a moving target." RSA Security, November 2001, www-106.ibm.com/developerworks/library/wi-sec1/index.html?dwzone=wireless.
2. Dylan Tweney, "Are You Broadcasting Secrets Over the Airwaves?" www.business2.com/articles/web/0,1653,36044,FF.html. December 06, 2001.
3. WEP is used at the two lowest layers of the Open Systems Interconnection (OSI) Reference Model; therefore, it does not offer end-to-end security that would be possible if the encryption took place, for example, at the application layer.
4. Nikita Borisov, Ian Goldberg, and David Wagner, "Security of the WEP Algorithm." www.isaac.cs.berkeley.edu/isaac/wep-faq.html, University of California at Berkeley, February 2001.
5. Mike McMurry, "Wireless Security." www.sans.org/infosecFAQ/wireless/wireless_sec.htm, January 22, 2001.
6. Nikita Borisov, Ian Goldberg, and David Wagner, "Intercepting mobile communications: The insecurity of 802.11." MOBICOM 2001 (2001).
7. Adam Stubblefield, John Ioannidis, and Aviel D. Rubin, "Using the Fluhrer, Mantin, and Shamir Attack to Break WEP." August 6, 2001.
8. Nikita Borisov, Ian Goldberg, and David Wagner, "Wired Equivalent Privacy (WEP)." www.isaac.cs.berkeley.edu/isaac/wep-faq.html, wep@isaac.cs.berkeley.edu, Summer 2001.
9. Dennis Fisher and Carmen Nobel, "Wireless LAN Holes." *eWeek*, www.zdnet.com/eweek/stories/general/0,11011,2684337,00.html, February 11, 2001.
10. Princy C. Mehta, "Wired Equivalent Privacy Vulnerability." http://rr.sans.org/wireless/equiv.php, April 4, 2001.
11. The rest of this section is taken from Princy C. Mehta. "Wired Equivalent Privacy Vulnerability." http://rr.sans.org/wireless/equiv.php, April 4, 2001.
12. S. Kent and R. Atkinson, "Security architecture for the Internet protocol." Request for Comments 2401, Internet Engineering Task Force, November 1998.
13. T. Ylonen, "SSH-secure login connection over the Internet." USENIX Security Conference VI (1996), pp. 37–42.

14. Ellen Zurko, "Listwatch: Items from Security-Related Mailing Lists." www.ieee-security.org/Cipher/Newsbriefs/2001/022001 .ListWatch.html, IEEE, February 16, 2001.

15. Robert Lemos, "Wireless networks wide open to hackers." CNET News.com, July 12, 2001.

16. Andrew Garcia, "Holes in Wireless Nets." *eWeek*, www.zdnet.com/eweek/stories/general/0,11011,2687518,00.html, February 26, 2001.

17. Andrew Garcia, "WEP Remains Vulnerable." *eWeek*, www.zdnet.com/eweek/stories/general/0,11011,2700806,00.html, March 26, 2001.

18. Richard Shim, "How to Fill Wi-Fi's Security Holes." www.zdnet.com/enterprise/stories/main/0,10228,2693864,00.html, ZDNet, March 8, 2001.

19. This section is taken from work by Kim Getgen, used with permission. Kim Getgen (kgetgen@rsasecurity.com) is a product marketing manager at RSA Security in San Mateo, CA. She is responsible for the marketing of the RSA BSAFE product line and the RSA Wireless Security Portfolio. Reference: "Securing the air: Don't let your wireless LAN be a moving target," www-106.ibm.com/developerworks/library/wi-sec1/index.html?dwzone=wireless, November 2001.

20. Matthew Peretz, "IEEE 802.1X Standard Used Successfully with Windows XP." www.80211-planet.com/news/article/0,,1481_905331,00.html.

21. Nick Petroni and Bryan D. Payne, "Opensource Implementation of IEEE 802.1X." Maryland Information and Systems Security Lab at the University of Maryland, www.missl.cs.umd.edu/1x/index.html.

22. RFC 2284.

23. P. Roshan, "802.1X Authenticates 802.11 Wireless." *Network World*, September 24, 2001.

24. IEEE 802 minutes, 802.1 Interim meeting, Raleigh, NC. January 17 and 18, 2002.

25. Paul Congdon, "IEEE 802.1X Overview: Port-Based Network Access Control IEEE Plenary." http://grouper.ieee.org/groups/802/1/pages/802.1X.html, March 2000.

26. University of Maryland Information Systems Security Lab. "Implementing 802.1X on Wireless Networks with Cisco and Microsoft." www.cs.umd.edu/%7Emvanopst/8021x/howto/.

27. This section include information from promotional material from Meetinghouse Data Communications on SecureSupplicant, http://www.mtghouse.com/supplicant.html.

28. www.mtghouse.com.

29. W. Simpson, " STD 51, RFC 1661: The Point-to-Point Protocol (PPP)." July 1994.

30. L. Blunk and J. Vollbrecht, "Request for Comments: 2284, PPP Extensible Authentication Protocol (EAP)." March 1998.

31. The rest of this section taken from Blunk and Vollbrecht's "Request for Comments: 2284, PPP Extensible Authentication Protocol (EAP)," March 1998.

32. F. Yergeau, "RFC 2044: UTF-8, a transformation format of Unicode and ISO 10646." October 1996.

33. W. Simpson, "RFC 1994: PPP Challenge Handshake Authentication Protocol (CHAP)." RFC 1994, August 1996.

34. N. Haller and C. Metz, "A One-Time Password System." RFC 1938, May 1996.

35. RFC 2284 is Copyright © The Internet Society (1998). This document and translations of it may be copied and furnished to others, and derivative works that comment on or otherwise explain it or assist in its implementation may be prepared, copied, published and distributed, in whole or in part, without restriction of any kind, provided that the above copyright notice and this paragraph are included on all such copies and derivative works.

36. S. Glass, T. Hiller, S. Jacobs, and C. Perkins, "RFC 2977, Mobile IP Authentication, Authorization, and Accounting Requirements." October 2000.

37. The rest of this section is based on S. Glass, T. Hiller, S. Jacobs, and C. Perkins. "RFC 2977, Mobile IP Authentication, Authorization, and Accounting Requirements." October 2000.

38. C. Perkins, "RFC 2003: IP Encapsulation within IP." October 1996.

39. ———. " RFC 2002: IP Mobility Support." October 1996.

40. ———. "RFC 2004: Minimal Encapsulation within IP." October 1996.

41. J. Solomon and S. Glass, "RFC 2290: Mobile-IPv4 Configuration Option for PPP IPCP." February 1998.

42. C. Rigney, A. Rubens, W. Simpson, and S. Willens, "RFC 2138: Remote Authentication Dial In User Service (RADIUS)." April 1997.

43. B. Aboba and M. Beadles, "RFC 2486: The Network Access Identifier." January 1999.

44. P. Calhoun and C. Perkins, "RFC 2794: Mobile IP Network Address Identifier Extension." March 2000.

45. Ramon Caceres and Liviu Iftode, "Improving the Performance of Reliable Transport Protocols in Mobile Computing Environments." *IEEE Journal on Selected Areas in Communications*, 13(5):850–857, June 1995.

46. D. Harkins and D. Carrel, "RFC 2409: The Internet Key Exchange (IKE)." November 1998.

47. G. Montenegro and V. Gupta, "RFC 2356: Sun's SKIP Firewall Traversal for Mobile IP." June 1998.

48. P. Mockapetris, "STD 13, RFC 1035: Domain names—implementation and specification." November 1987.

49. C. Perkins and D. Johnson, "Route Optimization in Mobile IP." Work in progress.

50. B. Aboba and G. Zorn, "RFC 2477: Criteria for Evaluating Roaming Protocols." December 1998.

51. B. Aboba and J. Vollbrecht, "RFC 2607: Proxy Chaining and Policy Implementation in Roaming." June 1999.

52. RFC 2977 is Copyright © The Internet Society (2000). This document and translations of it may be copied and furnished to others, and derivative works that comment on or otherwise explain it or assist in its implementation may be prepared, copied, published, and distributed, in whole or in part, without restriction of any kind, provided that the above copyright notice and this paragraph are included on all such copies and derivative works.

53. R. Rivest and S. Dusse, "The MD5 Message-Digest Algorithm," RFC 1321, MIT Laboratory for Computer Science, RSA Data Security Inc., April 1992.

54. This means the implementation discards the packet without further processing. The implementation should provide the capability of logging the error, including the contents of the silently discarded packet, and should record the event in a statistics counter.

55. J. Postel, "STD 6, RFC 768: User Datagram Protocol." USC/Information Sciences Institute, August 1980.

56. This section is taken from a pointed article by Dave Molta. "WLAN Security on the Rise." *Network Computing*, February 4, 2002; used with permission.

57. Such as Cisco Systems, Agere Systems, and Symbol Technologies.

CHAPTER 5

IEEE 802.11

This chapter is a partial redux of the IEEE standard "IEEE Standard for Information Technology—Telecommunications and information exchange between systems—Local and metropolitan area networks—Specific requirements—Part 11: Wireless LAN Medium Access Control (MAC) and Physical Layer (PHY) specifications, which was adopted by the ISO/IEC and redesignated as ISO/IEC 8802-11:1999(E)."[1] This material is for pedagogical, advocacy, and educational purposes. It is meant to acquaint the reader with the major concepts of the most important protocol for hotspot networking. Developers and readers interested in the complete specification and subtending details should refer directly to the full standard from IEEE.

The standard defines the protocol and compatible interconnection of data communication equipment via the air, radio, or infrared in a local area network (LAN) using the carrier sense multiple access with collision avoidance (CSMA/CA) medium-sharing mechanism. The MAC supports operation under control of an access point (AP) as well as between independent stations. The protocol includes authentication, association, and reassociation services, an optional encryption/decryption procedure, power management,[2] and a point coordination function (PCF) for the time-bounded transfer of data. The infrared (IR) implementation of the PHY supports 1 Mbps data rate with an optional 2 Mbps extension. The radio implementations of the PHY specify either a frequency-hopping spread spectrum (FHSS) supporting 1 Mbps and an optional 2 Mbps data rate or a direct sequence spread spectrum (DSSS) supporting both 1 and 2 Mbps data rates. For the basic protocol model, see Figure 5-1.

Figure 5-1 Basic protocol model, IEEE 802.11

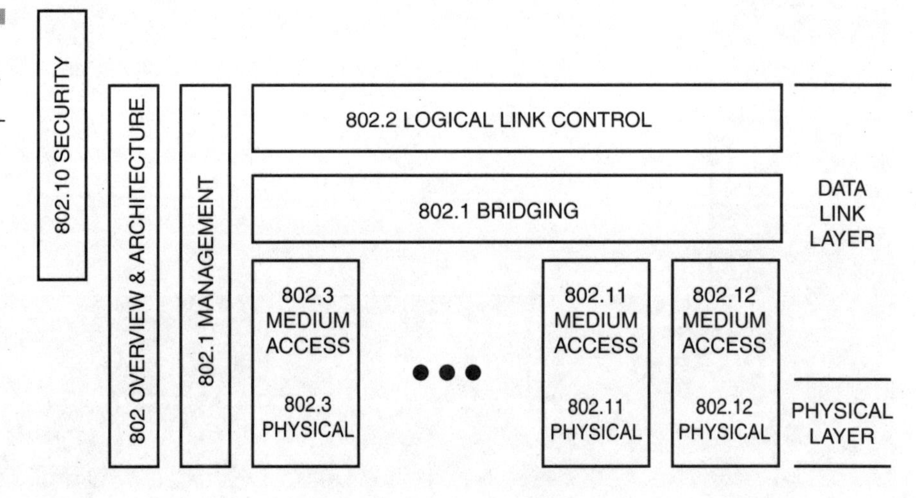

Specifically, the IEEE 802.11 standard

- Describes the functions and services required by an IEEE-802.11-compliant device to operate within ad hoc and infrastructure networks as well as the aspects of station mobility (transition) within those networks
- Defines the MAC procedures to support the asynchronous MAC service data unit (MSDU) delivery services
- Defines several PHY signaling techniques and interface functions that are controlled by the IEEE 802.11 MAC
- Permits the operation of an IEEE-802.11-conformant device within a wireless local area network (WLAN) that may coexist with multiple overlapping IEEE 802.11 WLANs
- Describes the requirements and procedures to provide privacy of user information being transferred over the wireless medium (WM) and authentication of IEEE 802.11 conformant devices

IEEE 802.11 uses an abstract architecture to describe functional components of an IEEE 802.11 LAN. The architectural descriptions are not intended to represent any specific physical implementation of IEEE 802.11.

802.11b and 802.11a are extensions of the basic standard and aim at supporting higher speeds. These are not further discussed here.

How WLAN Systems Are Different

Wireless networks have fundamental characteristics that make them significantly different from traditional wired LANs. Some countries impose specific requirements for radio equipment in addition to those specified in the 802.11 standard.

Destination Address Does Not Equal Destination Location

In wired LANs, an address is equivalent to a physical location. This is implicitly assumed in the design of wired LANs. In IEEE 802.11, the addressable unit is a station (STA). The STA is a message destination, but not generally a fixed location.

The Media Impact of the Design

The physical layers used in IEEE 802.11 are fundamentally different from wired media. Thus, IEEE 802.11 PHYs

- Use a medium that has neither absolute nor readily observable boundaries outside of which stations with conformant PHY transceivers are known to be unable to receive network frames.
- Are unprotected from outside signals.
- Communicate over a medium significantly less reliable than wired PHYs.
- Have dynamic topologies.
- Lack full connectivity, and therefore the assumption normally made that every STA can hear every other STA is invalid (that is, STAs may be hidden from each other).
- Have time-varying and asymmetric propagation properties.

Because of limitations on wireless PHY ranges, WLANs intended to cover reasonable geographic distances may be built from basic coverage building blocks.

The Impact of Handling Mobile Stations

One of the requirements of IEEE 802.11 is to handle mobile as well as portable stations. A portable station is one that is moved from location to location, but that is only used while at a fixed location. Mobile stations actually access the LAN while in motion.

For technical reasons, it is not sufficient to handle only portable stations. Propagation effects blur the distinction between portable and mobile stations; stationary stations often appear to be mobile due to propagation effects. Another aspect of mobile stations is that they may often be battery powered. Hence, power management is an important consideration. For example, it cannot be presumed that a station's receiver will always be powered on.

Interaction with Other IEEE 802 Layers

IEEE 802.11 is required to appear to higher layers (Logical Link Control [LLC]) as a current style IEEE 802 LAN. This requires that the IEEE 802.11 network handle station mobility within the MAC sublayer. To meet reliability assumptions (that LLC makes about lower layers), it is necessary for IEEE 802.11 to incorporate functionality that is untraditional for MAC sublayers.

Components of the IEEE 802.11 Architecture

The IEEE 802.11 architecture consists of several components that interact to provide a WLAN that supports station mobility transparently to upper layers. The basic service set (BSS) is the basic building block of an IEEE 802.11 LAN. Figure 5-2 shows two BSSs, each of which has two stations that are members of the specified BSS. It is useful to think of the ovals used to depict a BSS as the coverage area within which the member stations of the BSS may remain in communication. (The concept of area, although not precise, is often good enough.) If a station moves out of its BSS, it can no longer directly communicate with other members of the BSS.

The Independent BSS (IBSS) as an Ad Hoc Network

The IBSS is the most basic type of IEEE 802.11 LAN. A minimum IEEE 802.11 LAN may consist of only two stations that communicate directly with each other. Figure 5-2 shows two IBSSs (in this case, the BSS is

Figure 5-2 BSSs

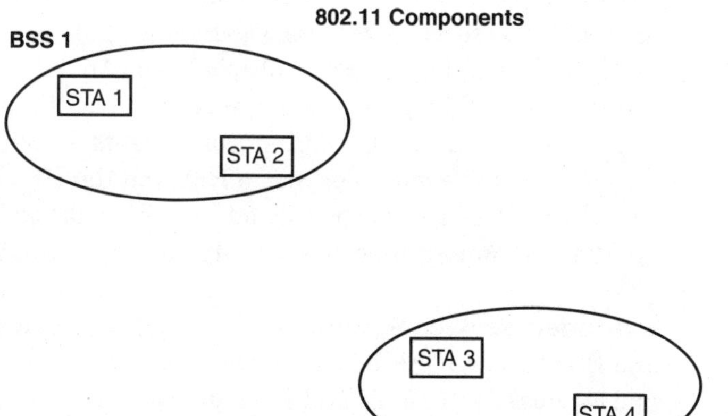

802.11 Components
BSS 1
STA 1
STA 2
STA 3
STA 4
BSS 2

actually an IBSS since there is no centralized device—an access point—that controls communcation). This mode of operation is possible when IEEE 802.11 stations are able to communicate directly. Because this type of IEEE 802.11 LAN is often formed without preplanning, for only as long as the LAN is needed, this type of operation is often referred to as an *ad hoc network*.

STA to BSS Association Is Dynamic The association between an STA and BSS is dynamic (STAs turn on, turn off, come within range, and go out of range). To become a member of an infrastructure BSS, a station needs to become associated. These associations are dynamic and involve the use of the distribution system service (DSS).

Distribution System (DS) Concepts

PHY limitations determine the direct station-to-station distance that may be supported. For some networks, this distance is sufficient; for other networks, increased coverage is required. Instead of existing independently, a BSS may also form a component of an extended form of network that is built with multiple BSSs. The architectural component used to interconnect BSSs is the distribution system (DS).

IEEE 802.11 logically separates the WM from the distribution system medium (DSM). Each logical medium is used for different purposes by a different component of the architecture. The IEEE 802.11 definitions neither preclude, nor demand, that the multiple media be either the same or different.

Recognizing that the multiple media are logically different is key to understanding the flexibility of the architecture. The IEEE 802.11 LAN architecture is specified independently of the physical characteristics of any specific implementation. The DS enables mobile device support by providing the logical services necessary to handle address-to-destination mapping and seamless integration of multiple BSSs. An AP is an STA that provides access to the DS by providing DS services in addition to acting as an STA.

Figure 5-3 adds the DS and AP components to the IEEE 802.11 architecture picture. Data move between a BSS and the DS via an AP. Note that all APs are also STAs; thus, they are addressable entities. The addresses used by an AP for communication on the WM and DSM are not necessarily the same.

Extended Service Set (ESS): The Large Coverage Network The DS and BSSs allow IEEE 802.11 to create a wireless network of arbitrary size and complexity. IEEE 802.11 refers to this type of network as the extended service set (ESS) network.

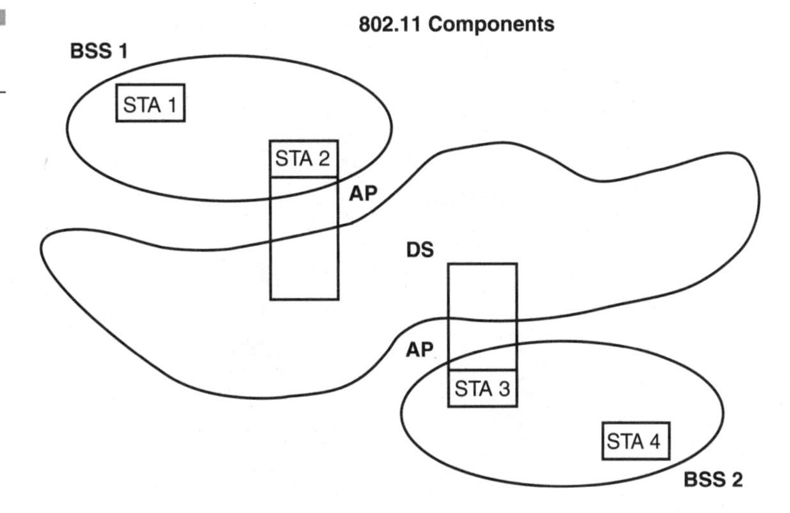

Figure 5-3
DSs and APs

The key concept is that the ESS network appears the same to an LLC layer as an IBSS network. Stations within an ESS may communicate and mobile stations may move from one BSS to another (within the same ESS) transparently to LLC.

802.11 assumes nothing about the relative physical locations of the BSSs in Figure 5-4. All of the following are possible:

- The BSSs may partially overlap. This is commonly used to arrange contiguous coverage within a physical volume.
- The BSSs could be physically disjointed. Logically, there is no limit to the distance between BSSs.
- The BSSs may be physically collocated. This may be done to provide redundancy.
- One (or more) IBSS or ESS networks may be physically present in the same space as one (or more) ESS networks. This may arise for a number of reasons. Two of the most common are when an ad hoc network is operating in a location that also has an ESS network and when physically overlapping IEEE 802.11 networks have been set up by different organizations.

Area Concepts

For wireless PHYs, well-defined coverage areas simply do not exist. Propagation characteristics are dynamic and unpredictable. Small changes in

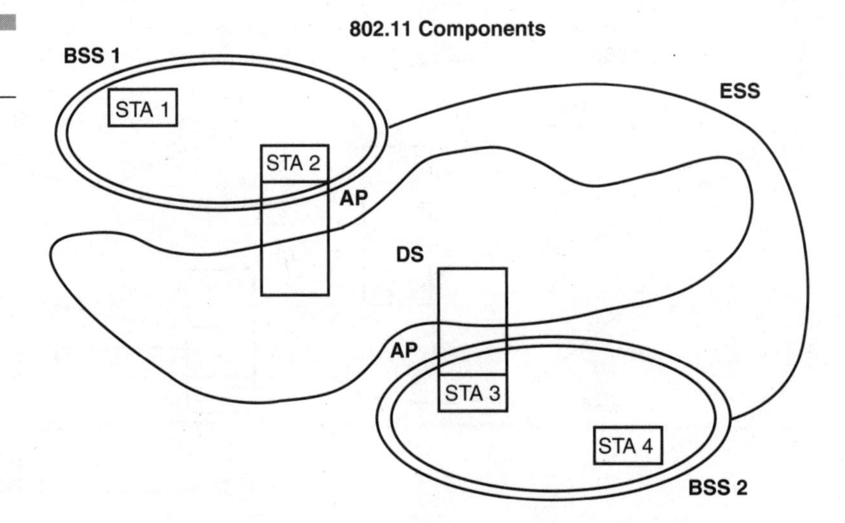

Figure 5-4
ESS

position or direction may result in dramatic differences in signal strength. Similar effects occur whether an STA is stationary or mobile (as moving objects may impact station-to-station propagation). Since dynamic three-dimensional field strength pictures are difficult to draw, well-defined two-dimensional shapes are used in IEEE 802.11 architectural diagrams to represent the coverage of a BSS.

Integration with Wired LANs

To integrate the IEEE 802.11 architecture with a traditional wired LAN, a final logical architectural component is introduced—a *portal*. A portal is the logical point at which MSDUs from an integrated non-IEEE 802.11 LAN enter the IEEE 802.11 DS. For example, the portal in Figure 5-5 is connecting to a wired IEEE 802 LAN. All data from non-IEEE 802.11 LANs enter the IEEE 802.11 architecture via a portal. The portal provides logical integration between the IEEE 802.11 architecture and existing wired LANs. It is possible for one device to offer the functions of an AP and a portal; this could be the case when a DS is implemented from IEEE 802 LAN components.

The ESS architecture (APs and the DS) provides traffic segmentation and range extension. Logical connections between 802.11 and other LANs are via the portal. Portals connect between the DSM and the LAN medium that is to be integrated.

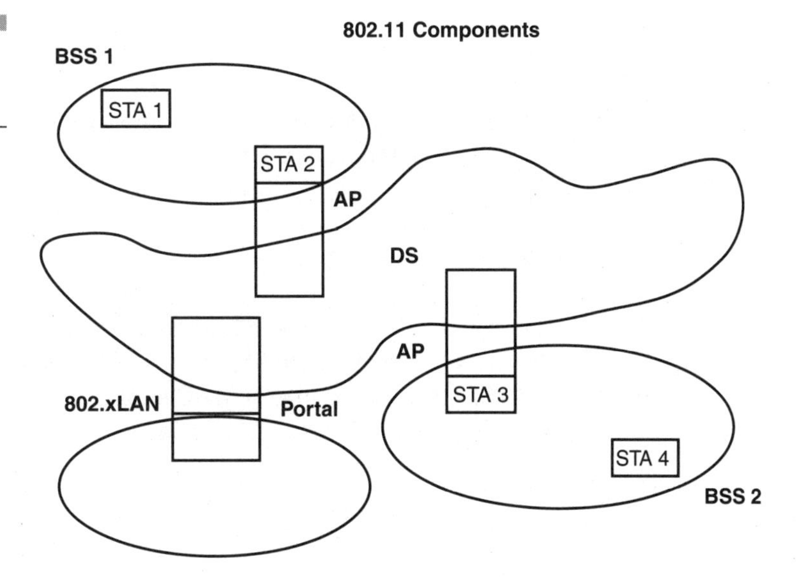

Figure 5-5 Connecting to other IEEE 802 LANs

Logical Service Interfaces

The IEEE 802.11 architecture allows for the possibility that the DS may not be identical to an existing wired LAN. A DS may be created from many different technologies including current IEEE 802 wired LANs. IEEE 802.11 does not constrain the DS to be either data link or network layer based. Nor does IEEE 802.11 constrain a DS to be either centralized or distributed in nature. IEEE 802.11 explicitly does not specify the details of DS implementations. Instead, IEEE 802.11 specifies *services*. The services are associated with different components of the architecture. There are two categories of IEEE 802.11 service—the station service (SS) and the DSS. Both categories of service are used by the IEEE 802.11 MAC sublayer.

The complete set of IEEE 802.11 architectural services are as follows:

- Authentication
- Association
- Deauthentication
- Disassociation
- Distribution
- Integration
- Privacy

- Reassociation
- MSDU delivery

This set of services is divided into two groups: those that are part of every STA and those that are part of a DS.

Station Service (SS)

The service provided by stations is known as the SS. The SS is present in every IEEE 802.11 station (including APs, as APs include station functionality). The SS is specified for use by MAC sublayer entities. All conformant stations provide SS. The SSs are as follows:

- Authentication
- Deauthentication
- Privacy
- MSDU delivery

Distribution System Service (DSS)

The service provided by the DS is known as the DSS. These services are represented in the IEEE 802.11 architecture by arrows within the APs, indicating that the services are used to cross media and address space logical boundaries. This is the convenient place to show the services in the picture. The physical embodiment of various services may or may not be within a physical AP. The DSSs are provided by the DS. They are accessed via an STA that also provides DSSs. An STA that provides DSS access is called an AP. The DSSs are as follows:

- Association
- Disassociation
- Distribution
- Integration
- Reassociation

DSSs are specified for use by MAC sublayer entities.

Figure 5-6 combines the components from previous figures with both types of services to show the complete IEEE 802.11 architecture.

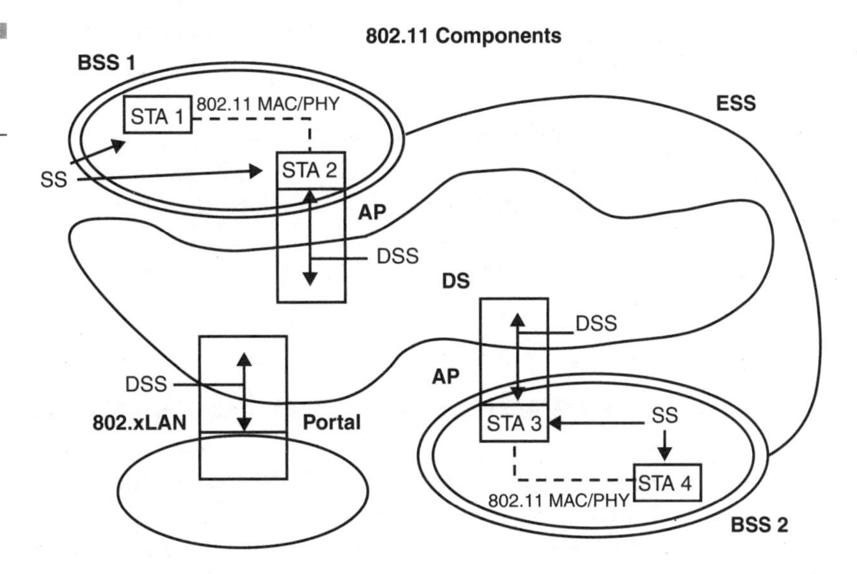

Figure 5-6 Complete IEEE 802.11 architecture

Multiple Logical Address Spaces

Just as the IEEE 802.11 architecture allows for the possibility that the WM, DSM, and an integrated wired LAN may all be different physical media, it also allows for the possibility that each of these components may be operating within different address spaces. IEEE 802.11 only uses and specifies the use of the WM address space.

Each IEEE 802.11 PHY operates in a single medium—the WM. The IEEE 802.11 MAC operates in a single address space. MAC addresses are used on the WM in the IEEE 802.11 architecture. Therefore, it is unnecessary for the standard to explicitly specify that its addresses are WM addresses. This is assumed throughout the 802.11 standard.

IEEE 802.11 has chosen to use the IEEE 802 48-bit address space. Thus, IEEE 802.11 addresses are compatible with the address space used by the IEEE 802 LAN family. The IEEE 802.11 choice of address space implies that for many instantiations of the IEEE 802.11 architecture, the wired LAN MAC address space and the IEEE 802.11 MAC address space may be the same. In those situations where a DS that uses MAC-level IEEE 802 addressing is appropriate, all three of the logical address spaces used within a system could be identical. Although this is a common case, it is not the only combination allowed by the architecture. The IEEE 802.11 architecture allows for all three logical address spaces to be distinct.

A multiple address space example is one in which the DS implementation uses network layer addressing. In this case, the WM address space and the DS address space would be different.

The capability of the architecture to handle multiple logical media and address spaces is key to IEEE 802.11's capability to be independent of the DS implementation and to interface cleanly with network layer mobility approaches. The implementation of the DS is unspecified and is beyond the scope of the 802.11 standard.

Overview of the Services

Nine services are specified by IEEE 802.11. Six of them are used to support MSDU delivery between STAs, and three are used to control IEEE 802.11 LAN access and confidentiality. This subsection defines the services, with an overview of how each service is used and a description of how each relates to other services and the 802.11 architecture. The services are presented in an order designed to help build an understanding of the operation of an IEEE 802.11 ESS network. As a result, the SSs and DSSs are intermixed in order (rather than being grouped by category). Each of the services is supported by one or more MAC frame types. Some of the services are supported by MAC management messages and some by MAC data messages. All of the messages gain access to the WM via the IEEE 802.11 MAC sublayer medium access method.

The IEEE 802.11 MAC sublayer uses three types of messages—data, management, and control. The data messages are handled via the MAC data service path.

MAC management messages are used to support the IEEE 802.11 services and are handled via the MAC management service data path. MAC control messages are used to support the delivery of IEEE 802.11 data and management messages. The examples here assume an ESS network environment.

Distribution of Messages within a DS

Distribution This is the primary service used by IEEE 802.11 STAs. It is conceptually invoked by every data message to or from an IEEE 802.11

STA operating in an ESS (when the frame is sent via the DS). Distribution is via a DSS. Refer to the ESS network in Figure 5-6 and consider a data message being sent from STA 1 to STA 4. The message is sent from STA 1 and received by STA 2 (the input AP). The AP gives the message to the distribution service of the DS. It is the job of the distribution service to deliver the message within the DS in such a way that it arrives at the appropriate DS destination for the intended recipient. In this example, the message is distributed to STA 3 (the output AP) and STA 3 accesses the WM to send the message to STA 4 (the intended destination). The specification does not determine how the message is distributed within the DS. All IEEE 802.11 is required to do is to provide the DS with enough information for the DS to be able to determine the output point that corresponds to the desired recipient. The necessary information is provided to the DS by the three association-related services (association, reassociation, and disassociation).

The previous example was a case in which the AP that invoked the distribution service was different from the AP that received the distributed message. If the message had been intended for a station that was a member of the same BSS as the sending station, then the input and output APs for the message would have been the same.

In both examples, the distribution service was logically invoked. Whether the message actually had to traverse the physical DSM or not is a DS implementation matter and is not specified by the 802.11 standard.

Although IEEE 802.11 does not specify DS implementations, it does recognize and support the use of the WM as the DSM. This is specifically supported by the IEEE 802.11 frame formats.

Integration If the distribution service determines that the intended recipient of a message is a member of an integrated LAN, the output point of the DS would be a portal instead of an AP. Messages that are distributed to a portal cause the DS to invoke the Integration function (conceptually after the distribution service). The Integration function is responsible for accomplishing whatever is needed to deliver a message from the DSM to the integrated LAN media (including any required media or address space translations). Integration is a DSS. Messages received from an integrated LAN (via a portal) by the DS for an IEEE 802.11 STA will invoke the Integration function before the message is distributed by the distribution service. The details of an Integration function are dependent on a specific DS implementation and are outside the scope of the 802.11 standard.

Services That Support the Distribution Service

The primary purpose of a MAC sublayer is to transfer MSDUs between MAC sublayer entities. The information required for the distribution service to operate is provided by the association services. Before a data message can be handled by the distribution service, an STA will be associated. To understand the concept of association, it is necessary first to understand the concept of mobility.

Mobility Types The three transition types of significance to the 802.11 standard that describe the mobility of stations within a network are as follows:

- **No-transition** In this type, two subclasses that are usually indistinguishable are identified:
  - **Static** No motion.
  - **Local movement** Movement within the PHY range of the communicating STAs (that is, movement within a basic service area [BSA]).
- **BSS-transition** This type is defined as a station movement from one BSS in one ESS to another BSS within the same ESS.
- **ESS-transition** This type is defined as station movement from a BSS in one ESS to a BSS in a different ESS. This case is supported only in the sense that the STA may move. 802.11 cannot guarantee maintenance of upper-layer connections; in fact, disruption of service is likely.

The different association services support the different categories of mobility.

Association To deliver a message within a DS, the distribution service needs to know which AP to access for the given IEEE 802.11 STA. This information is provided to the DS by the concept of association. Association is necessary, but not sufficient, to support BSS-transition mobility. Association is sufficient to support no-transition mobility. Association is a DSS. Before an STA is allowed to send a data message via an AP, it will first become associated with the AP. The act of becoming associated invokes the association service, which provides the STA-to-AP mapping to the DS. The DS uses this information to accomplish its message distribution service.

How the information provided by the association service is stored and managed within the DS is not specified by the standard. At any given instant, an STA may be associated with no more than one AP. This ensures that the DS may determine a unique answer to the question, "Which AP is serving STA X?" Once an association is completed, an STA may make full use of a DS (via the AP) to communicate. Association is always initiated by the mobile STA, not the AP. An AP may be associated with many STAs at one time. An STA learns what APs are present and then requests to establish an association by invoking the association service.

Reassociation Association is sufficient for no-transition message delivery between IEEE 802.11 stations. Additional functionality is needed to support BSS-transition mobility. The additional required functionality is provided by the reassociation service. Reassociation is a DSS. The reassociation service is invoked to move a current association from one AP to another. This keeps the DS informed of the current mapping between AP and STA as the station moves from BSS to BSS within an ESS. Reassociation also enables changing association attributes of an established association, while the STA remains associated with the same AP. It is always initiated by the mobile STA.

Disassociation The disassociation service is invoked whenever an existing association is to be terminated. Disassociation is a DSS. In an ESS, this tells the DS to void existing association information. Attempts to send messages via the DS to a disassociated STA will be unsuccessful. The disassociation service may be invoked by either party to an association (non-AP STA or AP). Disassociation is a notification, not a request. Disassociation cannot be refused by either party to the association. APs may need to disassociate STAs to enable the AP to be removed from a network for service or for other reasons. STAs will attempt to disassociate whenever they leave a network. However, the MAC protocol does not depend on STAs invoking the disassociation service. (MAC management is designed to accommodate loss of an associated STA.)

Access and Confidentiality Control Services

Two services are required for IEEE 802.11 to provide functionality equivalent to that inherent in wired LANs. The design of wired LANs assumes the physical attributes of wire. In particular, wired LAN design assumes the physically closed and controlled nature of wired media. The physically open medium nature of an IEEE 802.11 LAN violates those assumptions. Two services are provided to bring the IEEE 802.11 functionality in line with wired LAN assumptions: authentication and privacy. Authentication is used instead of the wired media physical connection. Privacy is used to provide the confidential aspects of closed wired media.

Authentication In wired LANs, physical security can be used to prevent unauthorized access by non-LAN resident parties. This is impractical in WLANs since they have a medium without precise bounds. IEEE 802.11 provides the ability to control LAN access via the authentication service. This service is used by all stations to establish their identity to stations with which they will communicate. This is true for both ESS and IBSS networks. If a mutually acceptable level of authentication has not been established between two stations, an association will not be established. Authentication is an SS. IEEE 802.11 supports several authentication processes. The IEEE 802.11 authentication mechanism also allows the expansion of the supported authentication schemes. *IEEE 802.11 does not mandate the use of any particular authentication scheme*. IEEE 802.11 provides link-level authentication between IEEE 802.11 STAs. *IEEE 802.11 does not provide either end-to-end (message-origin-to-message-destination) or user-to-user authentication*. IEEE 802.11 authentication is used simply to bring the wireless link up to the assumed physical standards of a wired link. (This use of authentication is independent of any authentication process that may be used in higher levels of a network protocol stack.) If authentication other than that described here is desired, it is recommended that IEEE Std 802.10-1992 be implemented.[3] If desired, an IEEE 802.11 network may be operated using open system authentication. This may violate implicit assumptions made by higher network layers. In an open system, any station may become authenticated. IEEE 802.11 also supports shared key authentication. Use of this authentication mechanism requires the implementation of the Wired Equivalent Privacy (WEP) option. In a shared key authentication system, identity is demonstrated by knowledge of a shared secret, the WEP encryption key. MIB functions are provided to support the standardized authentication schemes. IEEE 802.11 requires mutually acceptable, successful authentication. An STA may be authenticated with many other STAs at any given instant.

Preauthentication Because the authentication process could be time consuming (depending on the authentication protocol in use), the authentication service can be invoked independently of the association service. Preauthentication is typically done by an STA while it is already associated with an AP (with which it previously authenticated). IEEE 802.11 does not require that STAs preauthenticate with APs. However, authentication is required before an association can be established. If the authentication is left until reassociation time, this may impact the speed with which an STA can reassociate between APs, limiting BSS-transition mobility performance. The use of preauthentication takes the authentication service overhead out of the time-critical reassociation process.

Deauthentication The deauthentication service is invoked whenever an existing authentication is to be terminated. Deauthentication is an SS. In an ESS, since authentication is a prerequisite for association, the act of deauthentication will cause the station to be disassociated. The service may be invoked by either authenticated party (non-AP STA or AP). Deauthentication is not a request; it is a notification that cannot be refused by either party. When an AP sends a deauthentication notice to an associated STA, the association will also be terminated.

Privacy In a wired LAN, only those stations physically connected to the wire may hear LAN traffic. With a wireless shared medium, this is not the case. Any IEEE-802.11-compliant STA may hear all like-PHY IEEE 802.11 traffic that is within range. Thus, the connection of a single wireless link (without privacy) to an existing wired LAN may seriously degrade the security level of the wired LAN. To bring the functionality of the WLAN up to the level implicit in wired LAN design, IEEE 802.11 provides the ability to encrypt the contents of messages. This functionality is provided by the privacy service. Privacy is an SS for which 802.11 specifies an optional privacy algorithm, WEP, designed to satisfy the goal of wired-LAN-equivalent privacy. The algorithm is not, in other words, designed for complete security.

IEEE 802.11 uses the WEP mechanism to perform the actual encryption of messages. We have made reference to WEP in previous chapters. MIB functions are provided to support WEP. Note that privacy may only be invoked for data frames and some authentication management frames. All stations initially start in the clear in order to set up the authentication and privacy services. The default privacy state for all IEEE 802.11 STAs is in the clear. If the privacy service is not invoked, all messages will be sent unencrypted. If this default is not acceptable to one party or the other, data frames will not be successfully communicated between the LLC entities. Unencrypted data frames received at a station configured for mandatory privacy, as well as encrypted data frames using a key not available at the receiving station, are discarded without an indication to LLC (or without indication to distribution services in the case of "To DS" frames received at an AP). These frames are acknowledged on the WM (if received without frame check sequence [FCS] error) to avoid wasting WM bandwidth on retries.

Relationships Between Services

An STA keeps two state variables for each STA with which direct communication via the WM is needed:

- **Authentication state** The values are unauthenticated and authenticated.
- **Association state** The values are unassociated and associated.

These two variables create three local states for each remote STA:

- **State 1** Initial start state, unauthenticated, and unassociated
- **State 2** Authenticated, not associated
- **State 3** Authenticated and associated

The current state existing between the source and destination station determines the IEEE 802.11 frame types that may be exchanged between that pair of STAs. The allowed frame types are grouped into classes and the classes correspond to the station state. In State 1, only Class 1 frames are allowed. In State 2, either Class 1 or Class 2 frames are allowed. In State 3, all frames are allowed (Classes 1, 2, and 3). The frame classes are defined as follows:

1. Class 1 frames (permitted from within States 1, 2, and 3)
   a. Control frames
      i) Request to Send (RTS)
      ii) Clear to Send (CTS)
      iii) Acknowledgment (ACK)
      iv) Contention-free end (CF-End)+ACK
      v) CF-End
   b. Management frames
      i) Probe request/response.
      ii) Beacon.
      iii) Authentication: Successful authentication enables a station to exchange Class 2 frames. Unsuccessful authentication leaves the STA in State 1.

   iv) Deauthentication: Deauthentication notification when in State 2 or State 3 changes the STA's state to State 1. The STA will become authenticated again prior to sending Class 2 frames.
   v) Announcement traffic indication message (ATIM).
 c. Data frames
   i) Data: Data frames with frame control (FC) bits To DS and From DS are both false.

2. Class 2 frames (if and only if authenticated; allowed from within States 2 and 3 only)

 a. Management frames
   i) Association request/response
     - Successful association enables Class 3 frames.
     - Unsuccessful association leaves STA in State 2.
   ii) Reassociation request/response
     - Successful reassociation enables Class 3 frames.
     - Unsuccessful reassociation leaves the STA in State 2 (with respect to the STA that was sent the reassociation message). Reassociation frames will only be sent if the sending STA is already associated in the same ESS.
   iii) Disassociation
     - Disassociation notification when in State 3 changes a station's state to State 2. This station will become associated again if it wishes to utilize the DS. If STA A receives a Class 2 frame with a unicast address in the Address 1 field from STA B that is not authenticated with STA A, STA A will send a deauthentication frame to STA B.

3. Class 3 frames (if and only if associated; allowed only from within State 3)

 a. Data frames
   i) Data subtypes: Data frames allowed. That is, either the To DS or From DS FC bits may be set to true to utilize DSSs.
 b. Management frames
   i) Deauthentication: Deauthentication notification when in State 3 implies disassociation as well, changing the STA's state from 3 to 1. The station will become authenticated again prior to another association.
 c. Control frames
   i) Power-save poll (PS-Poll)

If STA A receives a Class 3 frame with a unicast address in the Address 1 field from STA B that is authenticated but not associated with STA A, STA A will send a disassociation frame to STA B. If STA A receives a Class 3 frame with a unicast address in the Address 1 field from STA B that is not authenticated with STA A, STA A will send a deauthentication frame to STA B.

Differences Between ESS and IBSS LANs

When the concept of the IBSS LAN was introduced, we noted that an IBSS is often used to support an ad hǫc network. In an IBSS network, an STA communicates directly with one or more other STAs. Consider the full IEEE 802.11 architecture, as shown in Figure 5-7. An IBSS consists of STAs that are directly connected. Thus, there is, by definition, only one BSS. Further, since there is no physical DS, there cannot be a portal, an integrated wired LAN, or the DSSs. The logical picture reduces to Figure 5-8. Only the minimum two stations are shown, although an IBSS may have an arbitrary number of members. In an IBSS, only Class 1 and 2 frames are allowed since there is no DS in an IBSS. The services that apply to an IBSS are the SSs.

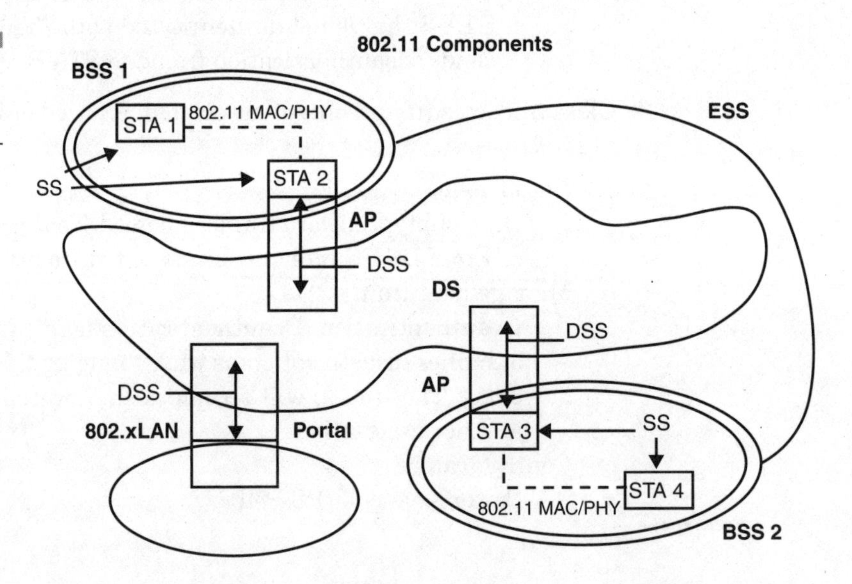

Figure 5-7 IEEE 802.11 architecture (again)

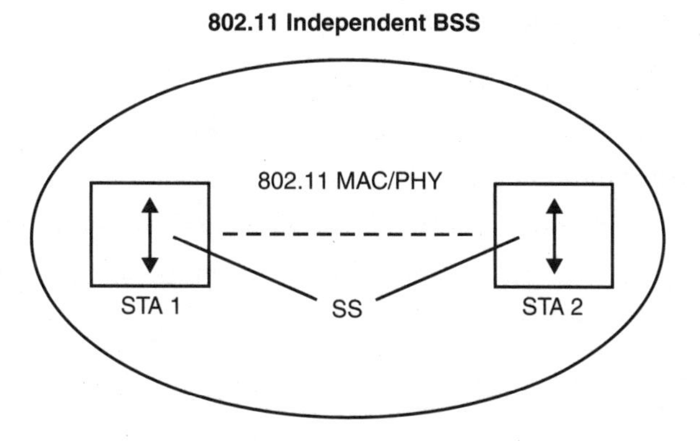

Figure 5-8
802.11 IBSS

Message Information Contents That Support Services

Each service is supported by one or more IEEE 802.11 messages. Information items are given by name.

Data

For an STA to send data to another STA, it sends a data message, as shown in the following:

- Message type: data
- Message subtype: data
- Information items:
  - IEEE source address of message
  - IEEE destination address of message
  - BSSID
- Direction of message: from STA to STA

Association

For an STA to associate, the association service causes the following messages to occur.

Association Request

- Message type: management
- Message subtype: association request
- Information items:
  - IEEE address of the STA initiating the association
  - IEEE address of the AP with which the initiating station will associate
  - ESSID
- Direction of message: from STA to AP

Association Response

- Message type: management
- Message subtype: association response
- Information items:
  - Result of the requested association. This is an item with the values successful and unsuccessful.
  - If the association is successful, the response will include the association identifier (AID).
- Direction of message: from AP to STA

Reassociation

For an STA to reassociate, the reassociation service causes the following message to occur.

Reassociation Request

- Message type: management
- Message subtype: reassociation request
- Information items:
  - IEEE address of the STA initiating the reassociation
  - IEEE address of the AP with which the initiating station will reassociate

  - IEEE address of the AP with which the initiating station is currently associated
  - ESSID
- Direction of message: from STA to AP (the AP with which the STA is requesting reassociation)

The address of the current AP is included for efficiency. The inclusion of the current AP address facilitates MAC reassociation to be independent of the DS implementation.

Reassociation Response

- Message type: management
- Message subtype: reassociation response
- Information items:
  - Result of the requested reassociation. This is an item with the values successful and unsuccessful.
  - If the reassociation is successful, the response will include the AID.
- Direction of message: from AP to STA

Disassociation

For an STA to terminate an active association, the disassociation service causes the following message to occur.

- Message type: management
- Message subtype: disassociation
- Information items:
  - IEEE address of the station that is being disassociated. This will be the broadcast address in the case of an AP disassociating with all associated stations.
  - IEEE address of the AP with which the station is currently associated.
- Direction of message: from STA to STA (for example, STA to AP or AP to STA)

Privacy

For an STA to invoke the WEP privacy algorithm (as controlled by the related MIB attributes), the privacy service causes MAC protocol data unit (MPDU) encryption and sets the WEP frame header bit appropriately.

Authentication

For an STA to authenticate with another STA, the authentication service causes one or more authentication management frames to be exchanged. The exact sequence of frames and their content is dependent on the authentication scheme invoked. For all authentication schemes, the authentication algorithm is identified within the management frame body. In an IBSS environment, either station may be the initiating STA (STA 1). In an ESS environment, STA 1 is the mobile STA, and STA 2 is the AP.

Authentication (First Frame of Sequence)

- Message type: management
- Message subtype: authentication
- Information items:
  - Authentication algorithm identification
  - Station identity assertion
  - Authentication transaction sequence number
  - Authentication algorithm-dependent information
- Direction of message: first frame in the transaction sequence is always from STA 1 to STA 2

The first frame in an authentication sequence will always be unencrypted.

Authentication (Intermediate Sequence Frames)

- Message type: management
- Message subtype: authentication
- Information items:
  - Authentication algorithm identification

- Authentication transaction sequence number
- Authentication algorithm-dependent information

- Direction of message:
  - Even transaction sequence numbers: from STA 2 to STA 1
  - Odd transaction sequence numbers: from STA 1 to STA 2

Authentication (Final Frame of Sequence)

- Message type: management
- Message subtype: authentication
- Information items:
  - Authentication algorithm identification.
  - Authentication transaction sequence number.
  - Authentication algorithm-dependent information.
  - The result of the requested authentication. This is an item with values successful and unsuccessful.
- Direction of message: from STA 2 to STA 1

Deauthentication

For an STA to invalidate an active authentication, the following message is sent:

- Message type: management
- Message subtype: deauthentication
- Information items:
  - IEEE address of the STA that is being deauthenticated.
  - IEEE address of the STA with which the STA is currently authenticated.
  - This will be the broadcast address in the case of an STA deauthenticating all STAs currently authenticated.
- Direction of message: from STA to STA

Reference Model

The 802.11 standard takes the architectural view, emphasizing the separation of the system into two major parts: the MAC of the data link layer and the PHY. These layers are intended to correspond closely to the lowest layers of the ISO/IEC basic reference model of OSI (ISO/IEC 7498-1: 1994 5). The layers and sublayers described in the 802.11 standard are shown in Figure 5-9.

MAC Service Definition

Asynchronous Data Service

This service provides peer LLC entities with the capability to exchange MSDUs. To support this service, the local MAC uses the underlying PHY-level services to transport an MSDU to a peer MAC entity, where it will be delivered to the peer LLC. Such asynchronous MSDU transport is performed on a best-effort connectionless basis. There are no guarantees that the submitted MSDU will be delivered successfully. Broadcast and multicast transport is part of the asynchronous data service provided by the MAC. Due

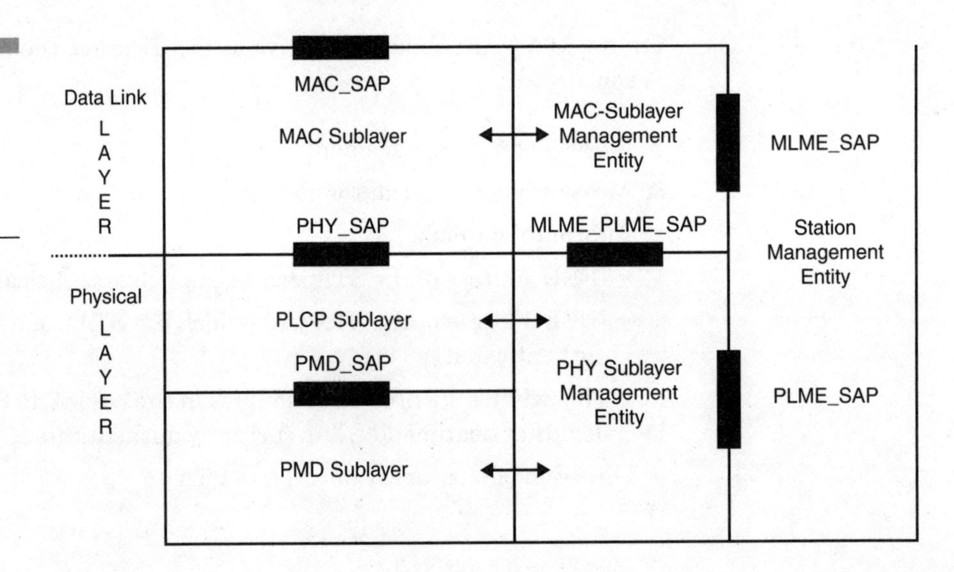

Figure 5-9 Portion of the ISO/IEC basic reference model covered in the 802.11 standard

to the characteristics of the WM, broadcast and multicast MSDUs may experience a lower quality of service compared to that of unicast MSDUs. All STAs will support the asynchronous data service. Because the operation of certain functions of the MAC may cause the reordering of some MSDUs, as discussed in more detail in the following section, there are two service classes within the asynchronous data service. By selecting the desired service class, each LLC entity initiating the transfer of MSDUs is able to control whether MAC entities are or are not allowed to reorder those MSDUs.

Security Services

Security services in IEEE 802.11 are provided by the authentication service and the WEP mechanism. The scope of the security services provided is limited to station-to-station data exchange. The privacy service offered by an IEEE 802.11 WEP implementation is the encryption of the MSDU. For the purposes of this standard, WEP is viewed as a logical service located within the MAC sublayer, as shown in the reference model Figure 5-9. Actual implementations of the WEP service are transparent to the LLC and other layers above the MAC sublayer. The security services provided by the WEP in IEEE 802.11 are as follows:

- Confidentiality
- Authentication
- Access control in conjunction with layer management

During the authentication exchange, parties A and B exchange authentication information. The MAC sublayer security services provided by WEP rely on information from non-layer-2 management or system entities. Management entities communicate information to WEP through a set of MIB attributes.

MSDU Ordering

The services provided by the MAC sublayer permit, and may in certain cases require, the reordering of MSDUs. The MAC does not intentionally reorder MSDUs except as may be necessary to improve the likelihood of successful delivery based on the current operational (power management) mode of the designated recipient station(s). The sole effect of this reordering (if any) for the set of MSDUs received at the MAC service interface of any single station is a change in the delivery order of broadcast and multicast MSDUs, relative to directed MSDUs, originating from a single source station address. If a higher-layer protocol using the asynchronous data service cannot tolerate this possible reordering, the optional StrictlyOrdered service class should be used. MSDUs transferred between any pair of stations using the StrictlyOrdered service class are not subject to the relative reordering that is possible when the ReorderableMulticast service class is used. However, the desire to receive MSDUs sent using the StrictlyOrdered service class at a station precludes simultaneous use of the MAC power management facilities at that station.

Frame Formats

The format of the MAC frames is specified in this clause. All stations will be able to properly construct frames for transmission and decode frames upon reception, as specified in this clause.

MAC Frame Formats

Each frame consists of the following basic components:

- MAC header, which comprises frame control, duration, address, and sequence control information
- A variable-length frame body, which contains information specific to the frame type
- An FCS, which contains an IEEE 32-bit cyclic redundancy code (CRC)

Conventions The MPDUs or frames in the MAC sublayer are described as a sequence of fields in specific order. Each of the following figures depicts the fields/subfields as they appear in the MAC frame and in the order in which they are passed to the Physical Layer Convergence Protocol (PLCP) from left to right. In figures, all bits within fields are numbered, from 0 to k, where the length of the field is $k + 1$ bit. The octet boundaries within a field can be obtained by taking the bit numbers of the field modulo 8. Octets within numeric fields that are longer than a single octet are depicted in increasing order of significance, from the lowest numbered bit to highest numbered bit. The octets in fields longer than a single octet are sent to the PLCP in order from the octet containing the lowest numbered bits to the octet containing the highest numbered bits. Any field containing a CRC is an exception to this convention and is transmitted commencing with the coefficient of the highest-order term.

MAC addresses are assigned as ordered sequences of bits. The Individual/Group bit is always transferred first and is bit 0 of the first octet. Values specified by a decimal are coded in natural binary unless otherwise stated. Reserved fields and subfields are set to 0 upon transmission and are ignored upon reception.

General Frame Format The MAC frame format comprises a set of fields that occur in a fixed order in all frames. Figure 5-10 depicts the general MAC frame format. The fields Address 2, Address 3, Sequence Control, Address 4, and Frame Body are only present in certain frame types.

Frame Fields

Frame Control Field The Frame Control field consists of the following subfields: Protocol Version, Type, Subtype, To DS, From DS, More Fragments, Retry, Power Management, More Data, WEP, and Order. The format of the Frame Control field is illustrated in Figure 5-11.

Duration/ID Field The Duration/ID field is 16 bits in length. The contents of this field are as follows:

- In control-type frames of subtype PS-Poll, the Duration/ID field carries the AID of the station that transmitted the frame in the 14 least significant bits with the 2 most significant bits both set to 1. The value of the AID is in the range 1 to 2,007.

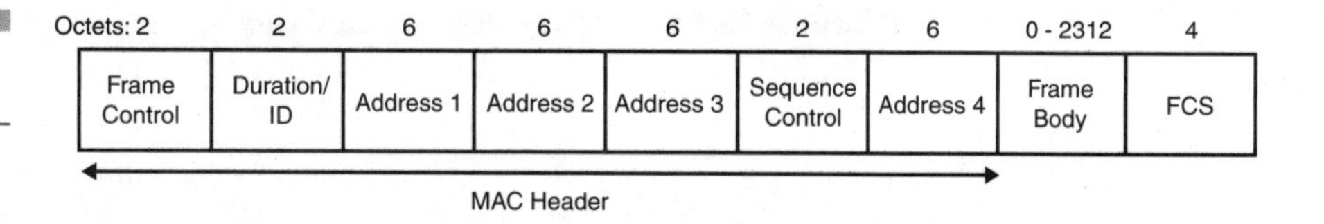

Figure 5-10 MAC frame format

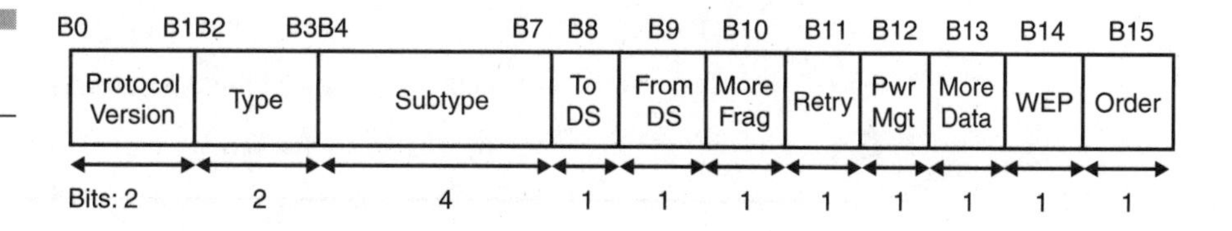

Figure 5-11 Frame Control field

- In all other frames, the Duration/ID field contains a duration value as defined for each frame type. For frames transmitted during the CF period (CFP), the Duration field is set to 32,768. Whenever the contents of the Duration/ID field are less than 32,768, the duration value is used to update the network allocation vector (NAV).

The encoding of the Duration/ID field is described in Table 5-1.

Address Fields There are four address fields in the MAC frame format. These fields are used to indicate the basic service set identifier (BSSID), source address, destination address, transmitting station address, and receiving station address. The usage of the four address fields in each frame type is indicated by the abbreviations BSSID, DA, SA, RA, and TA, BSSID, Destination Address, Source Address, Receiver Address, and Transmitter Address, respectively. Certain frames may not contain some of the address fields. Certain address field usage is specified by the relative position of the address field (1–4) within the MAC header, independent of the type of address present in that field. For example, receiver address matching is always performed on the contents of the Address 1 field in received frames, and the receiver address of CTS and ACK frames is always obtained from the Address 2 field in the corresponding RTS frame or from the frame being acknowledged.

Address Representation Each address field contains a 48-bit address, as defined in Clause 5.2 of IEEE 802-1990.

Table 5-1

Duration/ID field encoding

Bit 15	Bit 14	Bit 13–0	Usage
0	0–32,767		Duration
1	0	0	Fixed value within frames transmitted during the CFP
1	0	1–16,383	Reserved
1	1	0	Reserved
1	1	1–2,007	AID in PS-Poll frames
1	1	2,008–16,383	Reserved

Address Designation A MAC sublayer address is one of the following two types:

- **Individual address** The address associated with a particular station on the network
- **Group address** A multidestination address, associated with one or more stations on a given network

The two kinds of group addresses are as follows:

- **Multicast-group address** An address associated by a higher-level convention with a group of logically related stations.
- **Broadcast address** A distinguished, predefined multicast address that always denotes the set of all stations on a given LAN. All 1s in the Destination Address field are interpreted to be the broadcast address. This group is predefined for each communication medium to consist of all stations actively connected to that medium; it is used to broadcast to all the active stations on that medium. All stations are able to recognize the broadcast address. It is not necessary that a station be capable of generating the broadcast address.

The address space is also partitioned into locally administered and universal (globally administered) addresses. The nature of a body and the procedures by which it administers these universal (globally administered) addresses is beyond the scope of the 802.11 standard. See IEEE 802-1990 for more information.

BSSID Field The BSSID field is a 48-bit field of the same format as an IEEE 802 MAC address. This field uniquely identifies each BSS. The value of this field, in an infrastructure BSS, is the MAC address currently in use by the STA in the AP of the BSS. The value of this field in an IBSS is a locally administered IEEE MAC address formed from a 46-bit random number generated according to the procedure defined in the standard. The individual/group bit of the address is set to 0. The universal/local bit of the address is set to 1. This mechanism is used to provide a high probability of selecting a unique BSSID. The value of all 1s is used to indicate the broadcast BSSID. A broadcast BSSID may only be used in the BSSID field of management frames of subtype probe request.

Destination Address (DA) Field The DA field contains an IEEE MAC individual or group address that identifies the MAC entity or entities intended as the final recipient(s) of the MSDU (or fragment thereof) contained in the frame body field.

Source Address (SA) Field The SA field contains an IEEE MAC individual address that identifies the MAC entity from which the transfer of the MSDU (or fragment thereof) contained in the frame body field was initiated. The individual/group bit is always transmitted as a zero in the source address.

Receiver Address (RA) Field The RA field contains an IEEE MAC individual or group address that identifies the intended immediate recipient STA(s), on the WM, for the information contained in the frame body field.

Transmitter Address (TA) Field The TA field contains an IEEE MAC individual address that identifies the STA that has transmitted, onto the WM, the MPDU contained in the frame body field. The Individual/Group bit is always transmitted as a zero in the transmitter address.

Sequence Control Field The Sequence Control field is 16 bits in length and consists of two subfields: the Sequence Number and the Fragment Number. The format of the Sequence Control field is illustrated in Figure 5-12.

Sequence Number Field The Sequence Number field is a 12-bit field indicating the sequence number of an MSDU or MAC management protocol data unit (MMPDU). Each MSDU or MMPDU transmitted by an STA is assigned a sequence number. Sequence numbers are assigned from a single modulo 4096 counter, starting at 0 and incrementing by 1 for each MSDU or MMPDU. Each fragment of an MSDU or MMPDU contains the assigned sequence number. The sequence number remains constant in all retransmissions of an MSDU, MMPDU, or fragment thereof.

Fragment Number Field The Fragment Number field is a 4-bit field indicating the number of each fragment of an MSDU or MMPDU. The fragment number is set to zero in the first or only fragment of an MSDU or MMPDU and is incremented by one for each successive fragment of that MSDU or MMPDU. The fragment number remains constant in all retransmissions of the fragment.

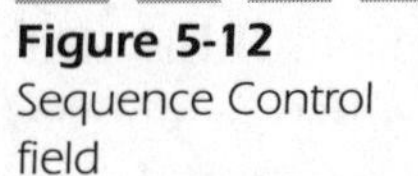

Figure 5-12
Sequence Control field

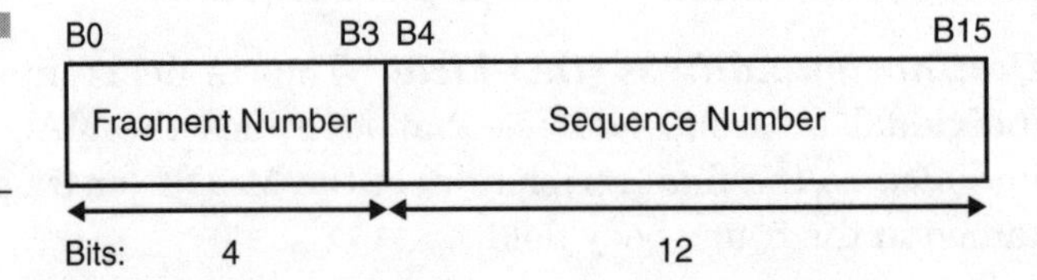

Frame Body Field The Frame Body is a variable-length field that contains information specific to individual frame types and subtypes. The minimum frame body is 0 octets. The maximum frame body is defined by the maximum length (MSDU + ICV + IV), where the integrity check value (ICV) and initialization vector (IV) are the WEP fields.

FCS Field The FCS field is a 32-bit field containing a 32-bit CRC. The FCS is calculated over all the fields of the MAC header and the Frame Body field.

Format of Individual Frame Types

Control Frames In the following descriptions, an immediately previous frame is a frame whose reception concluded within the prior short interframe space (SIFS) interval. The subfields within the Frame Control field of control frames are set, as illustrated in Figure 5-13.

RTS Frame Format The frame format for the RTS frame is defined in Figure 5-14.

The RA of the RTS frame is the address of the STA, on the WM, that is the intended immediate recipient of the pending directed data or management frame. The TA is the address of the STA transmitting the RTS frame.

The duration value is the time, in microseconds, required to transmit the pending data or management frame, plus one CTS frame, plus one ACK

Figure 5-13
Frame Control field subfield values within control frames

Protocol Version (B0)	Type	Subtype	To DS	From DS	More Frag	Retry	Pwr Mgt	More Data	WEP	Order (B15)
Protocol Version	Control	Subtype	0	0	0	0	Pwr Mgt	0	0	0
Bits: 2	2	4	1	1	1	1	1	1	1	1

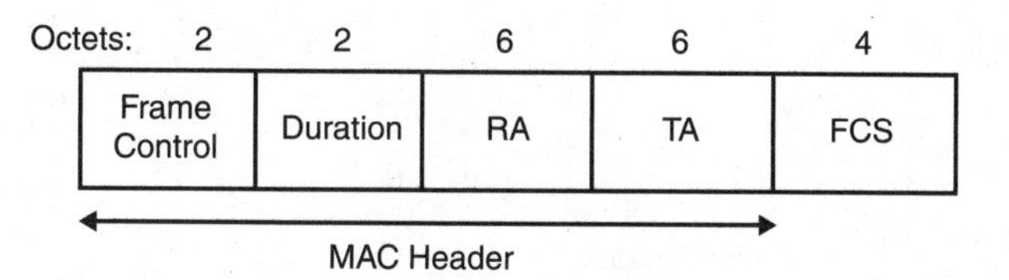

Figure 5-14
RTS frame

frame, plus three SIFS intervals. If the calculated duration includes a fractional microsecond, that value is rounded up to the next higher integer.

CTS Frame Format The frame format for the CTS frame is defined in Figure 5-15. The RA of the CTS frame is copied from the TA field of the previous RTS frame to which the CTS is a response. The duration value is the value obtained from the Duration field of the immediately previous RTS frame, minus the time, in microseconds, required to transmit the CTS frame and its SIFS interval. If the calculated duration includes a fractional microsecond, that value is rounded up to the next higher integer.

Acknowledgment (ACK) Frame Format The frame format for the ACK frame is defined in Figure 5-16. The RA of the ACK frame is copied from the Address 2 field of the immediately previous directed data, management, or PS-Poll control frame. If the More Fragment bit was set to 0 in the Frame Control field of the immediately previous directed data or management frame, the duration value is set to 0. If the More Fragment bit was set to 1 in the Frame Control field of the immediately previous directed data or management frame, the duration value is the value obtained from the Duration field of the immediately previous data or management frame, minus the time, in microseconds, required to transmit the ACK frame and its SIFS interval. If the calculated duration includes a fractional microsecond, that value is rounded up to the next higher integer.

Power-Save Poll (PS-Poll) Frame Format The frame format for the PS-Poll frame is defined in Figure 5-17. The BSSID is the address of the STA con-

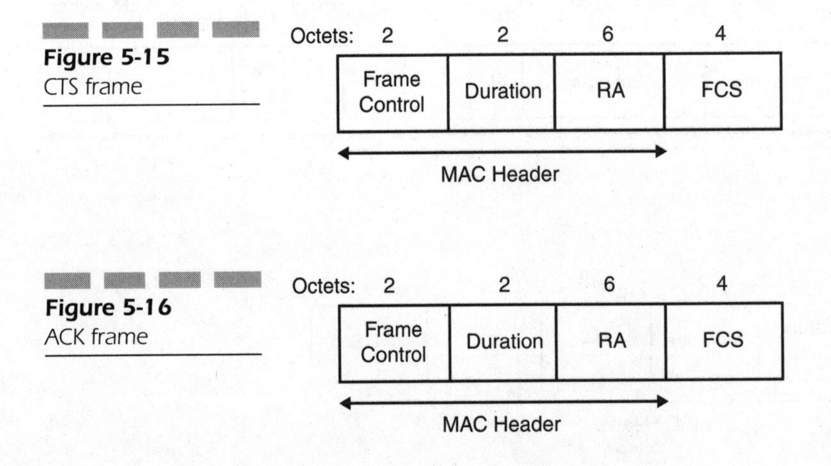

Figure 5-15 CTS frame

Figure 5-16 ACK frame

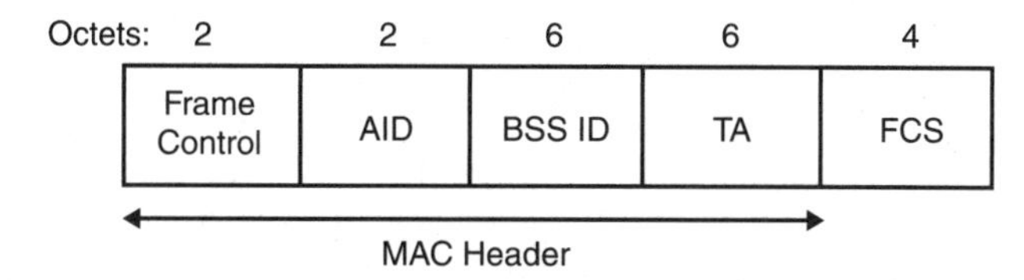

Figure 5-17
PS-Poll frame

tained in the AP. The TA is the address of the STA transmitting the frame. The AID is the value assigned to the STA transmitting the frame by the AP in the association response frame that established that STA's current association. The AID value always has its two most significant bits set to 1. All STAs, upon receipt of a PS-Poll frame, update their NAV settings as appropriate under the coordination function rules using a duration value equal to the time, in microseconds, required to transmit one ACK frame plus one SIFS interval.

CF-End Frame Format The frame format for the CF-End frame is defined in Figure 5-18.

The BSSID is the address of the STA contained in the AP. The RA is the broadcast group address. The Duration field is set to 0.

CF-End+CF-Ack Frame Format The frame format for the CF-End acknowledge (CF-End+CF-Ack) frame is defined in Figure 5-19.

Data Frames The frame format for a data frame is independent of subtype and is defined in Figure 5-20. The content of the address fields of the data frame is dependent upon the values of the "To DS" and "From DS" bits and is defined in Table 5-2. Where the content of a field is shown as not applicable (N/A), the field is omitted. Note that Address 1 always holds the receiver address of the intended receiver (or, in the case of multicast frames, receivers) and that Address 2 always holds the address of the station that is transmitting the frame.

A station uses the contents of the Address 1 field to perform address matching for receive decisions. In cases where the Address 1 field contains a group address, the BSSID also is validated to ensure that the broadcast or multicast originated in the same BSS. A station uses the contents of the Address 2 field to direct the acknowledgment if an acknowledgment is necessary. The DA is the destination of the MSDU (or fragment thereof) in the frame body field. The SA is the address of the MAC entity that initiated the MSDU (or fragment thereof) in the frame body field. The RA is the address of the STA contained in the AP in the wireless distribution system

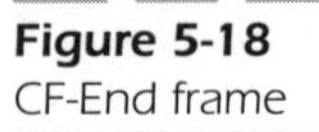

Figure 5-18
CF-End frame

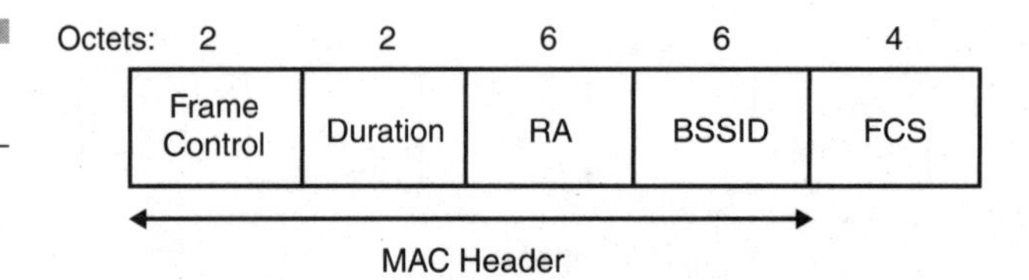

Figure 5-19
CF-End+CF-Ack frame

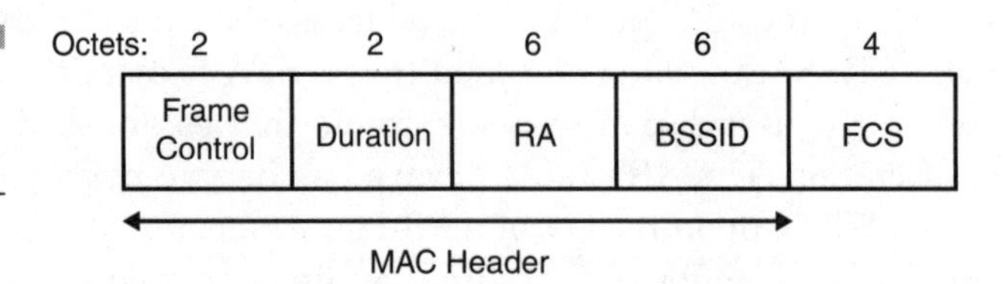

Figure 5-20
Data frame

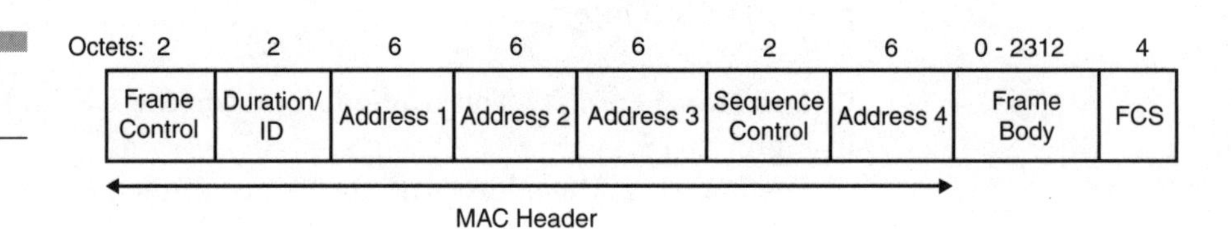

Table 5-2
Address field contents

To DS	From DS	Address 1	Address 2	Address 3	Address 4
0	0	DA	SA	BSSID	N/A
0	1	DA	BSSID	SA	N/A
1	0	BSSID	SA	DA	N/A
1	1	RA	TA	DA	SA

that is the next immediate intended recipient of the frame. The TA is the address of the STA contained in the AP in the wireless distribution system that is transmitting the frame. The BSSID of the data frame is determined as follows:

- If the station is an AP or is associated with an AP, the BSSID is the address currently in use by the STA contained in the AP.
- If the station is a member of an IBSS, the BSSID is the BSSID of the IBSS.

The frame body consists of the MSDU or a fragment thereof, and a WEP IV and ICV (if and only if the WEP subfield in the frame control field is set to 1). The frame body is null (0 octets in length) in data frames of Subtype Null function (no data), CF-Ack (no data), CF-Poll (no data), and CF-Ack+CF-Poll (no data). Within all data-type frames sent during the CFP, the Duration field is set to the value 32,768. Within all data-type frames sent during the contention period, the Duration field is set according to the following rules:

- If the Address 1 field contains a group address, the duration value is set to 0.
- If the More Fragments bit is set to 0 in the Frame Control field of a frame and the Address 1 field contains an individual address, the duration value is set to the time, in microseconds, required to transmit one ACK frame, plus one SIFS interval.
- If the More Fragments bit is set to 1 in the Frame Control field of a frame and the Address 1 field contains an individual address, the duration value is set to the time, in microseconds, required to transmit the next fragment of this data frame, plus two ACK frames, plus three SIFS intervals.

The duration value calculation for the data frame is based on rules that determine the data rate at which the control frames in the frame exchange sequence are transmitted. If the calculated duration includes a fractional microsecond, that value is rounded up to the next higher integer. All stations process Duration field values less than or equal to 32,767 from valid data frames to update their NAV settings as appropriate under the coordination function rules.

Management Frames The frame format for a management frame is independent of frame subtype and is defined in Figure 5-21. An STA uses the contents of the Address 1 field to perform the address matching for receive decisions. In the case where the Address 1 field contains a group address and the frame type is other than Beacon, the BSSID also is validated to ensure that the broadcast or multicast originated in the same BSS.

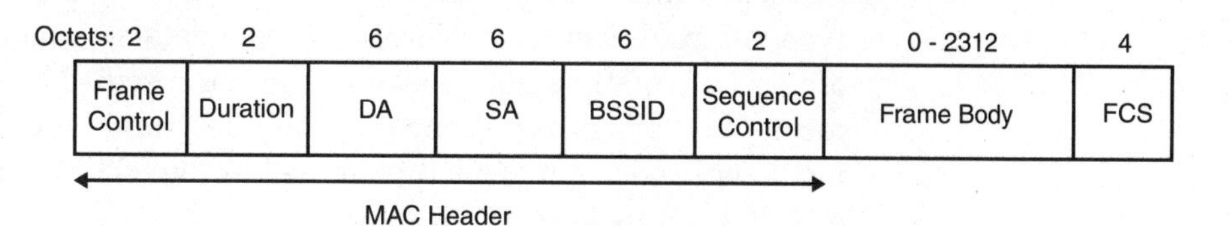

Figure 5-21
Management frame

If the frame type is Beacon, other address-matching rules apply. The address fields for management frames do not vary by frame subtype. The BSSID of the management frame is determined as follows:

- If the station is an AP or is associated with an AP, the BSSID is the address currently in use by the STA contained in the AP.
- If the station is a member of an IBSS, the BSSID is the BSSID of the IBSS.
- In management frames of subtype Probe Request, the BSSID is either a specific BSSID or the broadcast BSSID.

The DA is the destination of the frame. The SA is the address of the station transmitting the frame.

Within all management type frames sent during the CFP, the Duration field is set to the value 32,768. Within all management type frames sent during the contention period, the Duration field is set according to the following rules:

- If the DA field contains a group address, the duration value is set to 0.
- If the More Fragments bit is set to 0 in the Frame Control field of a frame and the DA contains an individual address, the duration value is set to the time, in microseconds, required to transmit one ACK frame plus one SIFS interval.
- If the More Fragments bit is set to 1 in the Frame Control field of a frame and the DA contains an individual address, the duration value is the time, in microseconds, required to transmit the next fragment of this management frame, plus two ACK frames, plus three SIFS intervals.

The duration value calculation for the management frame is based on rules that determine the data rate at which the control frames in the frame exchange sequence are transmitted. If the calculated duration includes a fractional microsecond, that value is rounded up to the next higher integer. All stations process Duration field values less than or equal to 32,767 from valid management frames to update their NAV settings as appropriate under the coordination function rules. The frame body consists of the fixed fields and information elements defined for each management frame subtype. All fixed fields and information elements are mandatory unless stated otherwise, and they can appear only in the specified order. Stations encountering an element type they do not understand ignore that element. Element type codes not explicitly defined in the 802.11 standard are reserved and do not appear in any frames.

Beacon Frame Format The frame body of a management frame of subtype Beacon contains the information shown in Table 5-3.

IBSS ATIM Frame Format The frame body of a management frame of subtype ATIM is null.

Disassociation Frame Format The frame body of a management frame of subtype Disassociation contains the information shown in Table 5-4.

Table 5-3

Beacon frame body

Order	Information	Notes
1	Timestamp	
2	Beacon interval	
3	Capability information	
4	SSID	
5	Supported rates	
6	FH Parameter Set	The FH Parameter Set information element is present within Beacon frames generated by STAs using frequency-hopping PHYs.
7	DS Parameter Set	The DS Parameter Set information element is present within Beacon frames generated by STAs using direct sequence PHYs.
8	CF Parameter Set	The CF Parameter Set information element is only present within Beacon frames generated by APs supporting a PCF.
9	IBSS Parameter Set	The IBSS Parameter Set information element is only present within Beacon frames generated by STAs in an IBSS.
10	TIM	The TIM information element is only present within Beacon frames generated by APs.

Table 5-4

Disassociation frame body

Order	Information
1	Reason code

Association Request Frame Format The frame body of a management frame of subtype Association Request contains the information shown in Table 5-5.

Association Response Frame Format The frame body of a management frame of subtype Association Response contains the information shown in Table 5-6.

Reassociation Request Frame Format The frame body of a management frame of subtype Reassociation Request contains the information shown in Table 5-7.

Table 5-5
Association Request frame body

Order	Information
1	Capability information
2	Listen interval
3	SSID
4	Supported rates

Table 5-6
Association Response frame body

Order	Information
1	Capability information
2	Status code
3	AID
4	Supported rates

Table 5-7
Reassociation Request frame body

Order	Information
1	Capability information
2	Listen interval
3	Current AP address
4	SSID
5	Supported rates

Reassociation Response Frame Format The frame body of a management frame of subtype Reassociation Response contains the information shown in Table 5-8.

Probe Request Frame Format The frame body of a management frame of subtype Probe Request contains the information shown in Table 5-9.

Probe Response Frame Format The frame body of a management frame of subtype Probe Response contains the information shown in Table 5-10.

Authentication Frame Format The frame body of a management frame of subtype Authentication contains the information shown in Table 5-11.

Deauthentication The frame body of a management frame of subtype Deauthentication contains the information shown in Table 5-13.

Management Frame Body Components

Within management frames, fixed-length mandatory frame body components are defined as fixed fields; variable-length mandatory and all optional frame body components are defined as information elements.

Table 5-8

Reassociation Response frame body

Order	Information
1	Capability information
2	Status code
3	AID
4	Supported rates

Table 5-9

Probe Request frame body

Order	Information
1	SSID
2	Supported rates

Table 5-10

Probe Response frame body

Order	Information	Notes
1	Timestamp	
2	Beacon interval	
3	Capability information	
4	SSID	
5	Supported rates	
6	FH Parameter Set	The FH Parameter Set information element is present within Probe Response frames generated by STAs using frequency-hopping PHYs.
7	DS Parameter Set	The DS Parameter Set information element is present within Probe Response frames generated by STAs using direct sequence PHYs.
8	CF Parameter Set	The CF Parameter Set information element is only present within Probe Response frames generated by APs supporting a PCF.
9	IBSS Parameter Set	The IBSS Parameter Set information element is only present within Probe Response frames generated by STAs in an IBSS.

Table 5-11

Authentication frame body

Order	Information	Notes
1	Authentication algorithm number	
2	Authentication transaction sequence number	
3	Status code	The status code information is reserved and set to 0 in certain Authentication frames, as defined in Table 5-12.
4	Challenge text	The challenge text information is only present in certain Authentication frames, as defined in Table 5-12.

Fixed Fields

Authentication Algorithm Number Field The Authentication Algorithm Number field indicates a single authentication algorithm. The length of the Authentication Algorithm Number field is 2 octets. The Authentication

Table 5-12

Presence of challenge text information

Authentication Algorithm	Authentication Transaction Sequence Number	Status Code	Challenge Text
Open system	1	Reserved	Not present
Open system	2	Status	Not present
Shared key	1	Reserved	Not present
Shared key	2	Status	Present
Shared key	3	Reserved	Present
Shared key	4	Status	Not present

Table 5-13

Deauthentication frame body

Order	Information
1	Reason code

Algorithm Number field is illustrated in Figure 5-22. The following values are defined for authentication algorithm number:

- Authentication algorithm number = 0: open system.
- Authentication algorithm number = 1: shared key.
- All other values of authentication number are reserved.

Authentication Transaction Sequence Number Field The Authentication Transaction Sequence Number field indicates the current state of progress through a multistep transaction. The length of the Authentication Transaction Sequence Number field is 2 octets. The Authentication Transaction Sequence Number field is illustrated in Figure 5-23.

Beacon Interval Field The Beacon Interval field represents the number of time units (TUs) between target beacon transmission times (TBTTs). The length of the Beacon Interval field is 2 octets. The Beacon Interval field is illustrated in Figure 5-24.

Capability Information Field The Capability Information field contains a number of subfields that are used to indicate requested or advertised capabilities. The length of the Capability Information field is 2 octets. The

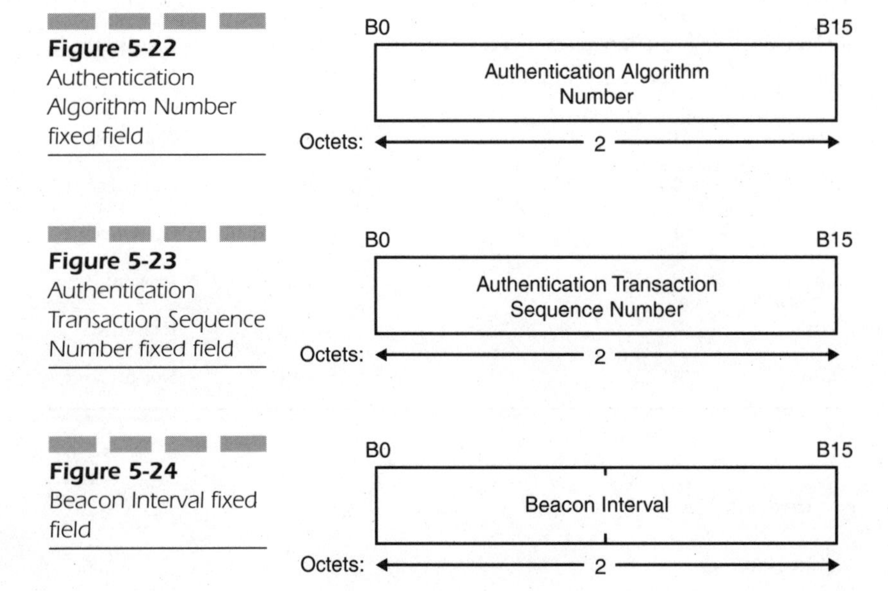

Figure 5-22 Authentication Algorithm Number fixed field

Figure 5-23 Authentication Transaction Sequence Number fixed field

Figure 5-24 Beacon Interval fixed field

Capability Information field consists of the following subfields: ESS, IBSS, CF-Pollable, CF-Poll Request, and Privacy. The remaining part of the Capability Information field is reserved. The format of the Capability Information field is illustrated in Figure 5-25.

Each Capability Information subfield is interpreted only in the management frame subtypes for which the transmission rules are defined. APs set the ESS subfield to 1 and the IBSS subfield to 0 within transmitted Beacon or Probe Response management frames. STAs within an IBSS set the ESS subfield to 0 and the IBSS subfield to 1 in transmitted Beacon or Probe Response management frames. STAs set the CF-Pollable and CF-Poll Request subfields in Association and Reassociation Request management frames according to Table 5-14.

APs set the CF-Pollable and CF-Poll Request subfields in Beacon, Probe Response, Association Response, and Reassociation Response management frames according to Table 5-15. An AP sets the CF-Pollable and CF-Poll Request subfield values in Association Response and Reassociation Response management frames equal to the values in the last Beacon or Probe Response frame that it transmitted.

APs set the Privacy subfield to 1 within transmitted Beacon, Probe Response, Association Response, and Reassociation Response management frames if WEP encryption is required for all data-type frames exchanged

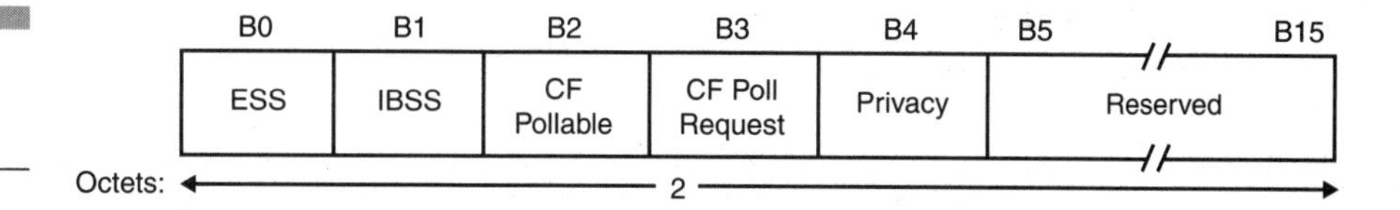

Figure 5-25 Capability Information field

Table 5-14 STA usage of CF-Pollable and CF-Poll Request

CF-Pollable	CF-Poll Request	Meaning
0	0	STA is not CF-Pollable.
0	1	STA is CF-Pollable, not requesting to be placed on the CF-Polling list.
1	0	STA is CF-Pollable, requesting to be placed on the CF-Polling list.
1	1	STA is CF-Pollable, requesting never to be polled.

Table 5-15 AP usage of CF-Pollable and CF-Poll Request

CF-Pollable	CF-Poll Request	Meaning
0	0	No PC at AP
0	1	PC at AP for delivery only (no polling)
1	0	PC at AP for delivery and polling
1	1	Reserved

within the BSS. If WEP encryption is not required, the Privacy subfield is set to 0.

STAs within an IBSS set the Privacy subfield to 1 in transmitted Beacon or Probe Response management frames if WEP encryption is required for all data-type frames exchanged within the IBSS. If WEP encryption is not required, the Privacy subfield is set to 0.

Current AP Address Field The Current AP Address field is the MAC address of the AP with which the station is currently associated. The length of the Current AP Address field is 6 octets. The Current AP Address field is illustrated in Figure 5-26.

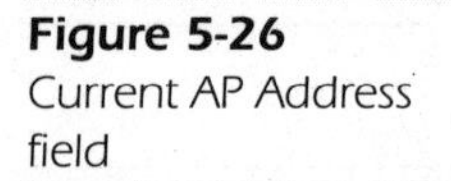

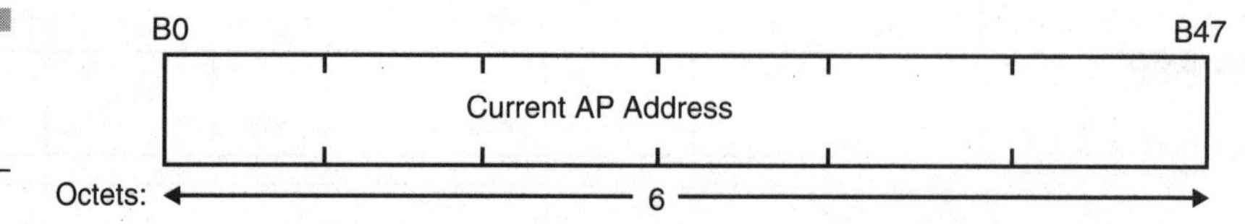

Figure 5-26 Current AP Address field

Listen Interval Field The Listen Interval field is used to indicate to the AP how often an STA wakes to listen to Beacon management frames. The value of this parameter is the STA's Listen Interval parameter of the MLME-Associate request primitive and is expressed in units of Beacon Interval. The length of the Listen Interval field is 2 octets. The Listen Interval field is illustrated in Figure 5-27. An AP may use the Listen Interval information in determining the lifetime of frames that it buffers for an STA.

Reason Code Field This Reason Code field is used to indicate the reason that an unsolicited notification management frame of type Disassociation or Deauthentication was generated. The length of the Reason Code field is 2 octets. The Reason Code field is illustrated in Figure 5-28.

AID Field The AID field is a value assigned by an AP during association that represents the 16-bit ID of an STA. The length of the AID field is 2 octets. The AID field is illustrated in Figure 5-29. The value assigned as the AID is in the range 1 to 2,007 and is placed in the 14 least significant bits of the AID field, with the two most significant bits of the AID field each set to 1.

Status Code Field The Status Code field is used in a response management frame to indicate the success or failure of a requested operation. The length of the Status Code field is 2 octets. The Status Code field is illustrated in Figure 5-30.

If an operation is successful, then the status code is set to 0. If an operation results in failure, the status code indicates a failure cause.

Timestamp Field This field represents the value of the TSFTIMER of a frame's source. The length of the Timestamp field is 8 octets. The Timestamp field is illustrated in Figure 5-31.

Information Elements Elements are defined to have a common general format consisting of a 1-octet Element ID field, a 1-octet length field, and

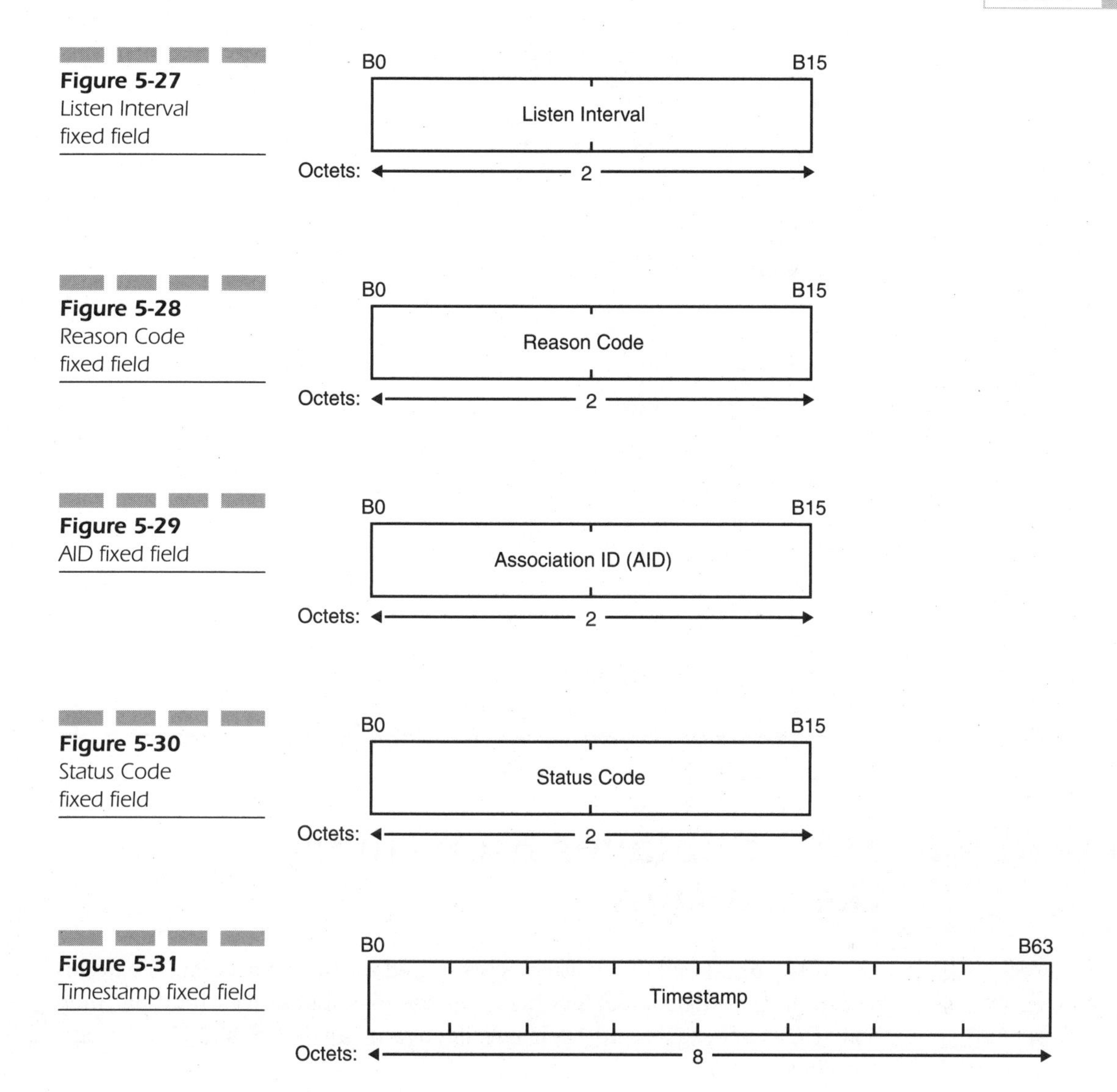

Figure 5-27 Listen Interval fixed field

Figure 5-28 Reason Code fixed field

Figure 5-29 AID fixed field

Figure 5-30 Status Code fixed field

Figure 5-31 Timestamp fixed field

a variable-length element-specific information field. Each element is assigned a unique Element ID, as defined in the 802.11 standard. The Length field specifies the number of octets in the Information field (see Figure 5-32). The set of valid elements is defined in Table 5-16.

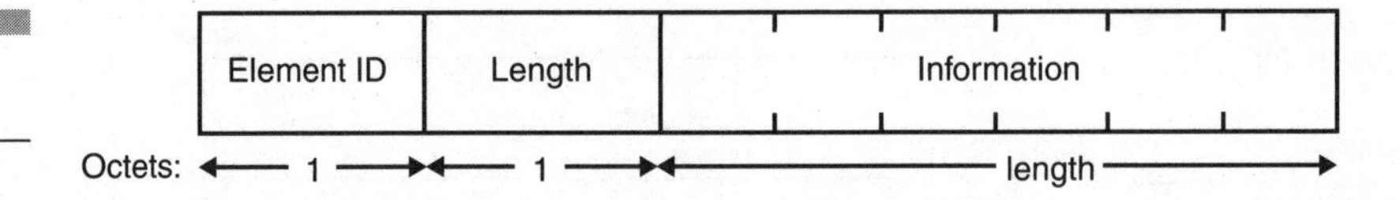

Figure 5-32 Element format

Table 5-16 Element IDs

Information Element	Element ID
SSID	0
Supported rates	1
FH Parameter Set	2
DS Parameter Set	3
CF Parameter Set	4
TIM	5
IBSS Parameter Set	6
Reserved	7–15
Challenge text	16
Reserved for challenge text extension	17–31
Reserved	32–255

MAC Sublayer Functional Description

The MAC functional description is presented in this subsection. The architecture of the MAC sublayer, including the distributed coordination function (DCF), the PCF, and their coexistence in an IEEE 802.11 LAN are introduced.

MAC Architecture

The MAC architecture can be described as providing the PCF through the services of the DCF, as shown in Figure 5-33.

Distributed Coordination Function (DCF) The fundamental access method of the IEEE 802.11 MAC is a DCF known as CSMA/CA. The DCF

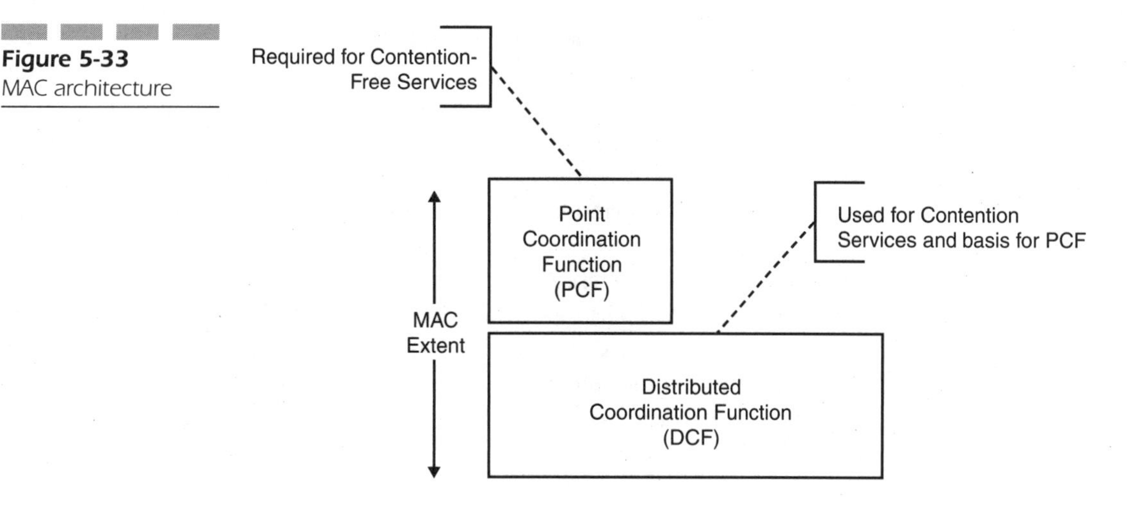

Figure 5-33 MAC architecture

will be implemented in all STAs, for use within both IBSS and infrastructure network configurations.

For an STA to transmit, it will sense the medium to determine if another STA is transmitting. If the medium is not determined to be busy, the transmission may proceed. The CSMA/CA distributed algorithm mandates that a gap of a minimum specified duration exist between contiguous frame sequences. A transmitting STA will ensure that the medium is idle for this required duration before attempting to transmit. If the medium is determined to be busy, the STA will defer until the end of the current transmission. After deferral, or prior to attempting to transmit again immediately after a successful transmission, the STA will select a random backoff interval and will decrement the backoff interval counter while the medium is idle. A refinement of the method may be used under various circumstances to further minimize collisions—here the transmitting and receiving STA exchange short control frames (RTS and CTS frames) after determining that the medium is idle and after any deferrals or backoffs prior to data transmission.

Point Coordination Function (PCF) The IEEE 802.11 MAC may also incorporate an optional access method called a PCF, which is only usable on infrastructure network configurations. This access method uses a point coordinator (PC), which will operate at the AP of the BSS, to determine which STA currently has the right to transmit. The operation is essentially that of polling, with the PC performing the role of the polling master. The operation of the PCF may require additional coordination, not specified in the 802.11 standard, to permit efficient operation in cases where multiple point-coordinated BSSs are operating on the same channel, in overlapping physical space. The PCF uses a virtual carrier sense mechanism aided by an access priority mechanism. The PCF will distribute information within Beacon management frames to gain control of the medium by setting the NAV in STAs. In addition, all frame transmissions under the PCF may use an interframe space (IFS) that is smaller than the IFS for frames transmitted via the DCF. Using a smaller IFS implies that point-coordinated traffic will have priority access to the medium over STAs in overlapping BSSs operating under the DCF access method. The access priority provided by a PCF may be utilized to create a CF access method. The PC controls the frame transmissions of the STAs so as to eliminate contention for a limited period of time.

Coexistence of DCF and PCF The DCF and the PCF will coexist in a manner that permits both to operate concurrently within the same BSS. When a PC is operating in a BSS, the two access methods alternate, with a CFCFP followed by a contention period (CP).

DSSS PHY Specification for the 2.4 GHz Band Designated for Industrial, Science, and Medical (ISM) Applications

Overview

The PHY for the DSSS system is described in this clause. The radio frequency (RF) LAN system is initially aimed for the 2.4 GHz band designated for ISM applications as provided in the United States according to FCC 15.247 and in Europe by ETS 300-328. The DSSS system provides a WLAN with both a 1 and 2 Mbps data payload communication capability. According to the FCC regulations, the DSSS system will provide a processing gain of at least 10 dB. This will be accomplished by chipping the baseband signal at 11 MHz with an 11-chip pseudonoise (PN) code. The DSSS system uses baseband modulations of differential binary phase-shift keying

(DBPSK) and differential quadrature phase-shift keying (DQPSK) to provide the 1 and 2 Mbps data rates, respectively.

Scope The PHY services provided to the IEEE 802.11 WLAN MAC by the 2.4 GHz DSSS system are described in this subsection. The DSSS PHY layer consists of two protocol functions:

- A physical layer convergence function, which adapts the capabilities of the physical medium dependent (PMD) system to the PHY service. This function will be supported by the PLCP, which defines a method of mapping the IEEE 802.11 MAC MPDUs into a framing format suitable for sending and receiving user data and management information between two or more STAs using the associated PMD system.
- A PMD system, whose function defines the characteristics of, and method of transmitting and receiving data through, a WM between two or more STAs each using the DSSS system.

DSSS PHY Functions The 2.4 GHz DSSS PHY architecture is depicted in the reference model in Figure 5-9. The DSSS PHY contains three functional entities: the PMD function, the physical layer convergence function, and the layer management function. Each of these functions is described in the following section. The DSSS PHY service will be provided to the MAC through the PHY service primitives.

PLCP Sublayer To allow the IEEE 802.11 MAC to operate with minimum dependence on the PMD sublayer, a physical layer convergence sublayer is defined. This function simplifies the PHY service interface to the IEEE 802.11 MAC services.

PMD Sublayer The PMD sublayer provides a means to send and receive data between two or more STAs. This clause is concerned with the 2.4 GHz ISM bands using direct sequence modulation.

Physical Layer Management Entity (PLME) The PLME performs management of the local PHY functions in conjunction with the MAC management entity. The models represented by figures and state diagrams are intended to be illustrations of functions provided. It is important to distinguish between a model and a real implementation. The models are optimized for simplicity and clarity of presentation; the actual method of implementation is left to the discretion of the IEEE 802.11 DSSS-PHY-compliant developer. The service of a layer or sublayer is a set of capabilities that it offers to a user in the next higher layer (or sublayer). Abstract services are specified here by describing the service primitives and parameters that characterize each service. This definition is independent of any particular implementation.

End Notes

1. The ANSI/IEEE Std 802.11, 1999 Edition [ISO/IEC 8802-11: 1999] standard is a revision of IEEE Std 802.11-1997. The management information base (MIB) according to Open Systems Interconnection (OSI) rules has been removed, redundant management items have been removed, and Annex D has been completed with the MIB according to Simple Network Management Protocol (SNMP).
2. This reduces power consumption in mobile stations.
3. IEEE 802.1X is now also applicable.

CHAPTER 6

IEEE 802.11b and IEEE 802.11a

This chapter is taken directly from IEEE Standard 802.11b-1999 and IEEE Standard 802.11a-1999. It is intended for pedagogical, advocacy, and educational purposes, so that readers have the core concepts at their command before making planning or implementation decisions. Designers and readers requiring detailed information are advised to consult the standards cited in their entirety.

IEEE 802.11b

IEEE Std 802.11b-1999 (a supplement to ANSI/IEEE Std 802.11, 1999 Edition) Supplement to IEEE Standard for Information technology—Telecommunications and information exchange between systems—Local and metropolitan area networks—Specific requirements—Part 11: Wireless LAN Medium Access Control (MAC) and Physical Layer (PHY) specifications: Higher-Speed Physical Layer Extension in the 2.4 GHz Band.

Figure 6-1 shows the reference model used in the protocol specification while Table 6-1 lists the modulation schemes supported by IEEE 802.11b.

Overview of the High Rate, Direct Sequence Spread Spectrum PHY Specification

The standard specifies the High Rate extension of the PHY for the Direct Sequence Spread Spectrum (DSSS) system. This is known as the High Rate

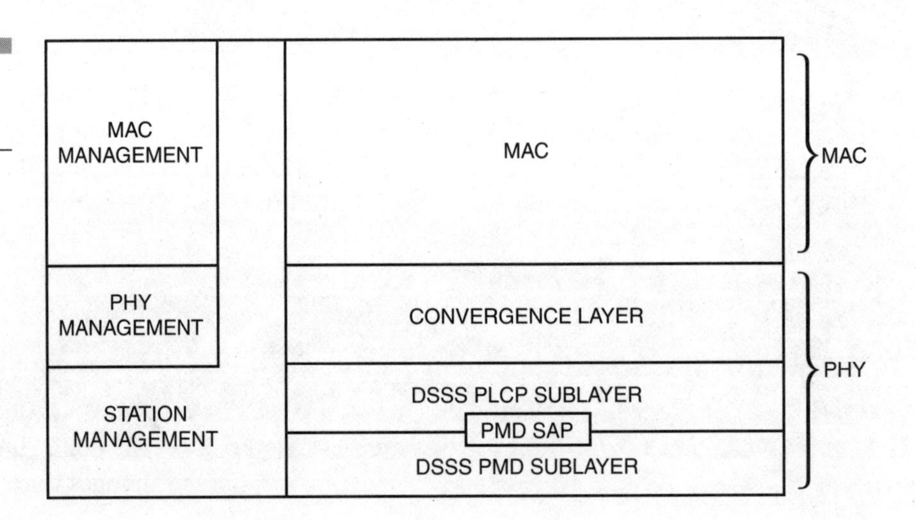

Figure 6-1 Layer reference model

Table 6-1

Modulation schemes supported by IEEE 802.11b

CCK	Complementary code keying
HR/DSSS	High Rate direct sequence spread spectrum using the Long Preamble and header
HR/DSSS/short	High Rate direct sequence spread spectrum using the optional Short Preamble and header mode
HR/DSSS/PBCC	High Rate direct sequence spread spectrum using the optional packet binary convolutional coding mode and the optional Short Preamble and header

PHY for the 2.4 GHz band designated for ISM applications. This extension of the DSSS system builds on the data rate capabilities, as described in Clause 15 of IEEE 802.11, 1999 Edition, to provide 5.5 Mbps and 11 Mbps payload data rates in addition to the 1 Mbps and 2 Mbps rates. To provide the higher rates, 8-chip complementary code keying (CCK) is employed as the modulation scheme. The chipping rate is 11 MHz, which is the same as the DSSS system described in Clause 15, thus providing the same occupied channel bandwidth. The basic new capability described in 802.11b is called High Rate Direct Sequence Spread Spectrum (HR/DSSS). The basic High Rate PHY uses the same PLCP preamble and header as the DSSS PHY, so that both PHYs can coexist in the same BSS and can use the rate switching mechanism as provided.

In addition to providing higher speed extensions to the DSSS system, a number of optional features included in IEEE 802.11b allow the performance of the radio frequency LAN system to be improved as technology allows the implementation of these options to become cost effective. An optional mode replacing the CCK modulation with packet binary convolutional coding (HR/DSSS/PBCC) is provided. Another optional mode is provided that allows data throughput at the higher rates (2, 5.5, and 11 Mbps) to be significantly increased by using a shorter PLCP preamble. This mode is called HR/DSSS/short, or HR/DSSS/PBCC/short. The Short Preamble mode can coexist with DSSS, HR/DSSS, or HR/DSSS/PBCC under limited circumstances, such as on different channels or with appropriate CCA mechanisms. An optional capability for Channel Agility is also provided. This option allows an implementation to overcome some inherent difficulty with static channel assignments (a tone jammer), without burdening all implementations with the added cost of this capability. This option can also be used to implement IEEE 802.11-compliant systems that are interoperable with both FH and DS modulations.

Scope of the Standard A clause in IEEE 802.11b specifies the PHY entity for the HR/DSSS extension and the changes that have to be made to the base standard (IEEE 802.11) to accommodate the High Rate PHY. The High Rate PHY layer consists of the following two protocol functions:

- A *PHY convergence function*, which adapts the capabilities of the physical medium dependent (PMD) system to the PHY service. This function is supported by the PHY convergence procedure (PLCP), which defines a method for mapping the MAC sublayer protocol data units (MPDU) into a framing format suitable for sending and receiving user data and management information between two or more STAs using the associated PMD system. The PHY exchanges PHY protocol data units (PPDU) that contain PLCP service data units (PSDU). The MAC uses the PHY service, so each MPDU corresponds to a PSDU that is carried in a PPDU.
- A *PMD system*, whose function defines the characteristics of, and method of transmitting and receiving data through, a wireless medium between two or more STAs, each using the High Rate PHY system.

High Rate PHY Functions The 2.4 GHz High Rate PHY architecture is depicted in the ISO/IEC basic reference model shown in Figure 6-1. The High Rate PHY contains three functional entities: (i) the PMD function, (ii) the PHY convergence function, and (iii) the layer management function.

High Rate PLCP Sublayer

Overview A convergence procedure is defined in the standard for the 2, 5.5, and 11 Mbps specification, in which PSDUs are converted to and from PPDUs. During transmission, the PSDU will be appended to a PLCP preamble and header to create the PPDU. Two different preambles and headers are defined: the mandatory supported Long Preamble and header, which interoperates with the original 1 Mbps and 2 Mbps DSSS specification (as described in IEEE Std 802.11, 1999 Edition), and an optional Short Preamble and header. At the receiver, the PLCP preamble and header are processed to aid in demodulation and delivery of the PSDU.

The optional Short Preamble and header is intended for applications where maximum throughput is desired and interoperability with legacy and nonshort-preamble capable equipment is not a consideration. That is, it is expected to be used only in networks of like equipment, which can all handle the optional mode.

PPDU Format Two different preambles and headers are defined: the mandatory supported Long Preamble and header, which is interoperable with the current 1 Mbps and 2 Mbps DSSS specification (as described in IEEE Std 802.11, 1999 Edition) and an optional Short Preamble and header. Below we only cover the Long PLCP PPDU format.

Figure 6-2 shows the format for the interoperable (long) PPDU, including the High Rate PLCP preamble, the High Rate PLCP header, and the PSDU. The PLCP preamble contains the following fields: synchronization (sync) and start frame delimiter (SFD). The PLCP header contains the following fields: signaling (SIGNAL), service (SERVICE), length (LENGTH), and CCITT/ITU CRC-16. The format for the PPDU, including the long High Rate PLCP preamble, the long High Rate PLCP header, and the PSDU, do not differ from 802.11, 1999 Edition for 1 Mbps and 2 Mbps. The only exceptions are

- The encoding of the rate in the SIGNAL field
- The use of a bit in the SERVICE field to resolve an ambiguity in PSDU length in octets, when the length is expressed in whole microseconds
- The use of a bit in the SERVICE field to indicate if the optional PBCC mode is being used
- The use of a bit in the SERVICE field to indicate that the transit frequency and bit clocks are locked

Figure 6-2 Long PLCP PDU format

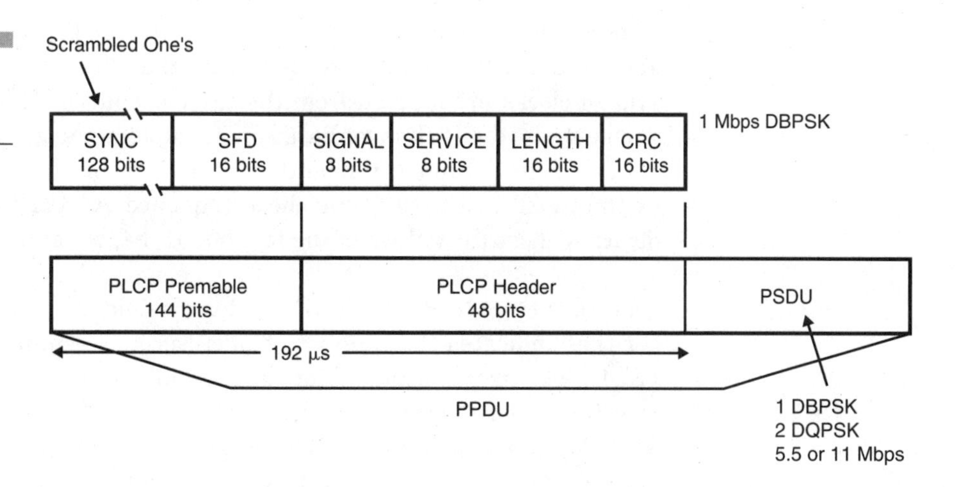

PLCP PPDU Field Definitions

Long PLCP SYNC Field The SYNC field consists of 128 bits of scrambled "1" bits. This field is provided so the receiver can perform the necessary synchronization operations.

Long PLCP SFD The SFD is provided to indicate the start of PHY-dependent parameters within the PLCP preamble. The SFD will be a 16-bit field, [1111 0011 1010 0000], where the rightmost bit will be transmitted first in time.

Long PLCP SIGNAL Field The 8-bit SIGNAL field indicates to the PHY the modulation that is used for transmission (and reception) of the PSDU. The data rate will be equal to the SIGNAL field value multiplied by 100 kbps. The High Rate PHY supports four mandatory rates given by the following 8-bit words, which represent the rate in units of 100 kbps, where the lsb will be transmitted first in time:

- X'0A' (msb to lsb) for 1 Mbps
- X'14' (msb to lsb) for 2 Mbps
- X'37' (msb to lsb) for 5.5 Mbps
- X'6E' (msb to lsb) for 11 Mbps

This field will be protected by the CCITT CRC-16 frame check sequence.

Long PLCP SERVICE Field Three bits have been defined in the SERVICE field to support the High Rate extension. The rightmost bit (bit 7) will be used to supplement the LENGTH field. Bit 3 will be used to indicate whether the modulation method is CCK <0> or PBCC <1>, as shown in Table 6-2. Bit 2 will be used to indicate that the transmit frequency and symbol clocks are derived from the same oscillator. This locked clocks bit will be set by the PHY layer based on its implementation configuration. The SERVICE field will be transmitted b0 first in time, and shall be protected by the CCITT CRC-16 frame check sequence. An IEEE 802.11-compliant device will set the values of the bits b0, b1, b4, b5, and b6 to 0.

Long PLCP LENGTH Field The PLCP length field is an unsigned 16-bit integer that indicates the number of microseconds required to transmit the PSDU. The transmitted value is determined from the LENGTH and DataRate parameters in the TXVECTOR issued with the PHY-TXSTART.request primitive. The length field provided in the TXVECTOR is in octets and is converted to microseconds for inclusion in the PLCP

Table 6-2
SERVICE field definitions

b0	b1	b2	b3	b4	b5	b6	b7
Reserved	Reserved	Locked clocks bit, 0 = not, 1 = locked	Mod. Selection bit, 0 = CCK, 1 = PBCC	Reserved	Reserved	Reserved	Length extension bit

LENGTH field. The LENGTH field is calculated as follows. Since there is an ambiguity in the number of octets that is described by a length in integer microseconds for any data rate over 8 Mbps, a length extension bit will be placed at bit position b7 in the SERVICE field to indicate when the smaller potential number of octets is correct.

- **5.5 Mbps CCK** Length = number of octets × 8/5.5, rounded up to the next integer.
- **11 Mbps CCK** Length = number of octets × 8/11, rounded up to the next integer; the service field (b7) bit will indicate a "0" if the rounding took less than 8/11 or a "1" if the rounding took more than or equal to 8/11.
- **5.5 Mbps PBCC** Length = (number of octets + 1) × 8/5.5, rounded up to the next integer.
- **11 Mbps PBCC** Length = (number of octets + 1) × 8/11, rounded up to the next integer; the service field (b7) bit will indicate a "0" if the rounding took less than 8/11 or a "1" if the rounding took more than or equal to 8/11.

At the receiver, the number of octets in the MPDU is calculated as follows:

- **5.5 Mbps CCK** Number of octets = Length × 5.5/8, rounded down to the next integer.
- **11 Mbps CCK** Number of octets = Length × 11/8, rounded down to the next integer, minus 1 if the service field (b7) bit is a "1."
- **5.5 Mbps PBCC** Number of octets = (Length × 5.5/8) − 1, rounded down to the next integer.
- **11 Mbps PBCC** Number of octets = (Length × 11/8) − 1, rounded down to the next integer, minus 1 if the service field (b7) bit is a "1."

PLCP CRC (CCITT CRC-16) Field The SIGNAL, SERVICE, and LENGTH fields will be protected with a CCITT CRC-16 frame check sequence (FCS).

Long PLCP Data Modulation and Modulation Rate Change The long PLCP preamble and header is transmitted using the 1 Mbps DBPSK modulation. The SIGNAL and SERVICE fields combined will indicate the modulation that will be used to transmit the PSDU. The SIGNAL field indicates the rate, and the SERVICE field indicates the modulation. The transmitter and receiver initiate the modulation and rate indicated by the SIGNAL and SERVICE fields, starting with the first octet of the PSDU. The PSDU transmission rate will be set by the DATARATE parameter in the TXVECTOR, issued with the PHY-TXSTART.request primitive.

PLCP Transmit Procedure The transmit procedures for a High Rate PHY using the long PLCP preamble and header are the same as those described in IEEE Std 802.11, 1999 Edition, and do not change apart from the ability to transmit 5.5 Mbps and 11 Mbps. The PLCP transmit procedure is shown in Figure 6-3.

PLCP Receive Procedure The receive procedures for receivers configured to receive the mandatory and optional PLCPs, rates, and modulations are described in this subclause. A receiver that supports this High Rate extension of the standard is capable of receiving 5.5 Mbps and 11 Mbps, in addition to 1 Mbps and 2 Mbps.

If the PHY implements the Short Preamble option, it will detect both Short and Long Preamble formats and indicate which type of preamble was received in the RXVECTOR. If the PHY implements the PBCC Modulation option, it will detect either CCK or PBCC Modulations, as indicated in the SIGNAL field, and will report the type of modulation used in the RXVECTOR.

High Rate PMD Sublayer

Scope and Field of Application Subclause 18.4 of the standard describes the PMD services provided to the PLCP for the High Rate PHY. Also defined here are the functional, electrical, and RF characteristics required for interoperability of implementations conforming to this specification. The relationship of this specification to the entire High Rate PHY is shown in Figure 6-1.

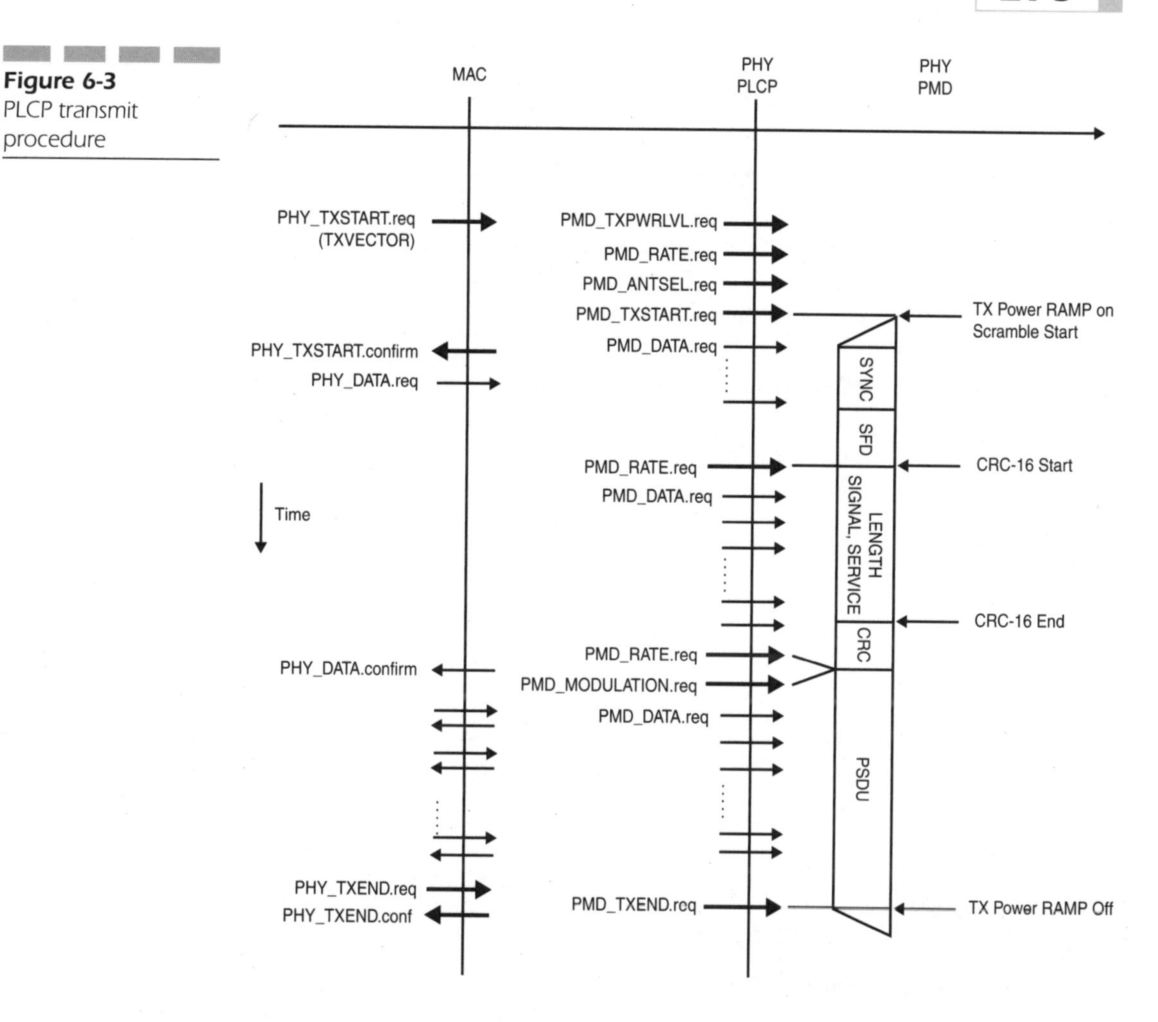

Figure 6-3 PLCP transmit procedure

Overview of Service The High Rate PMD sublayer accepts PLCP sublayer service primitives and provides the actual means by which data is transmitted or received from the medium. The combined functions of the High Rate PMD sublayer primitives and parameters for the receive function result in a data stream, timing information, and associated received signal parameters being delivered to the PLCP sublayer. A similar functionality is provided for data transmission.

Interactions The primitives associated with the PLCP sublayer to the High Rate PMD fall into two basic categories:

- Service primitives that support PLCP peer-to-peer interactions

- Service primitives that have local significance and that support sublayer-to-sublayer interactions

PMD_SAP Peer-to-Peer Service Primitives See Table 6-3.

PMD_SAP Sublayer-to-Sublayer Service Primitives See Table 6-4.

Refer to the IEEE 802.11b standard for a description of the services provided by each PMD primitive.

PMD Operating Specifications, General Subclauses 18.4.6.1 through 18.4.6.14 in the standard (summarized herewith) provide general specifications for the High Rate PMD sublayer. These specifications apply to both the receive and transmit functions and general operation of a High Rate PHY.

Table 6-3

PMD_SAP peer-to-peer service primitives

Primitive	Request	Indicative	Confirm	Response
PMD_DATA	X	X	—	—

Table 6-4

PMD_SAP sublayer-to-sublayer service primitives

Primitive	Request	Indicative	Confirm	Response
PMD_TXSTART	X	—	—	—
PMD_TXEND	X	—	—	—
PMD_ANTSEL	X	X	—	—
PMD_TXPWRLVL	X	—	—	—
PMD_MODULATION	X	X	—	—
PMD_PREAMBLE	X	X	—	—
PMD_RATE	X	X	—	—
PMD_RSSI	—	X	—	—
PMD_SQ	—	X	—	—
PMD_CS	—	X	—	—
PMD_ED	X	X	—	—

Operating Frequency Range The High Rate PHY will operate in the 2.4–2.4835 GHz frequency range, as allocated by regulatory bodies in the USA and Europe, or in the 2.471–2.497 GHz frequency range, as allocated by regulatory authority in Japan.

Number of Operating Channels The channel center frequencies and CHNL_ID numbers will be as shown in Table 6-5. The FCC (U.S.), IC (Canada), and ETSI (Europe) specify operation from 2.4–2.4835 GHz. For Japan, operation is specified as 2.471–2.497 GHz. France allows operation from 2.4465–2.4835 GHz, and Spain allows operation from 2.445–2.475 GHz. For each supported regulatory domain, all channels in Table 6-5 marked with an "X" will be supported.

Modulation and Channel Data Rates Four modulation formats and data rates are specified for the High Rate PHY. The basic access rate is based

Table 6-5

High Rate PHY frequency channel plan

		Regulatory domains					
CHNL_ID	Frequency (MHz)	X'10' FCC	X'20' IC	X'30' ETSI	X'31' Spain	X'32' France	X'40' MKK
1	2412	X	X	X	—	—	—
2	2417	X	X	X	—	—	—
3	2422	X	X	X	—	—	—
4	2427	X	X	X	—	—	—
5	2432	X	X	X	—	—	—
6	2437	X	X	X	—	—	—
7	2442	X	X	X	—	—	—
8	2447	X	X	X	—	—	—
9	2452	X	X	X	—	—	—
10	2457	X	X	X	X	X	—
11	2462	X	X	X	X	X	—
12	2467	—	—	X	—	X	—
13	2472	—	—	X	—	X	—
14	2484	—	—	—	—	—	X

on 1 Mbps DBPSK modulation. The enhanced access rate will be based on 2 Mbps DQPSK. The extended direct sequence specification defines two additional data rates. High Rate access rates are based on the CCK modulation scheme for 5.5 Mbps and 11 Mbps. An optional PBCC mode is also provided for potentially enhanced performance.

Spreading Sequences and Modulation for CCK Modulation at 5.5 Mbps and 11 Mbps For the CCK modulation modes, the spreading code length is 8 and is based on complementary codes. The chipping rate is 11 Mchip/s. The symbol duration will be exactly 8 complex chips long. The following formula is used to derive the CCK code words that will be used for spreading both 5.5 Mbps and 11 Mbps:

$$c = \{e^{j(\phi_1+\phi_2+\phi_3+\phi_4)}, e^{j(\phi_1+\phi_3+\phi_4)}, e^{j(\phi_1+\phi_2+\phi_4)},$$
$$-e^{j(\phi_1+\phi_4)}, e^{j(\phi_1+\phi_2+\phi_3)}, e^{j(\phi_1+\phi_3)}, -e^{j(\phi_1+\phi_2)}, e^{j\phi_1}\}$$

where C is the code word C = {c0 to c7}

CCK 5.5 Mbps Modulation At 5.5 Mbps 4 bits (d0 to d3; d0 first in time) are transmitted per symbol. The data bits d0 and d1 encode ϕ_1 based on DQPSK. The DQPSK encoder is specified in Table 6-6. (In the table, $+j\omega$ is defined as counterclockwise rotation.) The phase change for ϕ_1 is relative to the phase ϕ_1 of the preceding symbol. For the header to PSDU transition, the phase change for ϕ_1 is relative to the phase of the preceding DQPSK (2 Mbps) symbol. That is, the phase of the last symbol of the CRC-16 is the reference phase for the first symbol generated from the PSDU octets. A "+1" chip in the Barker code represents the same carrier phase as a "+1" chip in the CCK code.

All odd-numbered symbols generated from the PSDU octets will be given an extra 180 degree (π) rotation, in addition to the standard DQPSK modulation as shown in Table 6-6. Symbols of the PSDU are numbered from "0" for the first symbol, for the purposes of determining odd and even symbols. That is, the PSDU transmission starts on an even-numbered symbol.

The data dibits d2 and d3 CCK encode the basic symbol, as specified in Table 6-7. This table is derived from the formula above by setting $\phi_2 = (d2 \times \pi) + \pi/2$, $\phi_3 = 0$, and $\phi_4 = d3 \times \pi$. In this table, d2 and d3 are in the order

Table 6-6

DQPSK encoding table

Dibit pattern (d0, d1) (d0 is first in time)	Even symbols phase change (+jω)	Odd symbols phase change (+jω)
00	0	π
01	$\pi/2$	$3\pi/2$ $(-\pi/2)$
11	π	0
10	$3\pi/2$ $(-\pi/2)$	$\pi/2$

Table 6-7

5.5 Mbps CCK encoding table

d2, d3	c1	c2	c3	c4	c5	c6	c7	c8
00	1j	1	1j	−1	1j	1	−1j	1
01	−1j	−1	−1j	1	1j	1	−1j	1
10	−1j	1	−1j	−1	−1j	1	1j	1
11	1j	−1	1j	1	−1j	1	1j	1

shown, and the complex chips are shown c0 to c7 (left to right), with c0 transmitted first in time.

CCK 11 Mbps Modulation At 11 Mbps, 8 bits (d0 to d7; d0 first in time) are transmitted per symbol. The first dibit (d0, d1) encodes ϕ_1 based on DQPSK. The DQPSK encoder is specified in Table 6-6. The phase change for ϕ_1 is relative to the phase ϕ_1 of the preceding symbol. In the case of header to PSDU transition, the phase change for ϕ_1 is relative to the phase of the preceding DQPSK symbol. All odd-numbered symbols of the PSDU are given an extra 180 degree (π) rotation, in accordance with the DQPSK modulation shown in Table 6-6. Symbol numbering starts with "0" for the first symbol of the PSDU.

The data dibits (d2, d3), (d4, d5), and (d6, d7) encode ϕ_2, ϕ_3, and ϕ_4 respectively, based on QPSK as specified in Table 6-8 (this table is binary [not Grey] coded).

Table 6-8
QPSK encoding table

Dibit Patter [di, d(i + l)] (di is first in time)	Phase
00	0
01	$\pi/2$
10	π
11	$3\pi/2$ ($-\pi/2$)

IEEE 802.11a: OFDM PHY Specification for the 5 GHz Band

Supplementto IEEE Standard for Information technology—Telecommunications and information exchange between systems—Local and metropolitan area networks—Specific requirements Part 11: Wireless LAN Medium Access Control (MAC) and Physical Layer (PHY) specifications, High-speed Physical Layer in the 5 GHz Band, IEEE 802.11a-1999.

Concepts and Overview

IEEE 802.11a specifies the PHY entity for an orthogonal frequency division multiplexing (OFDM) system and the additions that have to be made to the base standard to accommodate the OFDM PHY. The radio frequency LAN system is initially aimed for the 5.15–5.25, 5.25–5.35, and 5.725–5.825 GHz unlicensed national information structure (U-NII) bands, as regulated in the United States by the Code of Federal Regulations, Title 47, Section 15.407. The OFDM system provides a wireless LAN with data payload communication capabilities of 6, 9, 12, 18, 24, 36, 48, and 54 Mbps. The support of transmitting and receiving at data rates of 6, 12, and 24 Mbps is mandatory. The system uses 52 subcarriers that are modulated using binary or quadrature phase shift keying (BPSK/QPSK), 16-quadrature amplitude modulation (QAM), or 64-QAM. Forward error correction coding (convolutional coding) is used with a coding rate of 1/2, 2/3, or 3/4.

Scope of the Standard This subclause describes the PHY services provided to the IEEE 802.11 wireless LAN MAC by the 5 GHz (bands) OFDM system. The OFDM PHY layer consists of two protocol functions, as follows (see Figure 6-4):

- A PHY convergence function, which adapts the capabilities of the physical medium dependent (PMD) system to the PHY service. This function is supported by the physical layer convergence procedure (PLCP), which defines a method of mapping the IEEE 802.11 PHY sublayer service data units (PSDU) into a framing format suitable for sending and receiving user data and management information between two or more stations using the associated PMD system.
- A PMD system whose function defines the characteristics and method of transmitting and receiving data through a wireless medium between two or more stations, each using the OFDM system.

OFDM PHY Functions The 5 GHz OFDM PHY contains three functional entities: the (i) PMD function, the (ii) PHY convergence function, and (iii) the layer management function. The OFDM PHY service is provided to the MAC through the PHY service primitives. The service of a layer or sublayer is the set of capabilities that it offers to a user in the next higher layer (or sublayer). Abstract services are specified here through the service primitives and parameters that characterize each service. This definition is independent of any particular implementation.

PLCP Sublayer In order to allow the IEEE 802.11 MAC to operate with minimum dependence on the PMD sublayer, a PHY convergence sublayer is

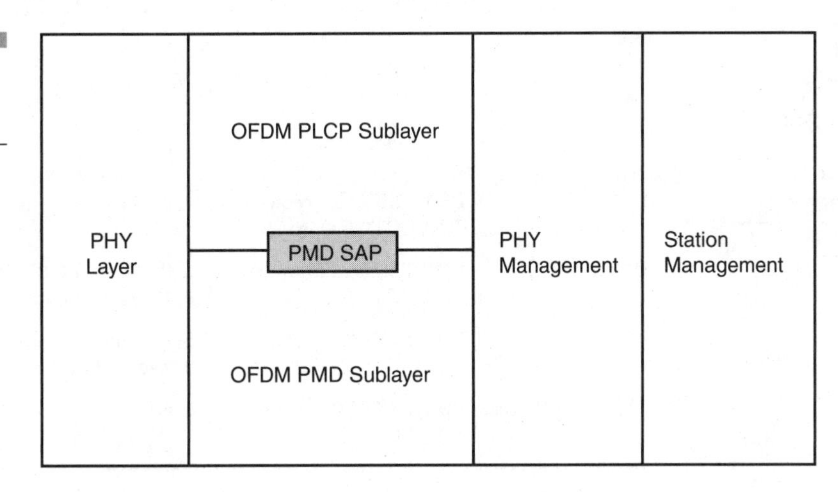

Figure 6-4 PMD layer reference model

defined. This function simplifies the PHY service interface to the IEEE 802.11 MAC services.

PMD Sublayer The PMD sublayer provides a means to send and receive data between two or more stations. This clause is concerned with the 5 GHz band using OFDM modulation.

PHY Management Entity (PLME) The PLME performs management of the local PHY functions in conjunction with the MAC management entity.

OFDM PHY Specific Service Parameter List

Introduction The architecture of the IEEE 802.11 MAC is intended to be PHYindependent. Some PHY implementations require medium management state machines running in the MAC sublayer in order to meet certain PMD requirements. These PHY-dependent MAC state machines reside in a sublayer defined as the MAC sublayer management entity (MLME). In certain PMD implementations, the MLME may need to interact with the PLME as part of the normal PHY SAP primitives. These interactions are defined by the PLME parameter list currently defined in the PHY service primitives as TXVECTOR and RXVECTOR. The list of these parameters, and the values they may represent, are defined in the specific PHY specifications for each PMD. This subsection addresses the TXVECTOR and RXVECTOR for the OFDM PHY.

TXVECTOR Parameters The parameters in Table 6-9 are defined as part of the TXVECTOR parameter list in the PHY-TXSTART.request service primitive.

Table 6-9

TXVECTPR parameters

Parameter	Associate primitive	Value
LENGTH	PHY-TXSTART.request (TXVECTOR)	1–4,095
DATA RATE	PHY-TXSTART.request (TXVECTOR)	6, 9, 12, 18, 24, 36, 48, and 54 (Support of 6, 12, and 24 data rates is mandatory.)
SERVICE	PHY-TXSTART.request (TXVECTOR)	Scrambler initialization: 7 null bits + 9 reserved null bits
TXPWR_LEVEL	PHY-TXSTART.request (TXVECTOR)	1–8

RXVECTOR Parameters The parameters listed in Table 6-10 are defined as part of the RXVECTOR parameter list in the PHY-RXSTART.indicate service primitive.

OFDM PLCP Sublayer

Introduction The mechanism described in this section provides a convergence procedure in which PSDUs are converted to and from PPDUs. During transmission, the PSDU is provided with a PLCP preamble and header to create the PPDU. At the receiver, the PLCP preamble and header are processed to aid in demodulation and delivery of the PSDU.

PLCP Frame Format Figure 6-5 shows the format for the PPDU including the OFDM PLCP preamble, OFDM PLCP header, PSDU, tail bits, and pad bits. The PLCP header contains the following fields: LENGTH, RATE, a reserved bit, an even parity bit, and the SERVICE field. In terms of modulation, the LENGTH, RATE, reserved bit, and parity bit (with six "zero" tail bits appended) constitute a separate single OFDM symbol, denoted SIGNAL, which is transmitted with the most robust combination of BPSK modulation, and a coding rate of R = 1/2. The SERVICE field of the PLCP header and the PSDU (with six "zero" tail bits and pad bits appended), denoted as DATA, are transmitted at the data rate described in the RATE field and may constitute multiple OFDM symbols. The tail bits in the SIGNAL symbol enable decoding of the RATE and LENGTH fields immediately after the reception of the tail bits. The RATE and LENGTH are required for decoding the DATA part of the packet. In addition, the CCA mechanism can be augmented by predicting the duration of the packet from the contents of the RATE and LENGTH fields, even if the data rate is not supported by the station.

Table 6-10

RXVECTOR parameters

Parameter	Associate primitive	Value
LENGTH	PHY-RXSTART.indicate	1–4,095
RSSI	PHY-RXSTART.indicate (RXVECTOR)	0–RSSI maximum
DATARATE	PHY-RXSTART.indicate (RXVECTOR)	6, 9, 12, 18, 24, 36, 48, and 54
SERVICE	PHY-RXSTART.indicate (RXVECTOR)	Null

Figure 6-5 PPDU frame format

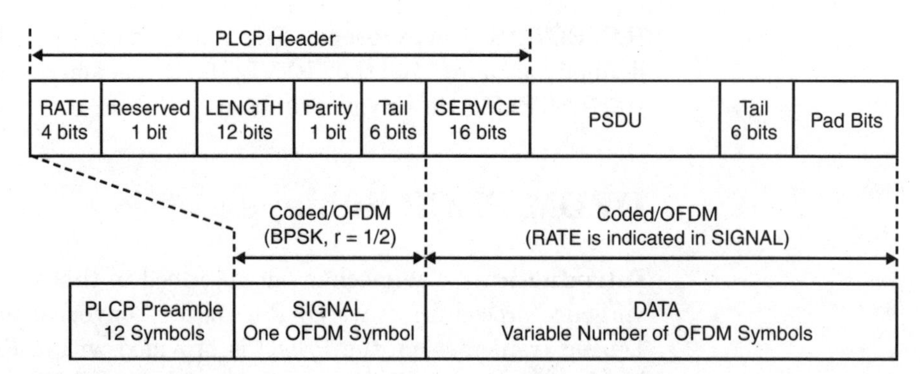

Overview of the PPDU Encoding Process The encoding process is composed of many detailed steps, which are summarized herewith.

- Produce the PLCP preamble field, composed of 10 repetitions of a "short training sequence" (used for AGC convergence, diversity selection, timing acquisition, and coarse frequency acquisition in the receiver) and two repetitions of a "long training sequence" (used for channel estimation and fine frequency acquisition in the receiver), preceded by a guard interval (GI). Refer to the standard for details.
- Produce the PLCP header field from the RATE, LENGTH, and SERVICE fields of the TXVECTOR by filling the appropriate bit fields. The RATE and LENGTH fields of the PLCP header are encoded by a convolutional code at a rate of R = 1/2, and are subsequently mapped onto a single BPSK encoded OFDM symbol, denoted as the SIGNAL symbol. In order to facilitate a reliable and timely detection of the RATE and LENGTH fields, six "zero" tail bits are inserted into the PLCP header. The encoding of the SIGNAL field into an OFDM symbol follows the same steps for convolutional encoding, interleaving, BPSK modulation, pilot insertion, Fourier transform, and prepending a GI as described subsequently for data transmission at 6 Mbps. The contents of the SIGNAL field are not scrambled. Refer to the standard for details.
- Calculate from the RATE field of the TXVECTOR the number of data bits per OFDM symbol (N DBPS), the coding rate (R), the number of bits in each OFDM subcarrier (N BPSC), and the number of coded bits per OFDM symbol (N CBPS). Refer to the standard for details.
- Append the PSDU to the SERVICE field of the TXVECTOR. Extend the resulting bit string with "zero" bits (at least six bits) so that the

resulting length will be a multiple of N DBPS. The resulting bit string constitutes the DATA part of the packet. Refer to the standard for details.

- Initiate the scrambler with a pseudorandom nonzero seed, generate a scrambling sequence, and XOR it with the extended string of data bits. Refer to the standard for details.
- Replace the six scrambled "zero" bits following the "data" with six nonscrambled "zero" bits.

  (Those bits return the convolutional encoder to the "zero state" and are denoted as "tail bits.") Refer to the standard for details.
- Encode the extended, scrambled data string with a convolutional encoder ($R = 1/2$). Omit (puncture) some of the encoder output string (chosen according to the "puncturing pattern") to reach the desired coding rate. Refer to the standard for details.
- Divide the encoded bit string into groups of N CBPS bits. Within each group, perform an "interleaving" (reordering) of the bits according to a rule corresponding to the desired RATE. Refer to the standard for details.
- Divide the resulting coded and interleaved data string into groups of N CBPS bits. For each of the bit groups, convert the bit group into a complex number according to the modulation encoding tables. Refer to the standard for details.
- Divide the complex number string into groups of 48 complex numbers. Each such group will be associated with one OFDM symbol. In each group, the complex numbers will be numbered 0 to 47 and mapped hereafter into OFDM subcarriers numbered −26 to −22, −20 to −8, −6 to −1, 1 to 6, 8 to 20, and 22 to 26. The subcarriers −21, −7, 7, and 21 are skipped and, subsequently, used for inserting pilot subcarriers. The "0" subcarrier, associated with center frequency, is omitted and filled with zero value. Refer to the standard for details.
- Four subcarriers are inserted as pilots into positions −21, −7, 7, and 21. The total number of the subcarriers is 52 (48 + 4). Refer to the standard for details.
- For each group of subcarriers −26 to 26, convert the subcarriers to a time domain using an inverse Fourier transform. Prepend to the Fourier-transformed waveform a circular extension of itself thus forming a GI, and truncate the resulting periodic waveform to a single OFDM symbol length by applying time domain windowing. Refer to the standard for details.

- Append the OFDM symbols one after another, starting after the SIGNAL symbol describing the RATE and LENGTH. Refer to the standard for details.
- Up-convert the resulting complex baseband waveform to an RF frequency according to the center frequency of the desired channel and transmit. Refer to the standard for details.

RATE-Dependent Parameters The modulation parameters dependent on the data rate used is set according to Table 6-11.

Timing Related Parameters Table 6-12 is the list of timing parameters associated with the OFDM PLCP.

Mathematical Conventions in the Signal Descriptions The transmitted signals will be described in a complex baseband signal notation. The signal actually transmitted relates to the complex baseband signal as follows:

$$r_{(RF)^{(t)}} = Re\{r\langle t\rangle \exp\langle j2\pi f_c t\rangle\} \quad (1)$$

where

$Re(.)$ represents the real part of a cokplex variable.
f_c denotes the carrier center frequency.

Table 6-11

Rate-dependent parameters

Data rate (Mbps)	Modulation	Coding rate (R)	Coded bits per subcarrier (N_{BPSC})	Coded bits per OFDM symbol (N_{CBPS})	Data bits per OFDM symbol (N_{DBPS})
6	BPSK	1/2	1	48	24
9	BPSK	3/4	1	48	36
12	QPSK	1/2	2	96	48
18	QPSK	3/4	2	96	72
24	16-QAM	1/2	4	192	96
36	16-QAM	3/4	4	192	144
48	64-QAM	2/3	6	288	192
54	64-QAM	3/4	6	288	216

Table 6-12

Timing-related parameters

Parameter	Value
N_{SD}: Number of data subcarriers	48
N_{SP}: Number of pilot subcarriers	4
N_{ST}: Number of subcarriers, total	52 ($N_{SD} + N_{SP}$)
Δ_F: Subcarrier frequency spacing	0.3125 MHz (= 20 MHz/64)
T_{FFT}: IFFT/FFT period	3.2 μs ($1/\Delta_F$)
$T_{PREAMBLE}$: PLCP preamble duration	16 μs ($T_{SHORT} + T_{LONG}$)
T_{SIGNAL}: Duration of the SIGNAL PBSK-OFDM symbol	4.0 μs ($T_{GI} + T_{FFT}$)
T_{GI}: GI duration	0.8 μs ($T_{FFT}/4$)
T_{GI2}: Training symbol GI duration	1.6 μs ($T_{FFT}/2$)
T_{SYM}: Symbol interval	4 μs ($T_{GI} + T_{FFT}$)
T_{SHORT}: Short training sequence duration	8 μs ($10 \times T_{FFT}/4$)
T_{LONG}: Long training sequence duration	8 μs ($T_{GI2} + 2 \times T_{FFT}$)

The transmitted baseband signal is composed of contributions from several OFDM symbols.

$$r_{PACKET}(t) = r_{PREAMBLE}(t) + r_{SIGNAL}(t - t_{SIGNAL}) + r_{DATA}(t - t_{DATA}) \quad (2)$$

PLCP Preamble (SYNC) The PLCP preamble field is used for synchronization. It consists of 10 short symbols and two long symbols that are shown in Figure 6-6.

Figure 6-6 shows the OFDM training structure (PLCP preamble), where t1 to t10 denote short training symbols and T1 and T2 denote long training symbols. The PLCP preamble is followed by the SIGNAL field and DATA. The total training length is 16 μs. The dashed boundaries in the figure denote repetitions due to the periodicity of the inverse Fourier transform. A short OFDM training symbol consists of 12 subcarriers.

Signal Field (SIGNAL) The OFDM training symbol is followed by the SIGNAL field, which contains the RATE and the LENGTH fields of the TXVECTOR. The RATE field conveys information about the type of modulation and the coding rate as used in the rest of the packet. The encoding

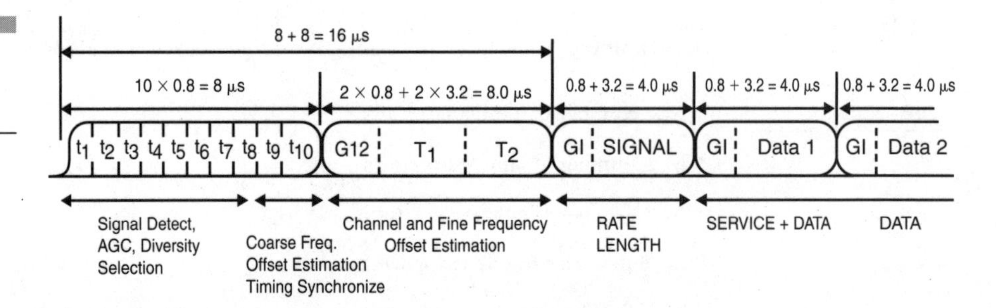

Figure 6-6 OFDM training structure

of the SIGNAL single OFDM symbol is performed with BPSK modulation of the subcarriers and using convolutional coding at R = 1/2. The encoding procedure includes convolutional encoding, interleaving, modulation mapping, pilot insertion, and OFDM modulation. The contents of the SIGNAL field are not scrambled.

The SIGNAL field is composed of 24 bits, as illustrated in Figure 6-7. The four bits 0 to 3 encode the RATE. Bit 4 is reserved for future use. Bits 5–16 encode the LENGTH field of the TXVECTOR, with the least significant bit (LSB) being transmitted first.

DATA Field The DATA field contains the SERVICE field, the PSDU, the TAIL bits, and the PAD bits, if needed.

Service Field (SERVICE) The IEEE 802.11 SERVICE field has 16 bits, which are denoted as bits 0–15. The bit 0 is transmitted first in time. The bits from 0–6 of the SERVICE field, which are transmitted first, are set to zeros and are used to synchronize the descrambler in the receiver. The remaining nine bits (7–15) of the SERVICE field are reserved for future use. All reserved bits are set to zero. Refer to Figure 6-8.

PPDU Tail Bit Field (TAIL) The PPDU tail bit field is six bits of "0," which are required to return the convolutional encoder to the "zero state." This procedure improves the error probability of the convolutional decoder, which relies on future bits when decoding and which may be not be available past the end of the message. The PLCP tail bit field is produced by replacing six scrambled "zero" bits following the message end with six nonscrambled zero bits.

Pad Bits (PAD) The number of bits in the DATA field is a multiple of N CBPS, the number of coded bits in an OFDM symbol (48, 96, 192, or 288 bits). To achieve that, the length of the message is extended so that it

RATE (4 bits)					LENGTH (12 bits)													SIGNAL TAIL (6 bits)					
R1	R2	R3	R4	R	LSB											MSB	P	"0"	"0"	"0"	"0"	"0"	"0"
0	1	2	3	4	5	6	7	8	9	10	11	12	13	14	15	16	17	18	19	20	21	22	23

Transmit Order →

Figure 6-7 SIGNAL field bit assignment

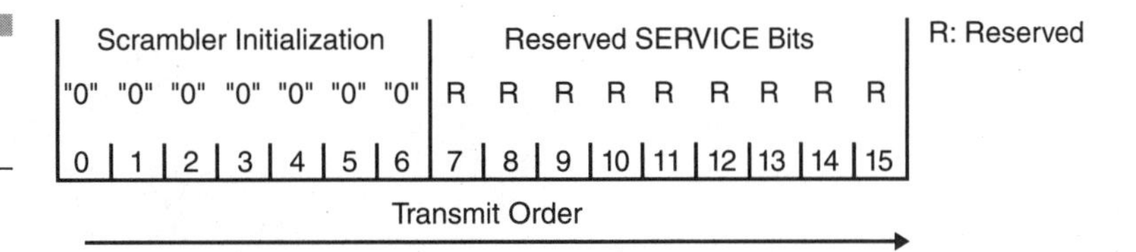

Figure 6-8 SERVICE field bit assignment

becomes a multiple of N DBPS, the number of data bits per OFDM symbol. At least six bits are appended to the message, in order to accommodate the TAIL bits.

Subcarrier Modulation Mapping OFDM subcarriers are modulated by using BPSK, QPSK, 16-QAM, or 64-QAM modulation, depending on the RATE requested. The encoded and interleaved binary serial input data is divided into groups of N BPSC (1, 2, 4, or 6) bits and converted into complex numbers representing BPSK, QPSK, 16-QAM, or 64-QAM constellation points. The conversion is performed according to Gray-coded constellation mappings, illustrated in Figure 6-9. Once again, please refer to the standard for details.

PMD Operating Specifications (General) Subclauses 17.3.8.1 through 17.3.8.8 in the standard provide general specifications for the BPSK OFDM, QPSK OFDM, 16-QAM OFDM, and 64-QAM OFDM PMD sublayers. These specifications apply to both the receive and transmit functions and general operation of the OFDM PHY.

Outline Description The general block diagram of the transmitter and receiver for the OFDM PHY is shown in Figure 6-10. Major specifications for the OFDM PHY are listed in Table 6-13.

Operating Channel Frequencies

Operating Frequency Range The OFDM PHY operates in the 5 GHz band, as allocated by a regulatory body in its operational region. Spectrum

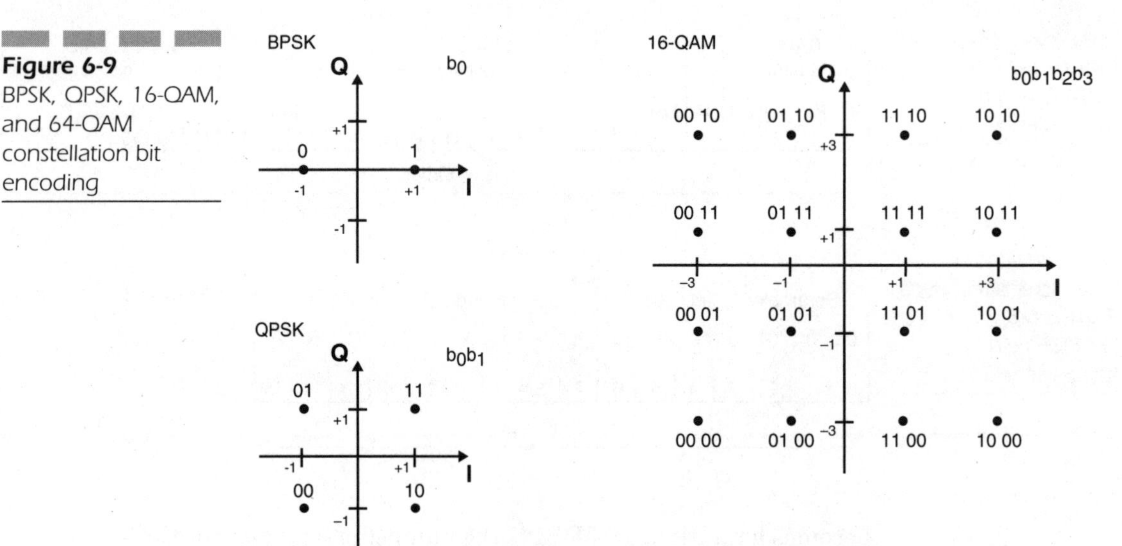

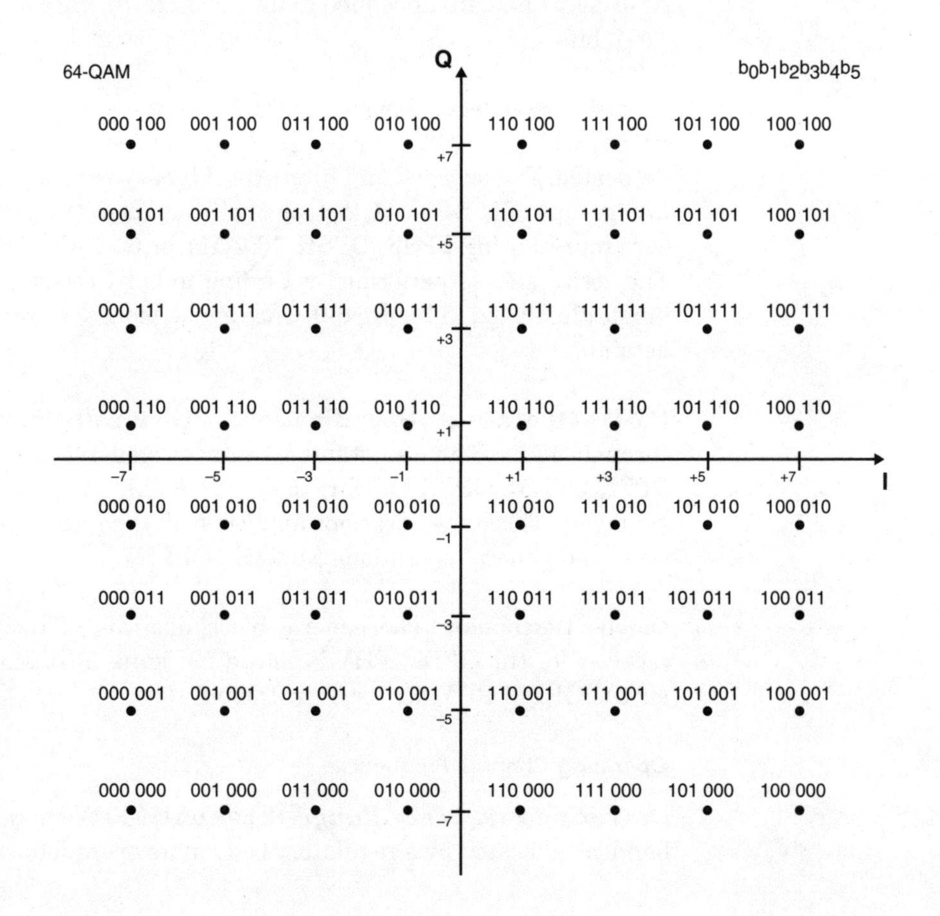

Figure 6-9 BPSK, QPSK, 16-QAM, and 64-QAM constellation bit encoding

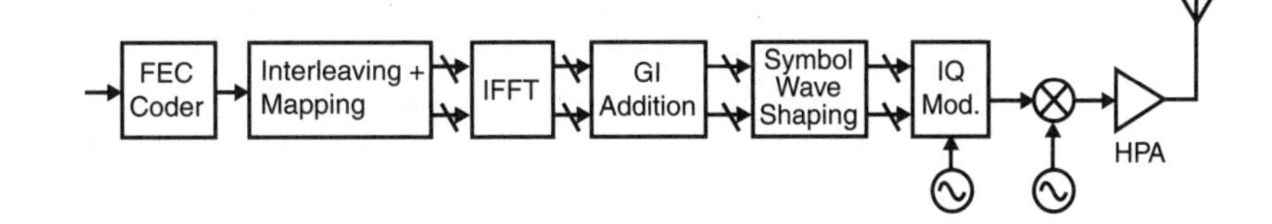

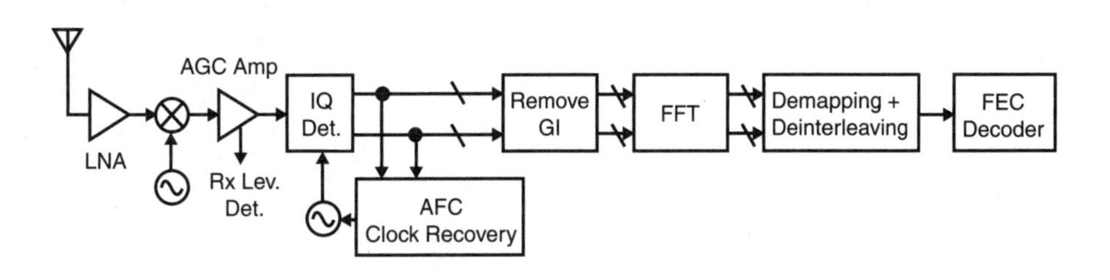

Figure 6-10 Transmitter and receiver block diagram for the OFDM PHY

Table 6-13 Major parameters of the OFDM PHY

Parameter	Value
Information data rate	6, 9, 12, 18, 24, 36, 48, and 54 Mbps (6, 12, and 24 Mbps are mandatory)
Modulation	BPSK OFDM
	QPSK OFDM
	16-QAM OFDM
	64-QAM OFDM
Error correcting code	K = 7 (64 states) convolutional code
Coding rate	1/2, 2/3, 3/4
Number of subcarriers	52
OFDM symbol duration	4.0 μs
Guard interval	0.8 μs (T_{GI})
Occupied bandwidth	16.6 MHz

allocation in the 5 GHz band is subject to authorities responsible for geographic-specific regulatory domains (global, regional, and national). The particular channelization to be used for this standard is dependent on such allocation as well as the associated regulations for use of the allocations. These regulations are subject to revision or may be superseded. In the United States, the FCC is the agency responsible for the allocation of the 5 GHz U-NII bands.

In some regulatory domains, several frequency bands may be available for OFDM PHY-based wireless LANs. These bands may be contiguous or not, and different regulatory limits may be applicable. A compliant OFDM PHY supports at least one frequency band in at least one regulatory domain.

Channel Numbering Channel center frequencies are defined at every integral multiple of 5 MHz above 5 GHz. The relationship between center frequency and channel number is given by the following equation:

$$\text{Channel center frequency} = 5000 + 5 \times n_{ch} \text{ (MHz)}$$

where $n_{ch} = 0,1, \ldots, 200$.

This definition provides a unique numbering system for all channels with 5 MHz spacing from 5 GHz to 6 GHz as well as the flexibility to define channelization sets for all current and future regulatory domains.

Channelization The set of valid operating channel numbers by regulatory domain is defined in Table 6-14.

Figure 6-11 shows the channelization scheme for this standard, which is used with the FCC U-NII frequency allocation. The lower and middle U-NII subbands accommodate eight channels in a total bandwidth of 200 MHz. The upper U-NII band accommodates four channels in a 100 MHz band-

Table 6-14

Valid operating channel numbers by regulatory domain and band

Regulatory domain	Band (GHz)	Operating channel numbers	Channel center frequencies (MHz)
United States	U-NII lower band (5.15–5.25)	36	5,180
		40	5,200
		44	5,220
		48	5,240
United States	U-NII middle band (5.25–5.35)	52	5,260
		56	5,280
		60	5,300
		64	5,320
United States	U-NII upper band (5.725–5.825)	149	5,745
		153	5,765
		157	5,785
		161	5,805

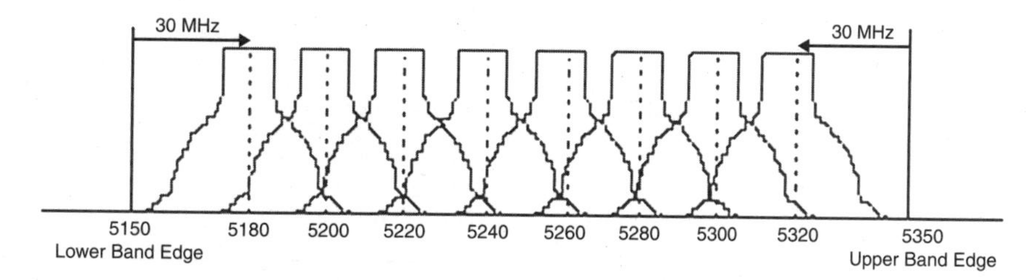

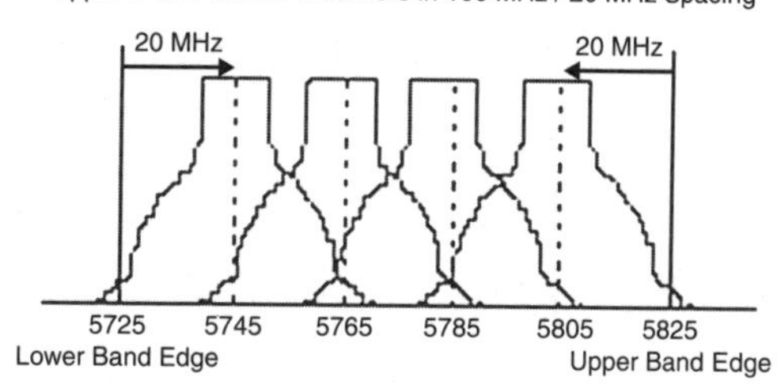

Figure 6-11 OFDM PHY frequency channel plan for the United States

width. The centers of the outermost channels are at a distance of 30 MHz from the band's edges for the lower and middle U-NII bands, and 20 MHz for the upper U-NII band. The OFDM PHY operates in the 5 GHz band, as allocated by a regulatory body in its operational region.

Transmit Power Levels The maximum output power allowable under FCC regulations is shown in Table 6-15.

PLCP Transmit Procedure See Figure 6-12.

OFDM PMD Sublayer

Scope and Field of Application This section of the standard describes the PMD services provided to the PLCP for the OFDM PHY. Also defined in this section are the functional, electrical, and RF characteristics required for the interoperability of implementations conforming to this specification (see Table 6-16). The relationship of this specification to the entire OFDM PHY was shown in Figure 6-4.

Table 6-15 Transmit power levels for the United States

Frequency band (GHz)	Maximum output power with up to 6 dBi antenna gain (mW)
5.15–5.25	40 (2.5 mW/MHz)
5.25–5.35	200 (12.5 mW/MHz)
5.725–5.825	800 (50 mW/MHz)

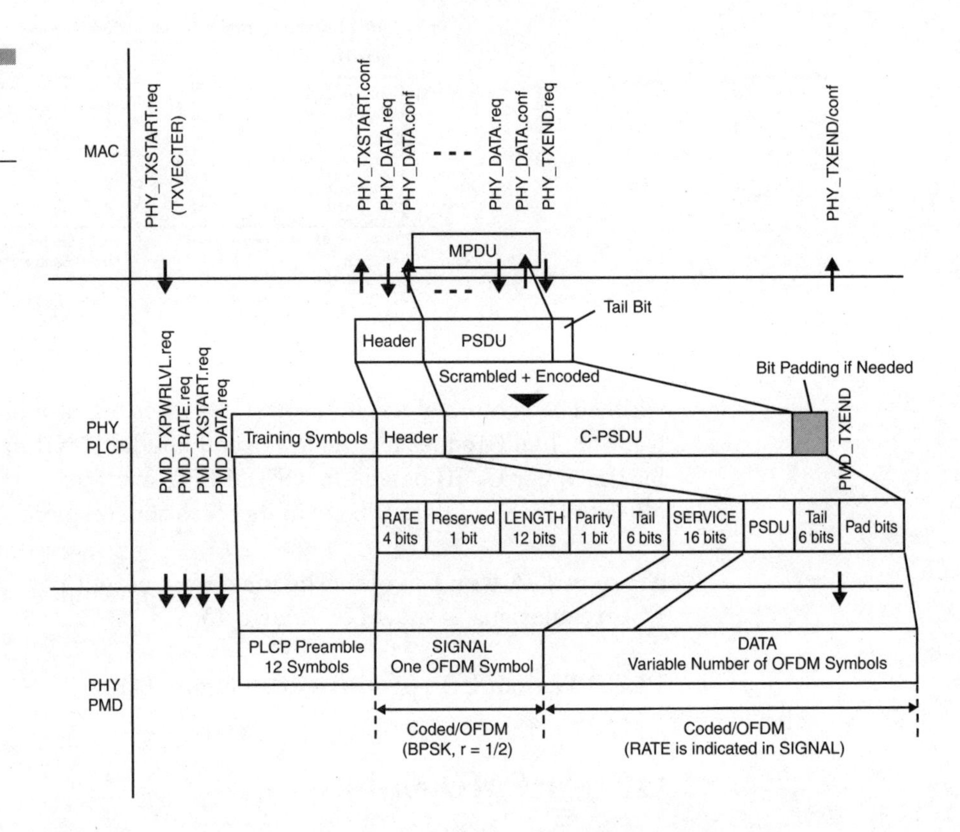

Figure 6-12 PLCP transmit procedure

Overview of Service The OFDM PMD sublayer accepts PLCP sublayer service primitives and provides the actual means by which data is transmitted or received from the medium. The combined function of the OFDM PMD sublayer primitives and parameters for the receive function results in a data stream, timing information, and associated received signal parameters being delivered to the PLCP sublayer. A similar functionality is provided for data transmission.

Table 6-16

OFDM PHY characteristics (partial)

Characteristics	Value
aSlotTime	9 μs
aSIFSTime	16 μs
aCCATime	<4 μs
aRxTxTurnaroundTime	<2 μs
aAirPropagationTime	<<1 μs
aMACProcessingDelay	<2 μs
aPreambleLength	20 μs
aPLCPHeaderLength	4 μs
aMPDUMaxLength	4,095
aCWmin	15
aCWmax	1,023

Overview of Interactions The primitives associated with the IEEE 802.11 PLCP sublayer to the OFDM PMD fall into two basic categories (see Figure 6-13):

- Service primitives that support PLCP peer-to-peer interactions
- Service primitives that have local significance and support sublayer-to-sublayer interactions

Example In this section we'll consider an example frame encoded for the OFDM PHY. Encoding goes through the following stages (refer to the standard for a detailed example of the encoding):

1. Generating the short training sequence section of the preamble
2. Generating the long preamble sequence section of the preamble
3. Generating the SIGNAL field bits
4. Coding and interleaving the SIGNAL field bits
5. Mapping the SIGNAL field into a frequency domain
6. Pilot insertion
7. Transforming into a time domain

Figure 6-13 Primitives

PMD_SAP peer-to-peer service primitives

Primitive	Request	Indicate	Confirm	Response
PMD_DATA	X	X	—	—

PMD_SAP sublayer-to-sublayer service primitives

Primitive	Request	Indicate	Confirm	Response
PMD_TXSTART	X	—	—	—
PMD_TXEND	X	—	—	—
PMD_TXPWRLVL	X	—	—	—
PMD_RATE	X	—	—	—
PMD_RSSI	—	X	—	—

8. Delineating the data octet stream into a bit stream
9. Prepending the SERVICE field and adding the pad bits, thus forming the DATA
10. Scrambling and zeroing the tail bits
11. Encoding the DATA with a convolutional encoder and puncturing
12. Mapping into complex 16-QAM symbols
13. Pilot insertion
14. Transforming from frequency to time and adding a circular prefix
15. Concatenating the OFDM symbols into a single, time-domain signal

CHAPTER 7

Wireless Application Protocol (WAP)

The Wireless Application Protocol (WAP) is an open specification that enables mobile users with wireless devices to easily access and interact with information and services instantly. The purpose of WAP is to enable the delivery of relevant information and services to mobile users. Handheld digital wireless devices such as mobile phones, pagers, two-way radios, smart phones and communicators—from low-end to high-end—all can utilize WAP.

WAP is a communications protocol and application environment. It can be built on any operating system including PalmOS, EPOC, Windows CE, FLEXOS, OS/9, JavaOS, and so on. It provides service interoperability even between different device families. WAP is also applicable in micro-browser environments. Micro-browsers are client software designed to address the challenges of mobile handheld devices that enable wireless access to Internet information (however, a suitable network server has to be deployed to use micro-browsers). WAP is designed to work with most wireless networks such as Cellular Digital Packet Data (CDPD), code division multiple access (CDMA), Global System for Mobile communications (GSM), personal digital cellular (PDC), PHS, Time Division Multiple Access (TDMA), FLEX, ReFLEX, iDEN, TETRA, Digital Enhanced Cordless Telecommunications (DECT), DataTAC, and Mobitex.

In comparing Bluetooth with WAP, recall that Bluetooth is a local area, low-power radio link between devices. Many of the usage scenarios for Bluetooth will also involve one of the devices communicating over the air using WAP. Although many Bluetooth member companies are also WAP member companies, and it is expected that many future handheld wireless devices will deploy both Bluetooth and WAP technology, the two technologies fundamentally address different problems.

Wireless Application Protocol (WAP) Fundamentals

WAP's advocacy comes from the WAP Forum, an industry association comprising over 500 members that has developed the de facto standard for wireless information and telephony services on digital mobile phones and other wireless terminals. The WAP Forum is an industry association that supports, gives input, and contributes specifications to the existing standards bodies.[1] The goal of the WAP Forum is to bring together companies from all segments of the wireless industry value chain to ensure product interoperability and the growth of the wireless market. WAP Forum members represent over 90 percent of the global handset market, including carriers with more than 100 million subscribers, leading infrastructure providers, software developers, and other organizations providing solutions to the wireless industry.[2]

Skeptics inquire, "Wireless data services have been talked about for a decade. Why will WAP succeed where previous attempts to reach mass markets have failed?" Proponents make the pitch that technology adoption has been slowed by the industry's lack of standards to make handheld products Internet compatible. WAP provides the de facto open standard for wireless data services. According to proponents, WAP has generated the critical mass for manufacturers, opening up new product and marketing opportunities in the wireless industry. Service providers are behind WAP because with relatively minimal risk and investment, WAP enables operators to decrease churn (people leaving the system to go to another provider) and increase revenues by improving existing, value-added services and offering exciting new informational services.[3]

WAP has been designed to be as independent as possible from the underlying network technology. Therefore, it will comply with evolving third-generation (3G) standards. WAP will continue to be necessary even with the higher-bandwidth 3G networks. Even as bandwidths increase, the cost of that bandwidth does not fall to zero. These costs result from higher power usage in the terminals, higher costs in the radio sections, greater use of the radio frequency (RF) spectrum, and increased network loading. In addition, the original constraints WAP was designed for—intermittent coverage, small screens, low power consumption, wide scalability over bearers and devices, and one-handed operation—are still valid in 3G networks. Finally, one can expect the bandwidth required by applications to steadily increase. Therefore, there is still a need to optimize the device and network resources for wireless environments. One can expect WAP to optimize support for multimedia applications that continue to be relevant. If WAP is successful in mass markets on 2.5G networks, 3G networks may be needed purely for capacity relief.

The following material is taken from WAP Architecture, Version 12-July-2001, Wireless Application Protocol, Architecture Specification, WAP-210-WAPArch-20010712.[4]

Scope

WAP is a result of continuous work to define an industry wide specification for developing applications that operate over wireless communication networks. The scope for the WAP Forum is to define a set of specifications to be used by service applications. The wireless market is growing quickly, reaching new customers and providing new services. To enable operators and manufacturers to meet the challenges in advanced services, differentiation, and fast/flexible service creation, WAP selects and defines a set of open, extensible protocols and content formats as a basis for interoperable implementations.

The objectives of the WAP Forum are:

- To bring Internet content and advanced data services to digital cellular phones and other wireless terminals.
- To create a global wireless protocol specification that will work across differing wireless network technologies.
- To enable the creation of content and applications that scale across a very wide range of bearer networks and device types.
- To embrace and extend existing standards and technology wherever appropriate.

The WAP Architecture Specification is intended to present the system and protocol architectures essential to achieving the objectives of the WAP Forum. The WAP Architecture Specification acts as the starting point for understanding the WAP technologies and the resulting specifications. As such, it provides an overview of the different technologies and references the appropriate specifications for further details.

This version of the WAP Architecture continues the themes and builds on the successes of the initial WAP Architecture. Network elements remain similar in function. For example, the architecture uses performance and feature-enhancing proxies to offload processing from constrained devices, to expose features and functions of the wireless network, and to provide for network and service management. Recent versions of the architecture have been enhanced to allow for a broader selection of connection paths between clients and origin servers as necessary, for example to provide end-to-end security.

The WAP Architecture Specification itself provides a framework for a variety of protocols, features, and services. It does not mandate any specific implementation, and is therefore be considered informative.

Definitions and Abbreviations

Definitions

Author An author is a person or program that writes or generates WML, WMLScript or other content.

Client A device (or application) that initiates a request for a connection with a server.

Content Subject matter (data) stored or generated at an origin server. Content is typically displayed or interpreted by a user agent in response to a user request.

Content Encoding When used as a verb, content encoding indicates the act of converting content from one format to another. Typically the resulting format requires less physical space than the original, is easier to process or store and/or is encrypted. When used as a noun, content encoding specifies a particular format or encoding standard or process.

Content Format Actual representation of content.

Device A network entity that is capable of sending and receiving packets of information and has a unique device address. A device can act as both a client or a server within a given context or across multiple contexts. For example, a device can service a number of clients (as a server) while being a client to another server.

JavaScript A *de facto* standard language that can be used to add dynamic behaviour to HTML documents. JavaScript is one of the originating technologies of ECMAScript.

Man-Machine Interface A synonym for user interface.

Origin Server The server on which a given resource resides or is to be created. Often referred to as a web server or an HTTP server.

Resource A network data object or service that can be identified by a URI or URL. Resources may be available in multiple representations (e.g., multiple languages, data formats, size and resolutions) or vary in other ways.

Server A device (or application) that passively waits for connection requests from one or more clients. A server may accept or reject a connection request from a client.

Terminal A device providing the user with user agent capabilities, including the ability to request and receive information. Also called a mobile terminal or mobile station.

User A user is a person who interacts with a user agent to view, hear, or otherwise use a resource.

User Agent A user agent is any software or device that interprets WML, WMLScript, WTAI or other resources. This may include textual browsers, voice browsers, search engines, etc.

WMLScript A scripting language used to program the mobile device. WMLScript is based on ECMAScript and loosely based on the JavaScript scripting languages.

Abbreviations

CGI	Common Gateway Interface
CPU	Central Processing Unit
DNS	Domain Name System
EFI	External Functionality Interface
HTML	HyperText Markup Language[5]
http	HyperText Transfer Protocol[6]
IP	Internet Protocol
MMI	Man-Machine Interface
MMS	Multimedia Message Service
OTA	Over The Air
PDA	Personal Digital Assistant
PICS	Protocol Implementation Conformance Statement
PKI	Public Key Infrastructure
RAM	Random Access Memory
RFC	Request For Comments
ROM	Read Only Memory
SCR	Static Conformance Requirement
SSL	Secure Sockets Layer
STD	Internet Standard
TCP	Transmission Control Protocol[7]

TLS	Transport Layer Security
UDP	User Datagram Protocol[8]
URI	Uniform Resource Identifier[9]
URL	Uniform Resource Locator[9]
W3C	World Wide Web Consortium
WAE	Wireless Application Environment[10]
WAP	Wireless Application Protocol
WDP	Wireless Datagram Protocol[11]
WIM	Wireless Identity Module[12]
WML	Wireless Markup Language[13]
WPKI	Wireless Public Key Infrastructure[14]
WSP	Wireless Session Protocol[15]
WTA	Wireless Telephony Application[16]
WTAI	Wireless Telephony Application Interface[17]
WTLS	Wireless Transport Layer Security[18]
WTP	Wireless Transaction Protocol[19]
WWW	World-Wide Web
XHTML	Extensible Hypertext Markup Language[20]
XML	Extensible Markup Language[21]

WAP Background

Motivation

WAP is positioned at the convergence of three rapidly evolving network technologies, wireless data, telephony, and the Internet. Both the wireless data market and the Internet are growing very quickly and are continuing to reach new customers. The growth of the Internet has fuelled the creation of new and exciting information services. Most of the original technology developed for the Internet has been designed for desktop and larger computers and medium to high bandwidth, generally reliable data networks. Mass-market, hand-held wireless devices present a more constrained computing environment compared to desktop computers. Because of fundamental limitations of power and form-factor, mass-market handheld devices tend to have:

- Less powerful CPUs,
- Less memory (ROM and RAM),
- Restricted power consumption,
- Smaller displays, and
- Different input devices (e.g., a phone keypad).

Similarly, wireless data networks present a more constrained communication environment compared to wired networks.

Because of fundamental limitations of power, available spectrum, and mobility, wireless data networks tend to have:

- Less bandwidth,
- More latency,
- Less connection stability, and
- Less predictable availability.

Mobile networks are growing in complexity and the cost of all aspects of providing value-added services is increasing. In order to meet the requirements of mobile network operators, solutions must be:

- **Interoperable** Terminals from different manufacturers communicate with services in the mobile network;
- **Scaleable** Mobile network operators are able to scale services to customer needs;
- **Efficient** Provides quality of service suited to the behaviour and characteristics of the mobile network;
- **Reliable** Provides a consistent and predictable platform for deploying services; and
- **Secure** Enables services to be extended over potentially unprotected mobile networks while still preserving the integrity of user data; protects the devices and services from security problems such as loss of confidentiality.

Many of the current mobile networks include advanced services that can be offered to end-users. Mobile network operators strive to provide advanced services in a useable and attractive way in order to promote increased usage of the mobile network services and to decrease the turnover rate of subscribers. Standard features, like call control, can be enhanced by using WAP technology to provide customised user interfaces. For example, services such as call forwarding may provide a user interface that prompts the user to make a choice between accepting a call, forwarding to another person, forwarding it to voice mail, etc. The nature of wireless devices is that they are inherently mobile. This mobility introduces new opportunities for services that are sensitive to mobility and can provide location-dependent information. The WAP specifications and architecture capitalize on this unique aspect of wireless devices by including mobility as part of the application model.

The WAP specifications address mobile network characteristics and operator needs by adapting existing network technology to the special requirements of mass-market, hand-held wireless data devices and by introducing new technology where appropriate. The WAP specifications will accommodate a range of devices, from devices that provide very basic functionality to devices that continue to expand their capabilities. This motivates an architecture where functionality may be moved to different locations within the network as appropriate, i.e. either to devices or to network servers as necessary.

Architectural Goals

The goals of the WAP Forum architecture are as follows. This summary is informative and non-exhaustive; the order of the items does not represent any priority or importance.

- Provide a web-centric application model for wireless data services that utilizes the telephony, mobility, and other unique functions of wireless devices and networks and allows maximum flexibility and ability for vendors to enhance the user experience.
- Enable the personalization and customization of the device, the content delivered to it, and the presentation of the content.
- Provide support for secure and private applications and communication in a manner that is consistent and interoperable with Internet security models.
- Enable wireless devices and networks that are currently or in the near future being deployed, including a wide variety of bearers from narrow-band to wide-band.
- Provide secure access to local handset functionality.
- Facilitate network-operator and third party service provisioning.

- Define a layered, scaleable and extensible architecture.
- Leverage existing standards where possible, especially existing and evolving Internet standards.

WAP Architecture

The World-Wide Web Model

The WWW architecture provides a very flexible and powerful programming model (Figure 7-1). Applications and content are presented in standard data formats, and are *browsed* by applications known as *web browsers*. The web browser is a networked application, i.e., it sends requests for named data objects to a network server and the network server responds with the data encoded using the standard formats.

WWW standards specify many of the mechanisms necessary to build a general-purpose application environment, including:

- **Standard naming model** All servers and content on the WWW are named with an Internet-standard *Uniform Resource Locator* (URL).[9]
- **Content typing** All content on the WWW is given a specific type thereby allowing web browsers to correctly process the content based on its type.[22,23]
- **Standard content formats** All web browsers support a set of standard content formats. These include the HyperText Markup Language (HTML),[5] scripting languages (ECMAScript, JavaScript),[24,25] and a large number of other formats.

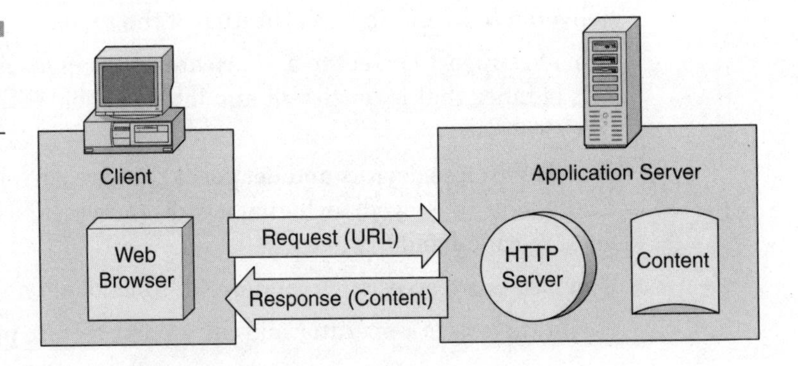

Figure 7-1 World-wide-web programming model

- **Standard Protocols** Standard networking protocols allow any web browser to communicate with any web server. The most commonly used protocol on the WWW is the HyperText Transport Protocol (HTTP),[6] operating on top of the TCP/IP protocol suite.[7]

This infrastructure allows users to easily reach a large number of third-party applications and content services. It also allows application developers to easily create applications and content services for a large community of clients.

The WAP Model

The WAP programming model (see Figure 7-2) is the WWW programming model with a few enhancements. Adopting the WWW programming model provides several benefits to the application developer community, including a familiar programming model, a proven architecture, and the ability to leverage existing tools (e.g., Web servers, XML tools, etc.). Optimizations and extensions have been made in order to match the characteristics of the wireless environment. Wherever possible, existing standards have been adopted or have been used as the starting point for the WAP technology.

The most significant enhancements WAP has added to the programming model are:

- Push
- Telephony Support (WTA)

Figure 7-2
WAP programming model

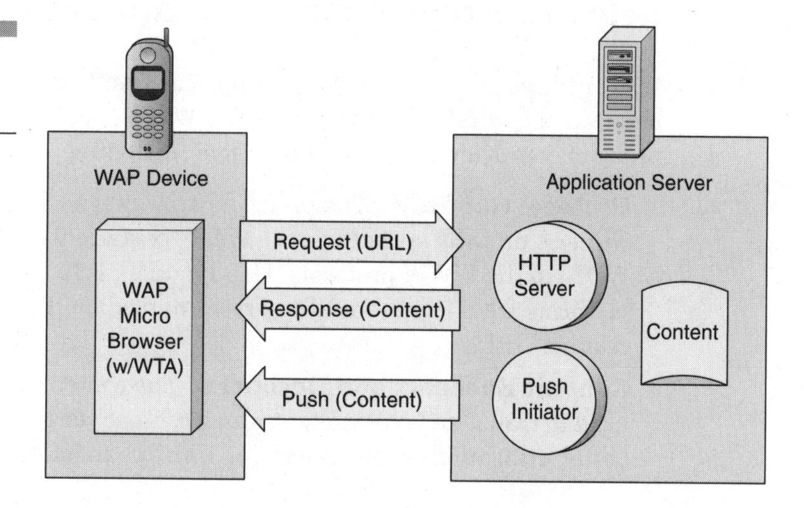

The classical request-response mechanism is commonly referred to as *pull* to contrast it with the *push* mechanism.

WAP content and applications are specified in a set of well-known content formats based on the familiar WWW content formats. Content is transported using a set of standard communication protocols based on the WWW communication protocols. The WAP *microbrowser* in the wireless terminal coordinates the user-interface and is analogous to a standard web browser.

WAP defines a set of standard components that enable communication between mobile terminals and network servers, including:

- **Standard naming model** WWW-standard URLs are used to identify WAP content on origin servers. WWW-standard URIs are used to identify local resources in a device, e.g. call control functions.
- **Content typing** All WAP content is given a specific type consistent with WWW typing. This allows WAP user agents to correctly process the content based on its type.
- **Standard content formats** WAP content formats are based on WWW technology and include display markup, calendar information, electronic business card objects, images and scripting language.
- **Standard communication protocols** WAP communication protocols enable the communication of browser requests from the mobile terminal to the network web server.

The WAP content types and protocols have been optimized for mass-market, hand-held wireless devices.

Feature/Performance-Enhancing Proxies

WAP utilizes proxy technology to optimize and enhance the connection between the wireless domain and the WWW (see Figure 7-3). The WAP proxy may provide a variety of functions, including:

- **Protocol Gateway** The protocol gateway translates requests from a wireless protocol stack (e.g., the WAP 1.x stack-WSP, WTP, WTLS, and WDP) to the WWW protocols (HTTP and TCP/IP). The gateway also performs DNS lookups of the servers named by the client in the request URLs.
- **Content Encoders and Decoders** The content encoders can be used to translate WAP content into a compact format that allows for better utilization of the underlying link due to its reduced size.

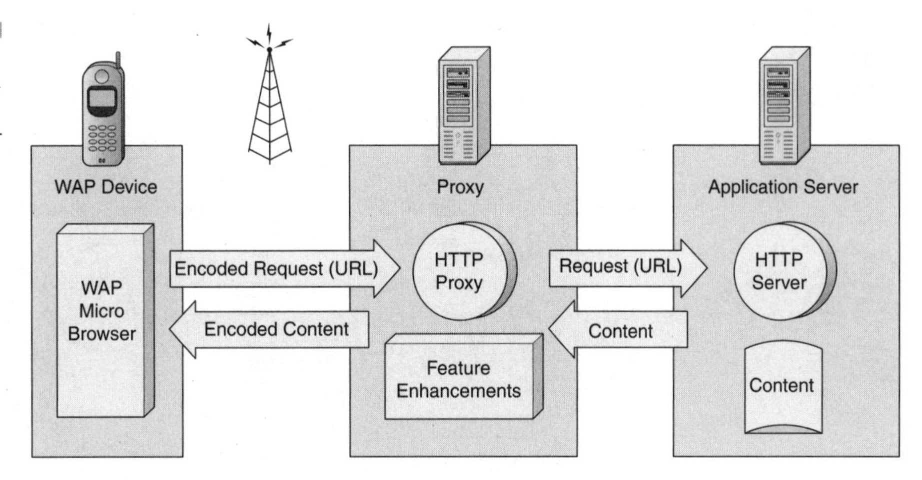

Figure 7-3 Feature/performance-enhancing proxy

- **User Agent Profile Management** User agent profiles describing client capabilities and personal preferences[26] are composed and presented to the applications.
- **Caching Proxy** A caching proxy can improve perceived performance and network utilization by maintaining a cache of frequently accessed resources.

This infrastructure ensures that mobile terminal users can access a wide variety of Internet content and applications, and that application authors are able to build content services and applications that run on a large base of mobile terminals.

The WAP proxy allows content and applications to be hosted on standard WWW servers and to be developed using proven WWW technologies such as CGI scripting. While the nominal use of WAP will include a web server, WAP proxy and WAP client, the WAP architecture can quite easily support other configurations.

Supporting Servers

The WAP Architecture also includes supporting servers (see Figure 7-4), which provide services to devices, proxies, and applications as needed. These services are often specific in function, but are of general use to a wide variety of applications.

Figure 7-4
Supporting services

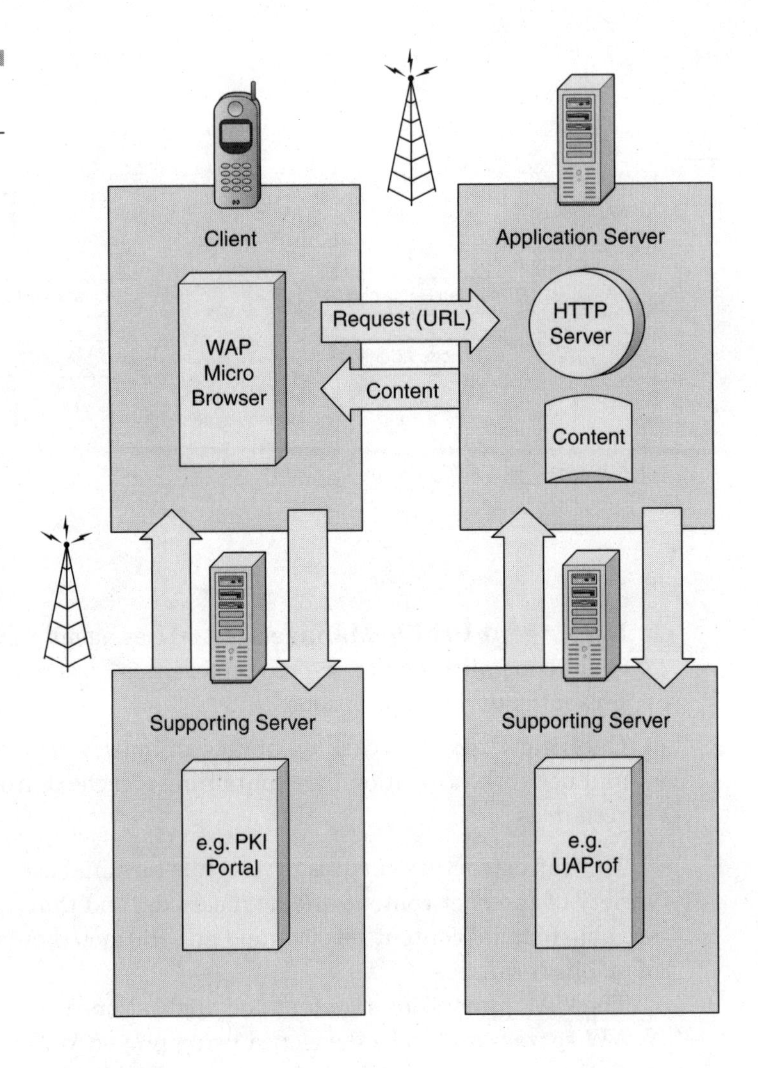

The supporting servers defined by the WAP Forum include, but are not limited to:

- **PKI Portal** The PKI Portal (shown in Figure 7-4)[14] allows devices to initiate the creation of new public key certificates.
- **UAProf Server** The UAProf Server[26] allows applications to retrieve the client capabilities and personal profiles of user agents and individual users.
- **Provisioning Server** The Provisioning Server[26] is trusted by the WAP device to provide its provisioning information.

WAP Network Elements

A typical WAP network is shown in Figure 7-5.

WAP clients communicate with application servers through a number of different proxies or directly. WAP clients support the *proxy selection* mechanism that allows them to utilize the most appropriate proxy for a given service or to connect directly to that service as necessary. Proxies can be used to augment a request. They translate between WAP and WWW protocols (HTTP, TCP), thereby allowing the WAP client to submit requests to the origin server.

Proxies may be located in a number of places, including wireless carriers or independent service providers in order to provide feature enhancements coupled to the wireless network (e.g., telephony, location and provisioning) or to optimize the communication between device and application server (e.g., protocol translation and cookie caching).

Proxies may be located in a secure network to provide a secure channel between wireless device and the secure network. In some instances, the device might make direct connections to application servers, for example to

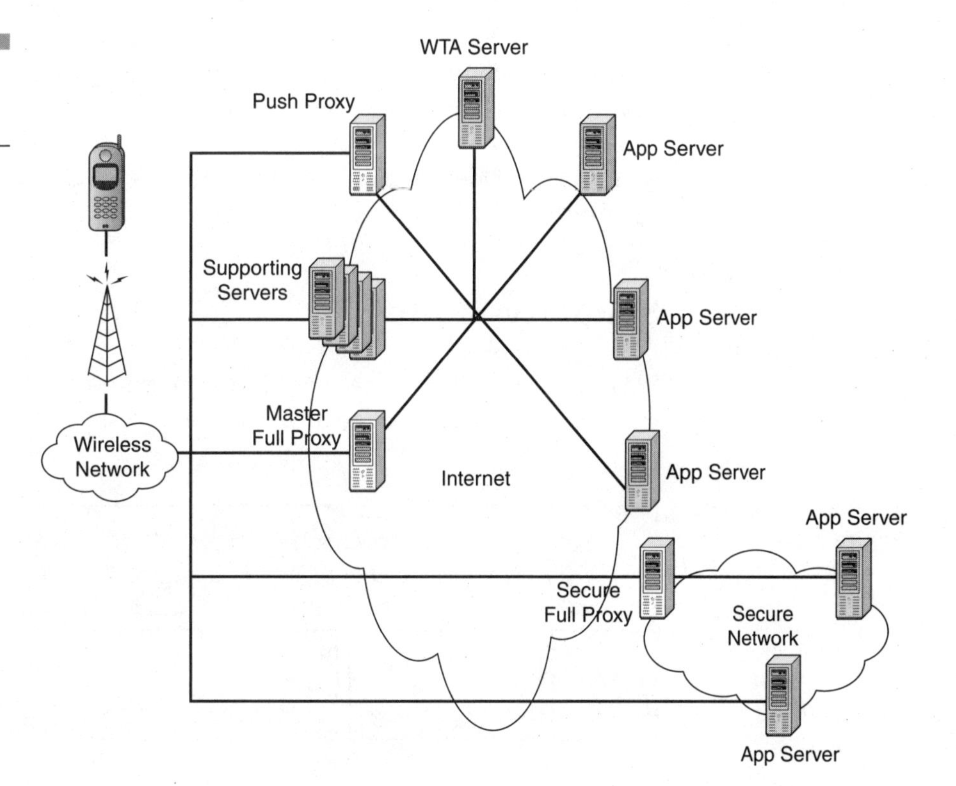

Figure 7-5
A WAP network example

provide a secure connection directly between the device and application server. The supporting servers provide support functions required by or generally useful to devices, proxies, and application servers. These functions include Provisioning, PKI, user agent profiles, etc.

Device Architecture

The architecture for WAP devices is shown in Figure 7-6. The Application Framework provides the device execution environment for WAP applications. WAP applications are comprised of markup, script, style sheets and multimedia content, all of which are rendered on the device. The WAP Application Environment (WAE) processing model defines the structure in which these various forms of executable and non-executable content interact.

The network protocols on the WAP client are shared between client and server. They are described in further detail below. Content renderers interpret specific forms of content and present them to the end user for perusal or interaction.

Common functions are defined to be utilized by the application framework, including persistence and data synchronisation. The Wireless Identity Module (WIM), as specified in "WAP Identity Module Specification,"[12] contains the identity of the device and the cryptographic means to mutually authenticate WAP devices and servers.

The architecture also provides a mechanism to access external functions that are embedded or attached to the devices via the External Functionality Interface (EFI).

Security Model

WAP enables a flexible security infrastructure that focuses on providing connection security between a WAP client and server. WAP can provide end-

Figure 7-6 WAP client architecture

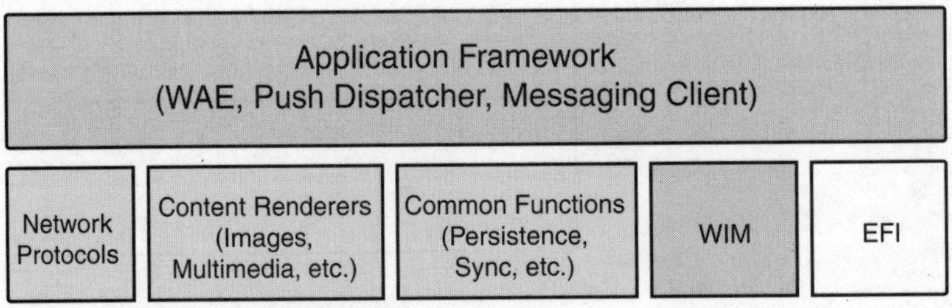

to-end security between protocol endpoints. If a browser and origin server desire end-to-end security, they can communicate directly using the security protocols. Moreover, the WAP specifications include support for application-level security, such as signed text.

Components of the WAP Architecture

WAP architecture provides a scaleable and extensible application development environment for mobile communication devices. This is achieved through a layered design of the protocol stack (Figure 7-7). Each layer provides a set of functions and/or services to other services and applications through a set of well-defined interfaces. Each of the layers of the

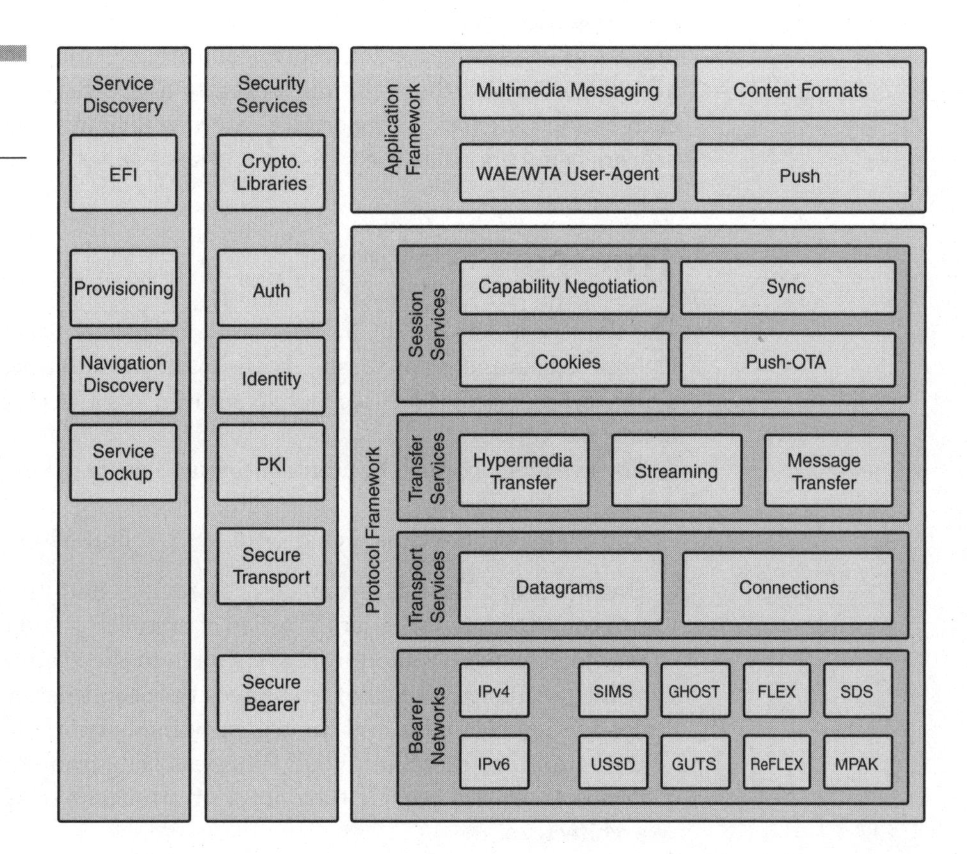

Figure 7-7 WAP stack architecture

architecture is accessible by the layers above, as well as by other services and applications.

The WAP architecture separates service interfaces from the protocols that provide those services to allow for evolution of the specifications and selection of the most appropriate protocol for a given context. Many of the services in the stack may be provided by more than one protocol. For example, either HTTP[6] or WSP[15] may provide the Hypermedia Transfer service.

Bearer Networks

Protocols have either been designed or selected to operate over a variety of different bearer services, including short message, circuit-switched data, and packet data. The bearers offer differing levels of quality of service with respect to throughput, error rate, and delays. The protocols are designed to compensate for or tolerate these varying levels of service.

Since the Transport Services layer provides the interface between the bearer service and the rest of the WAP stack, the transport specifications (e.g., "Wireless Datagram Protocol Specification"[11]) may list the bearers that are supported and the techniques used to allow the protocols to run over each bearer. The list of supported bearers will change over time with new bearers being added as the wireless market evolves.

Transport Services

The Transport Services layer offers a set of consistent services to the upper layer protocols and maps those services to the available bearer services. The Transport Services transport unstructured data across the underlying bearer networks.

These transport services create a common abstraction that is consistent across all the bearers.

The Transport Services include, but are not limited to:

- **Datagrams** The datagram service provides data transport in which self-contained, independent entities of data carry sufficient information to be routed from the source to the destination computer without reliance on earlier exchanges between this source and destination computer and the transporting network. UDP (User Datagram Protocol)[8] and WDP (Wireless Datagram Protocol)[11] are two protocols used to provide the datagram transport service in the WAP architecture.

- **Connections** The connection service provides data transport service in which communication proceeds in three well-defined phases: connection establishment, two-way reliable data transfer and connection release. TCP (Transmission Control Protocol)[7] is a protocol used to provide the connection transport service of IP[28] bearers for the WAP architecture. In order to cope with the wireless network characteristics, the TCP protocol can be profiled for its use.[29]

Transfer Services

The Transfer Services provide for the structured transfer of information between network elements. They include, but are not limited to:

- **Hypermedia Transfer** The hypermedia transfer services provide for the transfer of self-describing hypermedia resources. The combination of WSP (Wireless Session Protocol)[15] and WTP (Wireless Transaction Protocol)[19] provide the hypermedia transfer service over secure and non-secure datagram transports. The HTTP (Hypertext Transfer Protocol)[6] provides the hypermedia transfer service over secure and non-secure connection-oriented transports.
- **Streaming** The streaming services provide a means for transferring isochronous data such as audio and video.
- **Message Transfer** The message transfer services provide the means to transfer asynchronous multimedia messages such as email or instant messages. MMS Encapsulation[30] is a protocol used to transfer messages between WAP devices and MMS servers.

Session Services

The session services provide for the establishment of shared state between network elements that span multiple network requests or data transfers. For example, the Push session establishes that the WAP Device is ready and able to receive pushes from the Push Proxy. The Session Services include, but are not limited to:

- **Capability Negotiation** The WAP architecture includes specifications for describing, transmitting, and managing capabilities and preference information about the client, user, and network elements. See “User Agent Profile Specification” for more information.[26] This allows for customization of information and content returned by the origin server or pushed by the application.

- **Push-OTA** The Push-OTA (Over The Air) session service provides for network-initiated transactions to be delivered to wireless devices that are intermittently able to receive data (e.g., modal devices and devices with dynamically assigned addresses). The Push-OTA service operates over the connection-oriented transport service and datagram transport.[31]
- **Sync** The Sync service provides for the synchronization of replicated data.
- **Cookies** The Cookies service allows applications to establish state on the client or proxy that survives multiple hypermedia transfer transactions. See "HTTP State Management" for more information.[32]

Application Framework

The Application Framework provides a general-purpose application environment based on a combination of World Wide Web (WWW), Internet and Mobile Telephony technologies. The primary objective of the Application Framework is to establish an interoperable environment that will allow operators and service providers to build applications and services that can reach a wide variety of different wireless platforms in an efficient and useful manner.

The Application Frame work includes, but is not limited to:

- **WAE/WTA User-Agent** WAE is a micro-browser environment containing or allowing for markup (including WML and XHTML), scripting, style-sheet languages, and telephony services and programming interfaces, all optimized for use in hand-held mobile terminals. See "Wireless Application Environment Specification" for more information.[10]
- **Push** The Push service provides a general mechanism for the network to initiate the transmission of data to applications resident on WAP devices. See "WAP Push-Architectural Overview" for more information.[33]
- **Multimedia Messaging** The Multimedia Message Service (MMS) provides for the transfer and processing of multimedia messages such as email and instant messages to WAP devices.
- **Content Formats** The application framework includes support for a set of well-defined data formats, such as color images, audio, video, animation, phone book records, and calendar information.

Security Services

Security forms a fundamental part of the WAP Architecture, and its services can be found in many of its layers. In general, the following security facilities offered are:

- **Privacy** Facilities to ensure that communication is private and cannot be understood by any intermediate parties that may have intercepted it.
- **Authentication** Facilities to establish the authenticity of parties to the communication.
- **Integrity** Facilities to ensure that communication is unchanged and uncorrupted.
- **Non-Repudiation** Facilities to ensure parties to a communication cannot deny the communication took place.

The Security Services span all the various layers of the WAP Architecture. Some specific examples of the security services include:

- **Cryptographic Libraries** This application framework level library provides services for signing of data for integrity and non-repudiation purposes. See "WMLScript Crypto Library" for more information.[34]
- **Authentication** The Security Services provide various mechanisms for client and server authentication. At the Session Services layer, HTTP Client Authentication[35] may be used to authenticate clients to proxies and application servers. At the Transport Services layer, WTLS and TLS handshakes may be used to authenticate clients and servers.
- **Identity** WIM provides the functions that store and process information needed for user identification and authentication.[12]
- **PKI** The set of security services that enable the use and management of public-key cryptography and certificates.[14,36]
- **Secure Transport** The Transport Services layer protocols are defined for secure transport over datagrams and connections. WTLS is defined for secure transport over datagrams and TLS is defined for secure transport over connections (i.e. TCP). See "Wireless Transport Layer Security Protocol" and "WAPTLS Profiling and Tunneling" for more information.[18,37]
- **Secure Bearer** Some bearer networks provide bearer level security. For example, IP networks (especially in the context of IPv6) provide bearer-level security with IPSec[38].

Service Discovery

Service discovery forms a fundamental part of the WAP Architecture and its services can be found at many layers. Some specific examples of Service Discovery services include:

- **EFI** The External Functionality Interface (EFI) allows applications to discover what external functions/services are available on the device.
- **Provisioning** The Provisioning service allows a device to be provisioned with the parameters necessary to access network services. See "WAP Provisioning Architecture Overview" for more information.[27]
- **Navigation Discovery** The Navigation Discovery service allows a device to discover new network services (e.g. secure pull proxies) during the course of navigation such as when downloading resources from a hypermedia server. The WAP Transport-Level End-to-End Security specification[39] defines one navigation discovery protocol.
- **Service Lookup** The Service Lookup service provides for the discovery of a service's parameters through a directory lookup by name. One example of this is the Domain Name System (DNS).[40]

Other Services and Applications

The WAP layered architecture enables other services and applications to use the features of the WAP stack through a set of well-defined interfaces. External applications may access the various services directly. The WAP layered architecture builds upon an extensible set of protocols. This allows the WAP stack to be used for applications and services not currently specified by WAP, but deemed to be valuable for the wireless market. Such applications and services may benefit from adding protocols or particular protocol capabilities. For example, applications, such as electronic mail, calendar, phone book, notepad, and electronic commerce, or services, such as white and yellow pages, may be developed to use the WAP protocols.

Sample Configurations of WAP Technology

Because several of the services in the WAP stack can be provided using different protocols based on the circumstances, there are more than one possible stack configurations. The following figures depict several possible protocol stacks using WAP technology. These are for illustrative, informative purposes only and do not constitute a statement of conformance or interoperability, nor is this set of examples exhaustive.

Figure 7-8 depicts the protocol stacks for the original WAP Architecture. The WAP Gateway converts the hypermedia transfer service between the datagram-based protocols (WSP, WTP, WTLS, WDP) and connection-oriented protocols commonly used on the Internet (HTTP, SSL, TCP).

Figure 7-9 depicts a WAP HTTP proxy. The proxy configuration is widely used in the Internet for ordinary web access, multimedia data, e.g. music, video clip downloading and so on. This configuration locates the WAP Proxy between wireline and wireless networks to enhance performance by using the wireless profile of TCP (as shown with TCP*). In addition to TCP

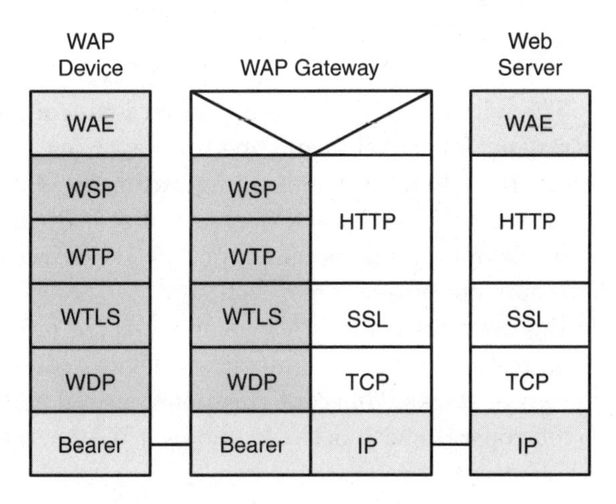

Figure 7-8 An example of WAP 1.x gateway

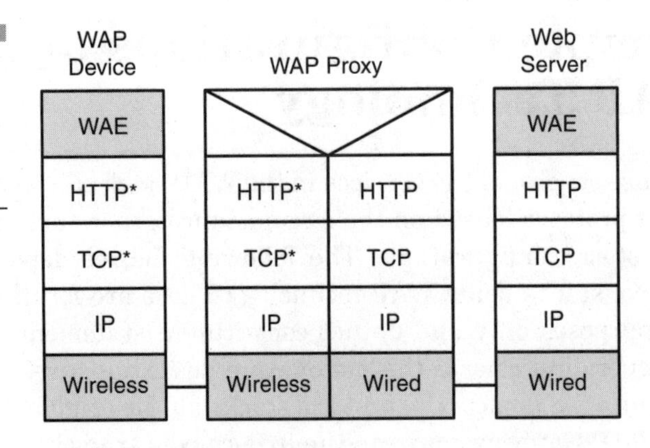

Figure 7-9 An example of WAP HTTP proxy with profiled TCP and HTTP

optimizations, the wireless profile of HTTP (as illustrated by HTTP*) allows for further performance enhancements. Both profiles comprise well-defined IETF options that provide for efficient operation over wireless networks as within the scope of WAP. The wireless profiled versions are interoperable with TCP and HTTP.

Figure 7-10 is a WAP HTTP proxy that has established a connection-oriented tunnel to the web server (e.g., in response to a CONNECT command). This configuration is used to allow TLS to provide end-to-end security between mobile terminal and origin server. E-commerce is a compelling use case for end-to-end security.

Figure 7-11 depicts a WAP device directly accessing a web server via the Internet. The wireless IP router is a standard part of an IP network that is used to transfer IP packets from one link layer (e.g., the wireless link) to another (e.g., the wired link). This configuration can apply to the case where bearer-level security (such as IPSec) is utilized. In the Direct Access scenario, wireless optimizations as defined by the Wireless Profiles for TCP and HTTP may not be available.

While the previous configurations show single protocol stacks for each WAP configuration, Figure 7-12 depicts a device that supports both the 1.x and 2.x protocol stacks. This configuration is useful in cases where a device needs to interoperate with both old and new WAP servers.

Conformance and Interoperability

The WAP Forum views vendor interoperability as an important element to the success of WAP products. In order to provide as high a probability as is

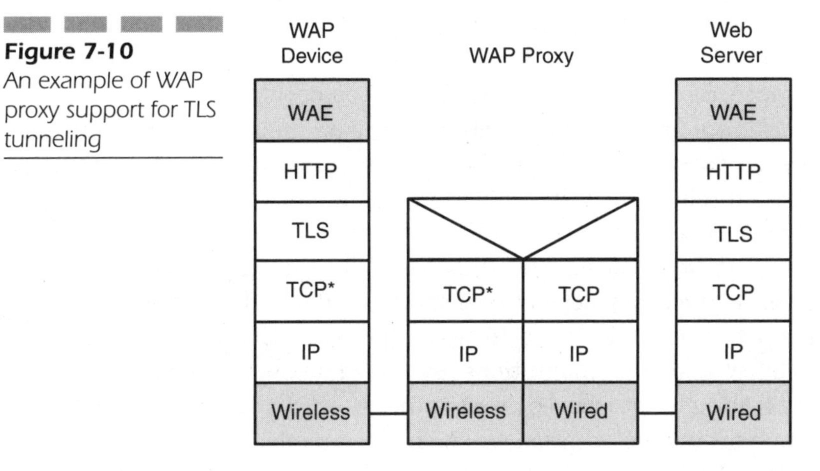

Figure 7-10 An example of WAP proxy support for TLS tunneling

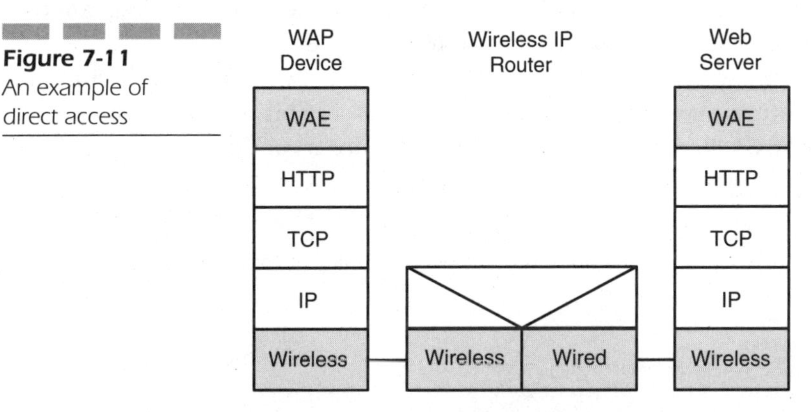

Figure 7-11 An example of direct access

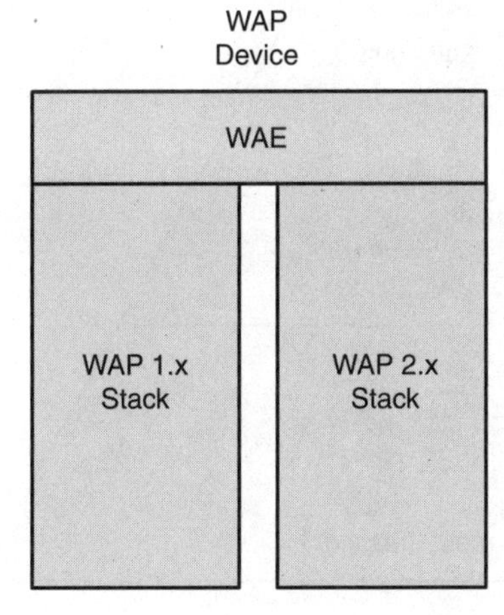

Figure 7-12 Dual stack support

technically possible that two WAP products developed independently by two different vendors will successfully interoperate, a rigorous definition of conformance, compliance, and testing has been developed.

Conformance answers the question, "Does an implementation meet the standard as written?" The WAP Forum charters a neutral third party to build a comprehensive test suite from its specifications. Usually, implementations are tested against known references.

Interoperability answers the question, "Will this implementation work with other products developed to the same standard?" Interoperability testing uses a test suite designed to test implementation-to-implementation compatibility, and implementations are tested against each other. Interoperability testing is not focused on compliance—two products with the same non-compliant implementation will be interoperable.

The WAP Forum Certification Program is focused on conformance, but offers some interoperability testing as well. The Certification Program covers the entire value chain as shown in Figure 7-13.

To improve interoperability at the authoring level, the WAP Forum provides authoring guidelines to improve the accessibility of WAP content. To certify WAP clients and servers, the WAP Forum conducts interoperability

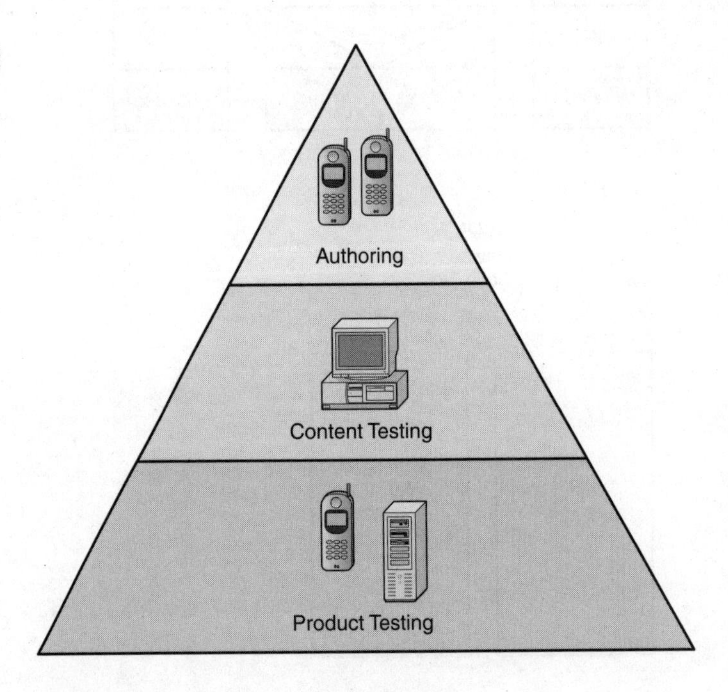

Figure 7-13 Interoperability and compliance

testing of an implementation against multiple reference implementations using a predefined suite of test cases.

The WAP Forum has defined a number of Class Conformance Profiles, e.g. Class A, Class B, and Class C. An implementation may be certified in one or more class. The class conformance requirements are specified in WAP Class Conformance Requirements.[41] Each WAP Specification includes static conformance requirements (SCRs) for that specification. These define which features are mandatory and optional and are the basis for the conformance test suite.

End Notes

1. The WAP Forum has relationships with the following standards bodies: The World Wide Web Consortium (W3C), European Telecommunication Standards Institute (ETSI), Telecommunications Industry Association (TIA), Internet Engineering Task Force (IETF), and the European Computer Manufacturers' Association (ECMA).
2. Promotional material from the WAP Forum.
3. Promotional material from the WAP Forum.
4. Copyright © 2000–2001 Wireless Application Protocol Forum Ltd. Terms and conditions of use are herewith met. One may use this document or any part of the document for educational purposes only, provided one does not modify, edit or take out of context the information in this document in any manner.
5. D. Raggett, et al., "HTML 4.0 Specification, W3C Recommendation, 18 December 1997, REC-HTML40-971218." www.w3.org/TR/REC-html40. September 17, 1997.
6. R. Fielding, J. Gettys, J. Mogul, H. Frystyk, L. Masinter, P. Leach, and T. Berners-Lee, "Hypertext Transfer Protocol—HTTP/1.1." www.rfc-editor.org/rfc/rfc2616.txt. June 1999.
7. J. Postel, "Transmission Control Protocol," www.rfc-editor.org/rfc/std/std7.txt. September 1981.
8. ——, "User Datagram Protocol." www.rfc-editor.org/rfc/std/std6.txt. August 1980.
9. T. Berners-Lee, R. Fielding, and L. Masinter, "Uniform Resource Identifiers (URI): Generic Syntax." www.rfc-editor.org/rfc/rfc2396.txt. August 1998.

10. ——, "Wireless Application Environment Specification." Version 2, WAP-236-WAESpec.
11. ——, "Wireless Datagram Protocol Specification."
12. ——, "WAP Identity Module Specification."
13. ——, "Wireless Markup Language."
14. ——, "WAP Public Key Infrastructure Definition."
15. ——, "Wireless Session Protocol."
16. ——, "Wireless Telephony Application Specification."
17. ——, "Wireless Telephony Application Interface."
18. ——, "Wireless Transport Layer Security Protocol."
19. ——, "Wireless Transaction Protocol Specification."
20. World Wide Web Consortium, "XHTML 1.1—Module Based XHTML." www.w3.org/TR/xhtml11/.
21. ——, "Extensible Markup Language (XML) 1.0." www.w3.org/TR/REC-xml/.
22. N. Freed, et al., "Multipurpose Internet Mail Extensions (MIME) Part One: Format of Internet Message Bodies." www.rfc -editor.org/rfc/rfc2045.txt. November 1996.
23. ——, "Multipurpose Internet Mail Extensions (MIME) Part Four: Registration Procedures." www.rfc-editor.org/rfc/rfc 2048.txt. November 1996.
24. ECMA, "Standard ECMA-262: ECMAScript Language Specification." June 1997.
25. David Flanagan, *JavaScript: The Definitive Guide*. New York: O'Reilly & Associates, Inc., 1997.
26. WAP Forum, "User Agent Profile Specification."
27. ——, "WAP Provisioning Architecture Overview."
28. The utilization of TCP connections over IP may require additional components of the TCP/IP protocol suite. One example for such a component is ICMP.
29. WAP Forum, "Wireless Profiled TCP Specification."
30. ——, "WAP Multimedia Messaging Service Message Encapsulation."
31. ——, "WAP Push OTA Protocol."
32. ——, "HTTP State Management." WAP-223-HTTPSM.
33. ——, "WAP Push Architectural Overview."

34. ——, "WMLScript Crypto Library."

35. J. Franks, P. Hallam-Baker, J. Hostetler, S. Lawrence, P. Leach, A. Luotonen, and L. Stewart, "HTTP Authentication: Basic and Digest Access Authentication." www.rfc-editor.org/rfc/rfc2617.txt. June 1999.

36. WAP Forum, "WAP Certificate and CRL Profiles."

37. ——, "WAPTLS Profile and Tunnelling." WAP-219-TLS-20010411.

38. S. Kent and R. Atkinson, "Security Architecture for the Internet Protocol." www.rfc-editor.org/rfc/rfc2401.txt. November 1998.

39. WAP Forum, "WAP Transport Layer End-to-End Security Specification."

40. P. Mockapetris, "Domain Name System," www.rfc-editor.org/rfc/std/std13.txt. November 1987.

41. WAP Forum, WAP Class Conformance Requirements.

CHAPTER 8

Designing Nomadic and Hotspot Networks

This chapter explores the key aspects of nomadic and hotspot network design. It is far from exhaustive but will convey some of the complexities and considerations that are particular to the design of such systems.

Introduction

A hotspot network supporting nomadic services will consist of remotely located low-end routers, access points (APs), wireless bridges, and so on, all connected via T1 lines to the Internet backbone. Figure 8-1 is one example of a hotspot application. Figure 8-2 shows a typical access point, and Figure 8-3 illustrates the basic configuration screen that an AP may present once installed, in particular the required Internet Protocol (IP) and Service Set Identifier (SSID) information.

Figure 8-1 Example of hotspot service configuration

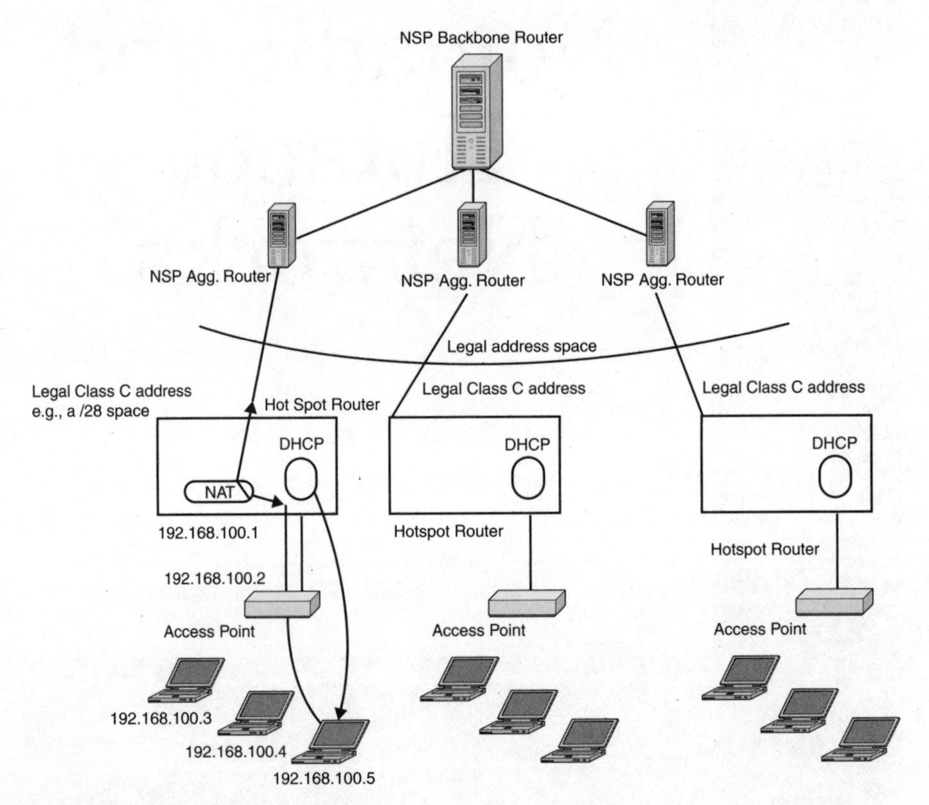

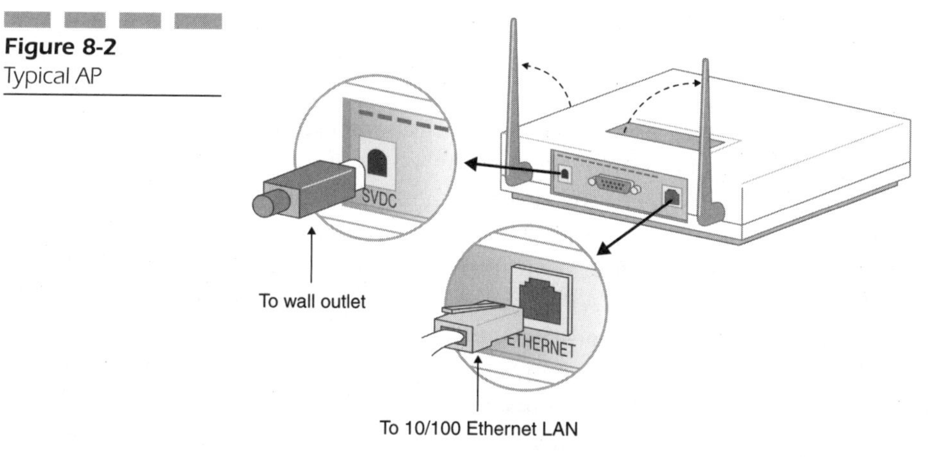

Figure 8-2
Typical AP

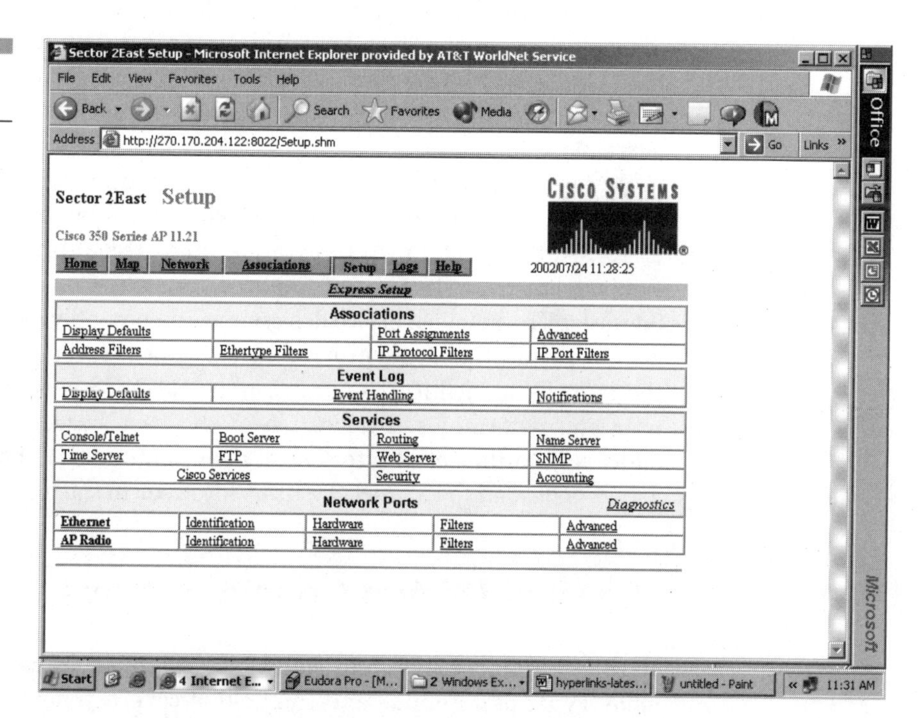

Figure 8-3
Typical setup screen

Logical Design Considerations

The first must-have is a cost-effective topology. The selection of hardware includes the router, the APs, and possibly wireless bridges. A terrestrial communication apparatus is required, such as T1-, DS3-, and/or Fast

Ethernet (FE)/Gigabit Ethernet (GbE)-based access to the Internet. A numbering plan is needed, and security considerations have to be taken into account (such as setting up the remote authentication dial-in user server [RADIUS] server). Other service-supporting servers (such as web servers) have to be deployed.

Once the data leaves the location's hotspot router, it needs to have a legal address. This implies that the hotspot router has to both run a local Dynamic Host Configuration Protocol (DHCP) as well as a Network Address Translation (NAT) function. Notice that the AP and the users that attach to it need to be on the same IP subnet. This implies that either a 192.168.0.0 address (the Internet Address Naming Authority's [IANA] recommended unregistered numbering plan) is used for both the clients and the AP, or that both the clients and the AP are assigned legal addresses. Therefore, Local Port Address Translation (PAT) plus on-router DHCP is the way to go. In the "Network Address Translation (NAT)" and "Dynamic Host Configuration Protocol (DHCP)" sections, we focus on NAT and DHCP support in the infrastructure behind the hotspot radios in the various dispersed locations.

Security is important, and it covers authentication, confidentiality, and integrity. This is the reason we covered Extensible Authentication Protocol (EAP); Authentication, Authorization, and Accounting (AAA); and Wired Equivalent Privacy (WEP) in Chapter 4, "Security Considerations for Hotspot Services." The hotspot service designer would probably like to deploy location-specific portals that could provide advertisement revenues from businesses in the immediate vicinity. Hence, after the RADIUS server authenticates the user and determines the location of the user based on the incoming IP address from the serial interface of the remote router, interplay with a web server is needed to redirect the user's browser to the location-specific URL.

Physical Design Considerations

In this section, we highlight some key design issues as graphically as possible. Types of antennas and the field of coverage for outdoor applications are shown in Figure 8-4, which illustrates omnidirectional antennas, sector antennas, and point-to-point patch antennas. Figure 8-5 further depicts the propagation characteristics of a typical omnidirectional antenna. Table 8-1 identifies typical antennas on the market for the APs, while Table 8-2 shows typical arrangements for the laptop. See Appendix A for a more inclusive listing. Figure 8-6 shows the use of wireless bridges to interconnect dispersed hotspots.

Figure 8-4
Antenna types

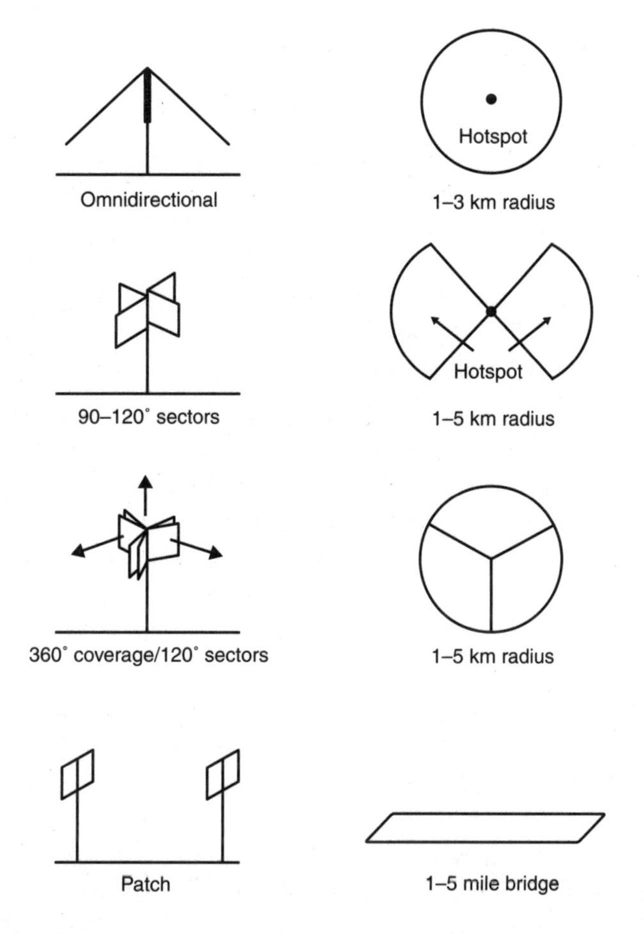

Figure 8-7 shows outdoor antenna coverage/designs for hotspot support. Figure 8-8 illustrates a number of possible hotspot designs along with some hotspot budgetary information. Figure 8-9 depicts an example of a prefabricated access point ready for deployment. Figure 8-10 depicts an indoor application. Figure 8-11 details indoors wireless local area network (WLAN) coverage options.

Some key considerations for the site survey are noted in the following boxed text.

Figures 8-12 through 8-15 illustrate examples of engineering packages that may be generated, based on the requirements identified via the site survey.

Figures 8-16 through 8-18 capture practical aspects of hotspot design. Figure 8-16 identifies what's needed within the Internet to support this

Site Surveys

Because differences exist in component configuration, placement, and physical environment, every network application is a unique installation. Before installing multiple APs, you should perform a site survey to determine the optimum utilization of networking components and to maximize range, coverage, and network performance.

Consider the following operating and environmental conditions when performing a site survey:

- **Data rates** Sensitivity and range are inversely proportional to data bit rates. The maximum radio range is achieved at the lowest workable data rate. A decrease in receiver threshold sensitivity occurs as the radio data increases.
- **Antenna type and placement** Proper antenna configuration is a critical factor in maximizing radio range. As a general rule, range increases in proportion to antenna height.
- **Physical environment** Clear or open areas provide better radio range than closed or filled areas. Also, the less cluttered the work environment, the greater the range.
- **Obstructions** A physical obstruction such as metal shelving or a steel pillar can hinder the performance of wireless devices. Avoid locating the devices in a location where there is a metal barrier between the sending and receiving antennas.
- **Building materials** Radio penetration is greatly influenced by the building material used in construction. For example, drywall construction allows a greater range than concrete blocks. Metal or steel construction is a barrier to radio signals.

Source: Cisco Systems

kind of service. Figure 8-17 shows a possible carrier installation process. Finally, Figure 8-18 depicts an actual hotspot designed by me for installation in the greater New York City area.

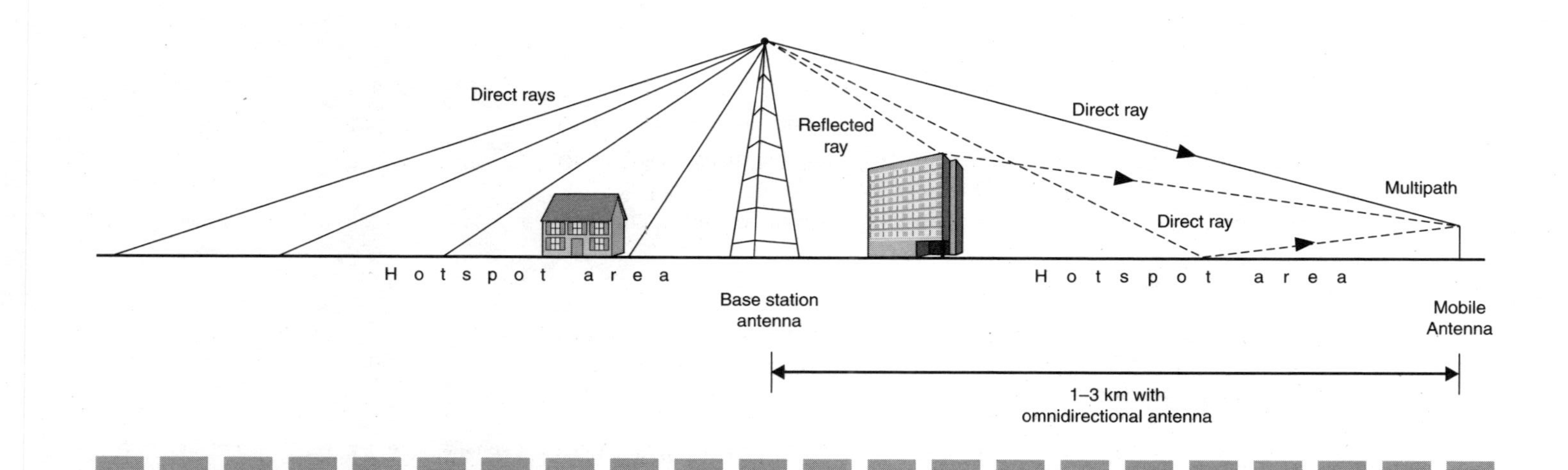

Figure 8-5 Example of propagation

Table 8-1

Typical antenna types

Type	dBi	Type	Application	Features/notes
Dipole "rubber duck"	2.2	Omnidirectional; two short rods	Indoors	Typically comes with AP.
Ceiling mount	2.2–5.2 (depends on model)	Omnidirectional; tubular	Indoors	Designed to be mounted to metal grid of suspended ceiling; aesthetically pleasing.
Mast mount vertical	5.14	Omnidirectional; tubular	Indoor or outdoor	To be clamped to a mast or pole; industrial applications; vertically polarized, must be installed perpendicular to ground.
Pillar mount	5.2	Omnidirectional; tubular; cloth-wrapped	Indoors	Aesthetically pleasing without looking like an antenna.
Ground plane	5.2	Omnidirectional; dish	Indoors	Useful for suspended ceilings.
Long range	12	Omnidirectional; tubular	Outdoors	+3.5 and -3.5 degree beam spread.
Patch antenna	3–6 (depends on model)	Directional, patch	Indoor or outdoor	Wide range.
Patch antenna	8.5–16	Directional, patch	Indoor or outdoor	Narrow beamwidth.
Sector antennas	8.5	90 or 120 degrees sector	Outdoor	For more complex designs.
Yagi	13.5	Directional	Long Distance	25 degrees beamwidth; 6.5 miles at 2 Mbps; 2 miles at 11 Mbps.
Parabolic	21	Directional	Long distance bridging	12 degrees beamwidth; 25 miles at 2 Mbps; 11.5 miles at 11 Mbps.

- Bridges are used to connect two or more wired LANs, usually located within separate buildings, to create one large LAN.
- A bridge can act as an AP in some applications by communicating with clients at the remote sites. This is accomplished with the Cisco Workgroup Bridge, PC Card, and PCI products.
- Cisco Bridges operate at the MAC address layer (data link layer), which means they have no routing capabilities.

Products: Wireless Bridges

Data Rate	Max. Distance	Optional Antenna (each side)	Standard Cable (6.7 dB / 100 ft) (each side)
11 Mbps	11.5 miles 18.75 km	21 dBi Dish	50 ft / 15.25 m
11 Mbps	18 miles 29.34 km	21 dBi Dish	20 ft / 6.1 m
5.5 Mbps	16 miles 26.08 km	21 dBi Dish	50 ft / 15.25 m
2 Mbps	25+ miles 40.75+ km	21 dBi Dish	50 ft / 15.25 m
1 Mbps	25+ miles 40.75+ km	21 dBi Dish	50 ft / 15.25 m

Point-to-Point Configuration

Building A
Optional Antenna
Bridge
Ethernet
0 to 25 miles or 1 to 40.75 km (line of sight)
Building B
Optional Antenna

Point-to-Multipoint Configuration

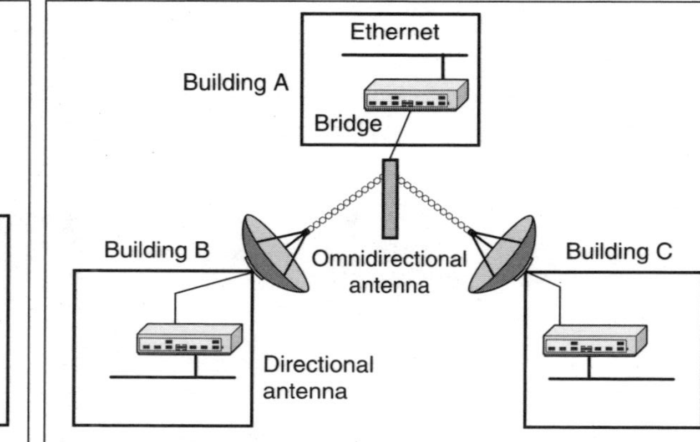

Figure 8-6 *Wireless bridging using Cisco products*

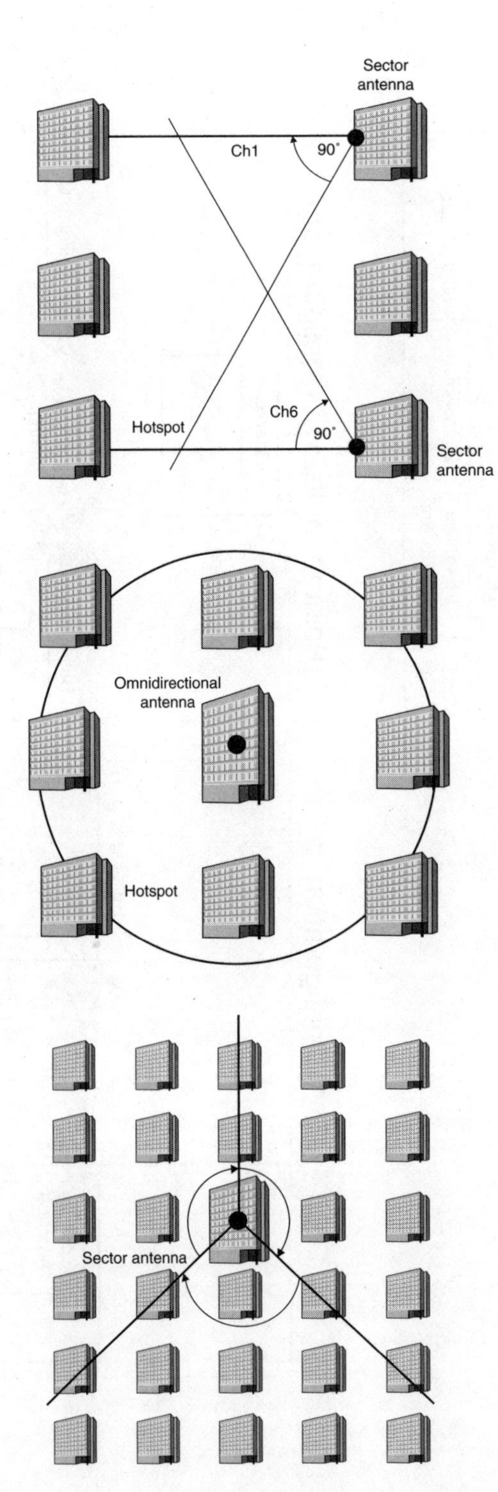

Figure 8-7
Coverage examples

Figure 8-8 Examples of hotspot outdoor topologies

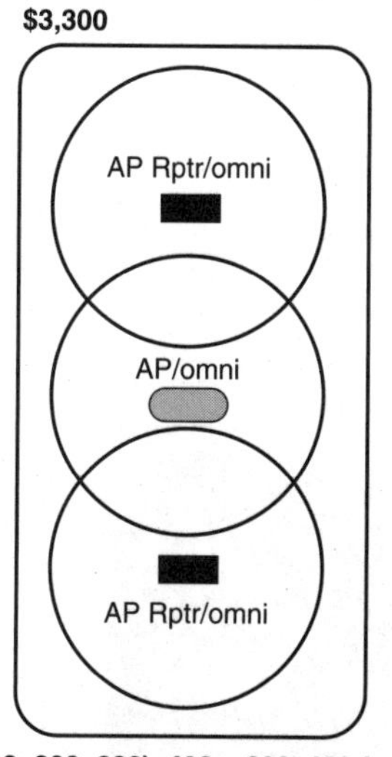

$4,200

AP Rptr/omni

AP Rptr/omni

AP/omni

AP Rptr/omni

(400+200+200) x (400+200+200) = 800x800 feet

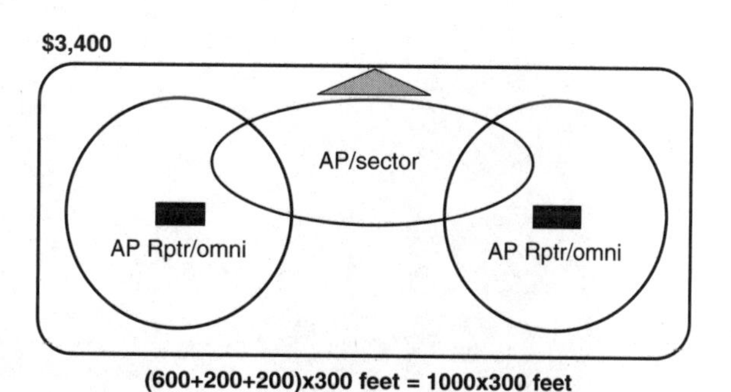

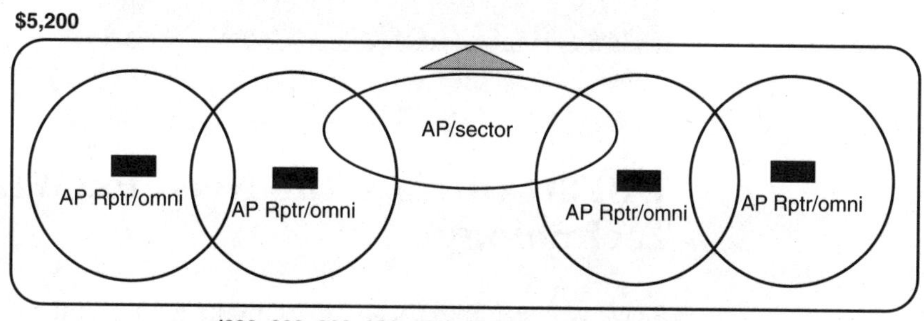

Figure 8-9 Examples of Prefabricated Hotspot node/repeater

Will 3G Obviate the Need for WLAN Hotspot Technology?

As we transition to these topics, make a note that *third generation (3G) will not replace the need for hotspot technologies based on WLANs or wireless personal area networks (WPANs).* 3G systems are wide area technologies, and although they have the advantage of seamless roaming, they offer much less bandwidth (2 Mbps at most and, more likely, at the practical deployment level, 384 Kbps for a long time to come.)

Figure 8-10 Indoor coverage

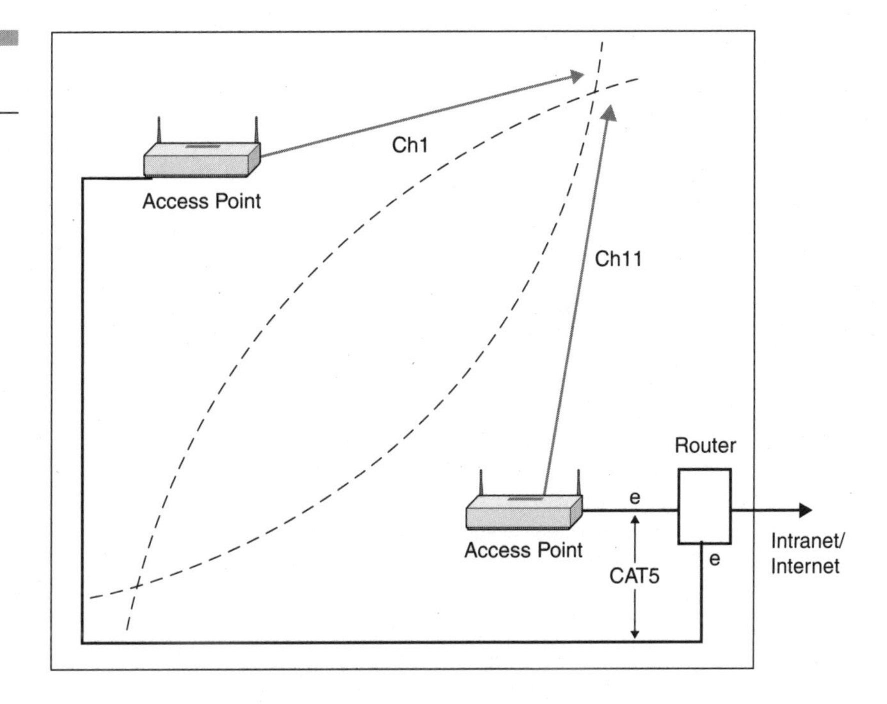

Indoors application (e.g. office, warehouse, etc.)
e = Ethernet

A wide area technology like 3G is likely to replace WLAN technology, just as Integrated Services Digital Network (ISDN) (a wide area technology, or even Asynchronous Transfer Mode [ATM]) replaced LANs in the mid- to late 1980s. Indeed, ISDN tried to play in the LAN game; one of the first "data" applications of ISDN was to attempt to replace IBM cluster controllers, which it totally failed to do. Eventually, incumbent carriers tried to deploy CO-LAN, again based on ISDN, but it also failed. And even with its high capacity of OC3, ATM could never catch up with FE, GbE, and 10GbE in the LAN.

The point is that a wide area network (WAN) technology has a hard time competing with a LAN technology when it attempts to provide a LAN service. Beyond that, there is the increased cost of a WAN/3G system compared with WLANs/WPANs that comes with paydowns on the licenses paid for the spectrum and with the major nationwide tower and transmission apparatus that has to be built.

Figure 8-11 Indoors WLAN coverage options (Source: Cisco Systems)

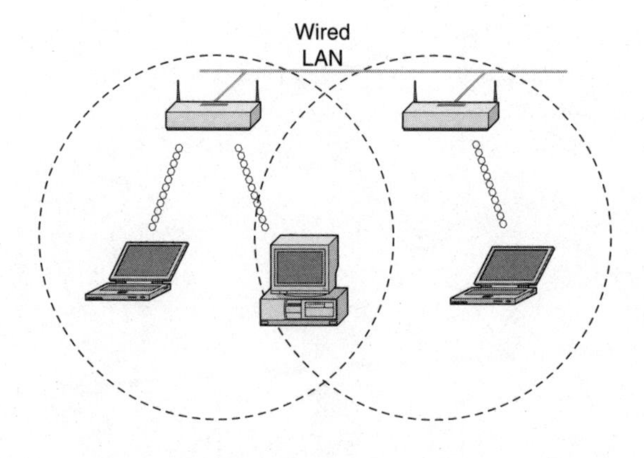

Multiple Overlapping Networks Coverage Option

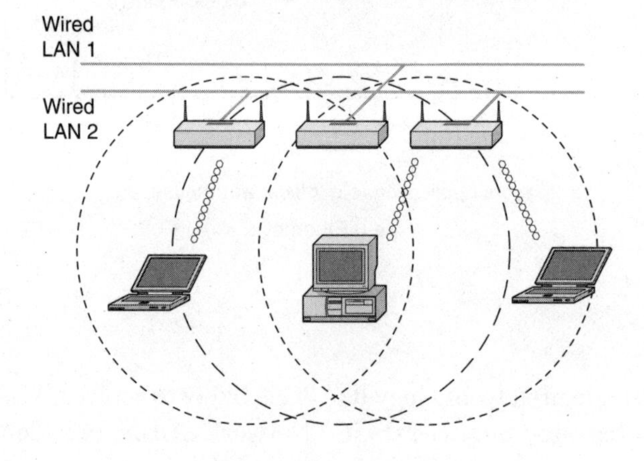

Heavy Overlap Coverage Option

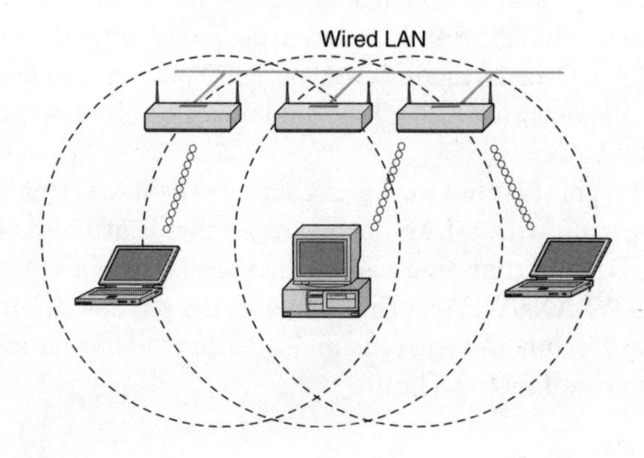

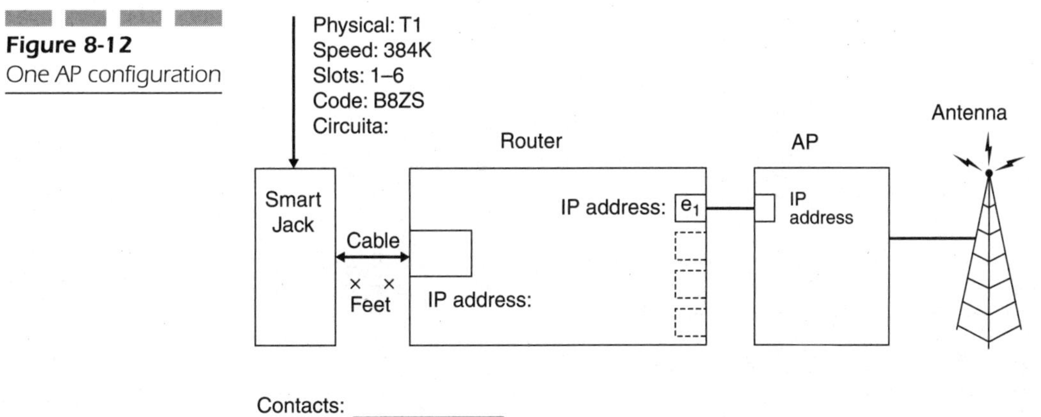

Figure 8-12 One AP configuration

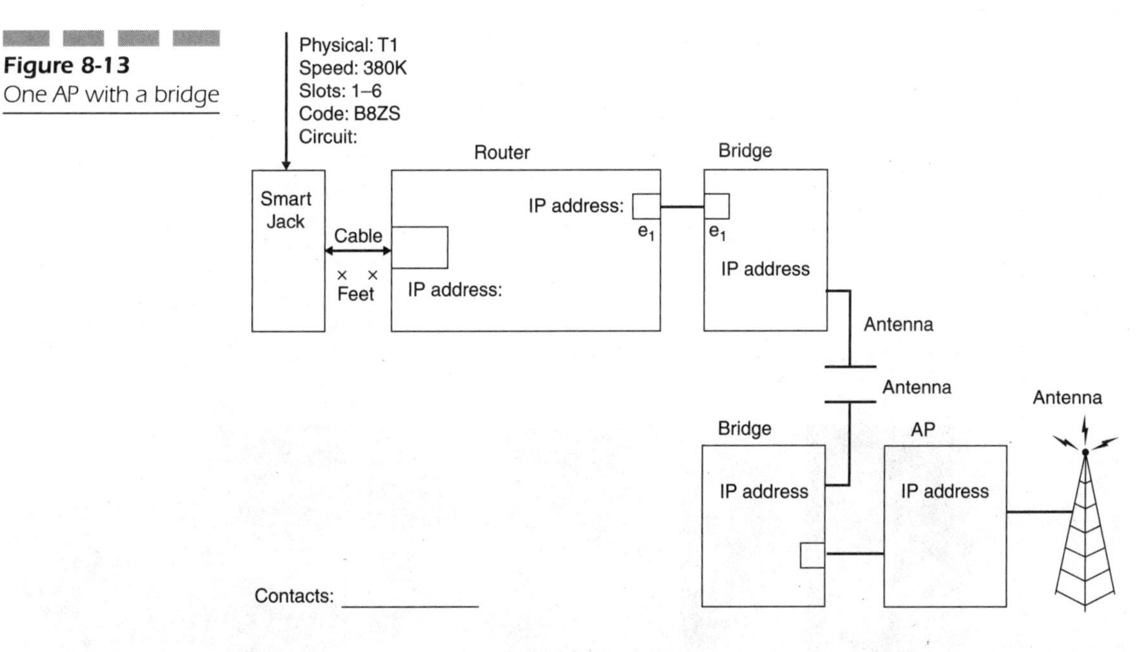

Figure 8-13 One AP with a bridge

Table 8-2

Solution	Picture	Budgetary Price	Application	Issues
PC PCMCIA Card		$175	Totally mobile PC	"No Gain"; could be limiting past 333 ft.
AIR-LMC352 card with "backcover antenna"	 	$275	Totally mobile PC	2.4-2.5 GHz: thin omnidirectional laptop computer antenna, 4.5 dBi gain, 18" pigtail with MMCX Connector (for Cisco LMC cards) "4.5 dbi of gain"; still limited by antenna positioning
AIR-LMC352 card with "backcover antenna" with 10 foot cable	 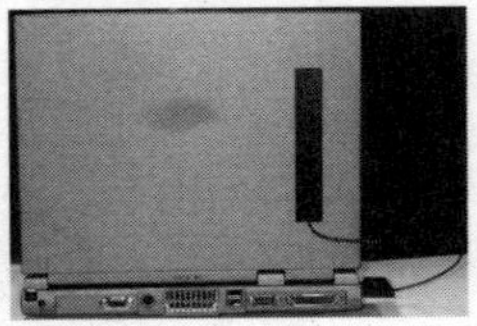 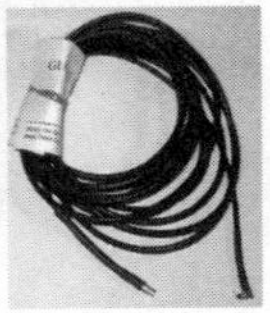	$300	Semimobile PC	2.4-2.5 GHz: thin omnidirectional laptop computer antenna, 4.5 dBi gain, 18" pigtail with MMCX Connector (for Cisco LMC cards) (10 ft RG-316 cable subtracts 6 db) "negative 2 gain"; but positioning can be helpful

AIR-LMC352 card with “mobile antenna” with 10 foot cable	$300	Semimobile PC	COMTELCO: 2.4-2.5 GHz, 7.5 dBi gain, 4.25” diameter patch antenna “net ~ 0 gain” (10 ft RG-316 cable subtracts 6 db); . . . but positioning can be helpful
AIR-LMC352 card with “monopole antenna” with 10 foot cable	$300	Semimobile PC	COMTELCO: 2.4-2.5 GHz, 8 dBi gain, 18” monopole / omnidirectional. Includes mounting hardware. “net ~ 2.5 gain” (10 ft RG-316 cable subtracts 6 db) ; . . . but positioning can be helpful
AIR-LMC352 card with “patch antenna” with 10 foot cable	$300	Semimobile PC	Conifer 2.4-2.5 GHz, 16 dBi gain, N-Female Connector, 27 degree BW “net ~10 gain” (10 ft RG-316 cable subtracts 6 db); No problems of any kind Antenna is about 9x8 inches . . . mount needed

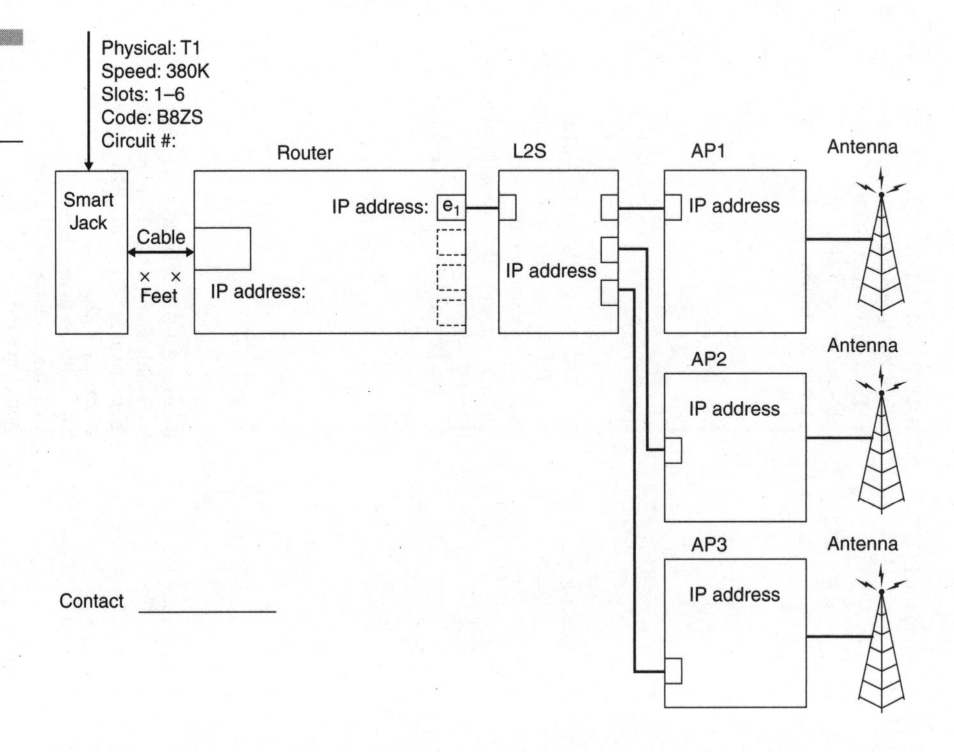

Figure 8-14 Multiple AP configuration

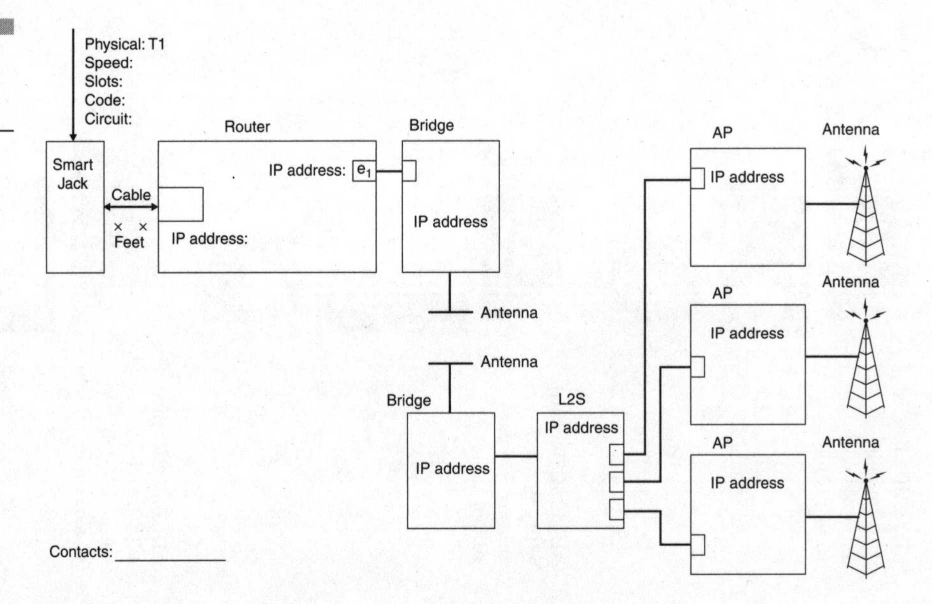

Figure 8-15 Multiple APs with a bridge

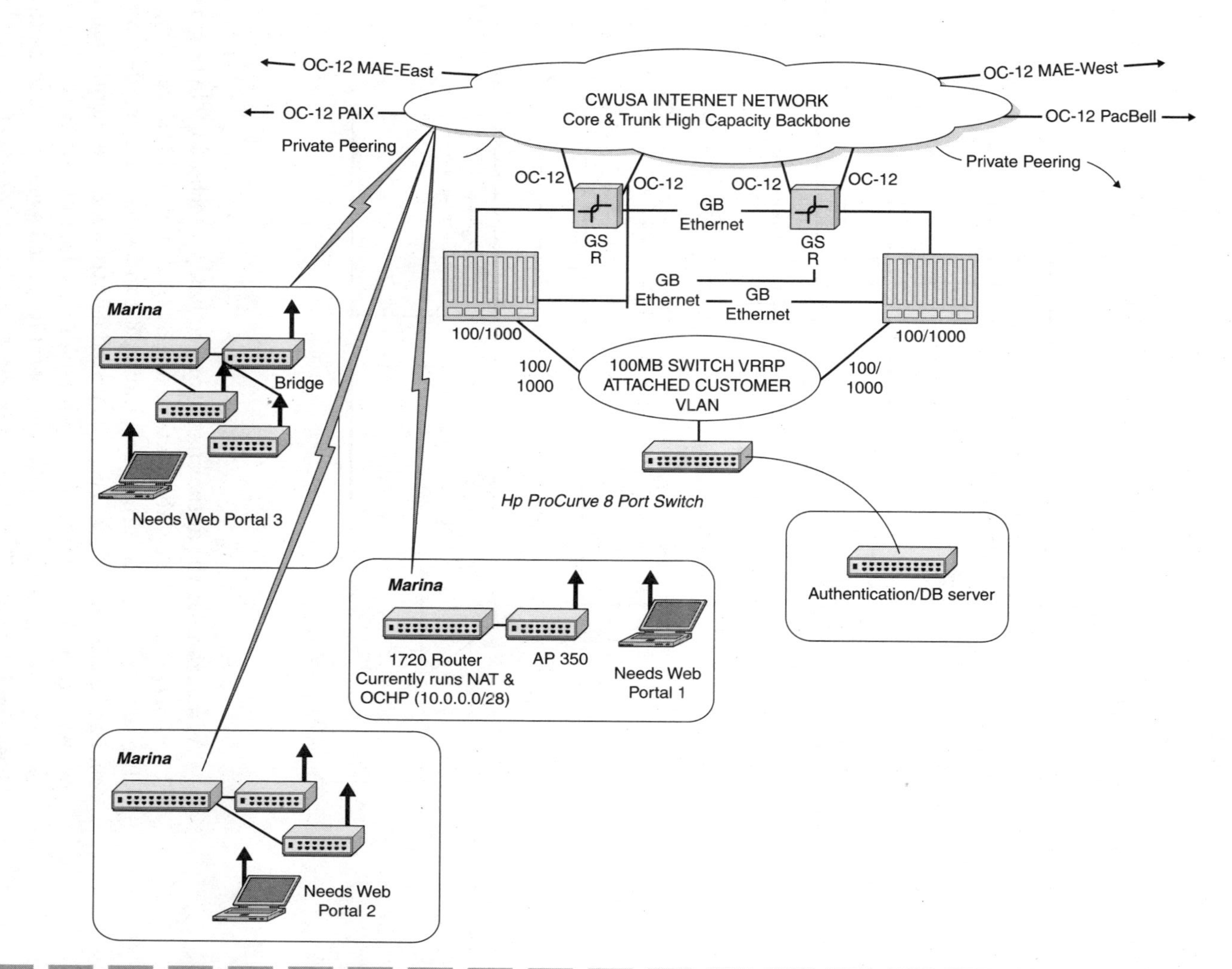

Figure 8-16 Connections back to Internet resources

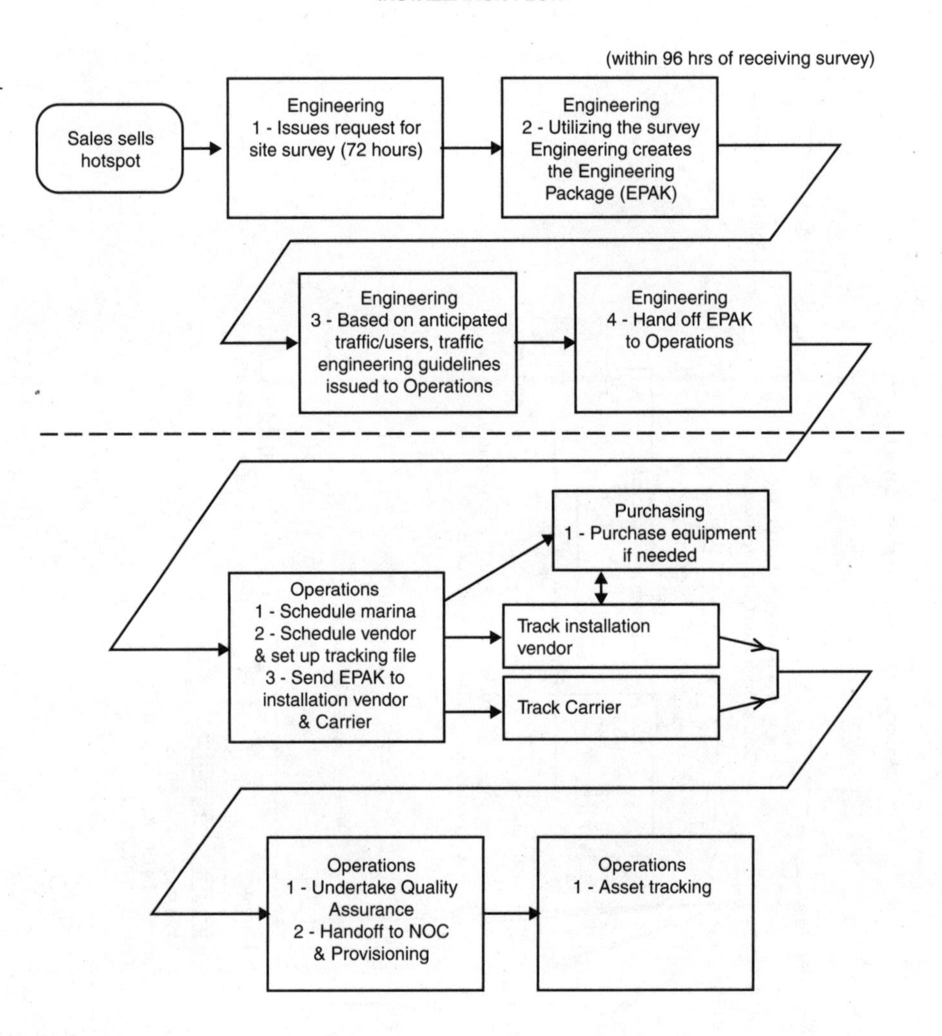

Figure 8-17 Example of hotspot installation process

As an example, the author has deployed hotspot technology in dozens of locations around the United States using 802.11b and 802.11a technologies. The service covers the areas for a radius of 1 to 2 miles or more, depending on antenna design. This service is priced as little as $40 a month (annual subscription) for unlimited use. Although this does not support a boater's access to the Internet while at sea, satellite-based systems (such as those based on INMARSAT) can cost between $2 to $8 per minute of usage, which

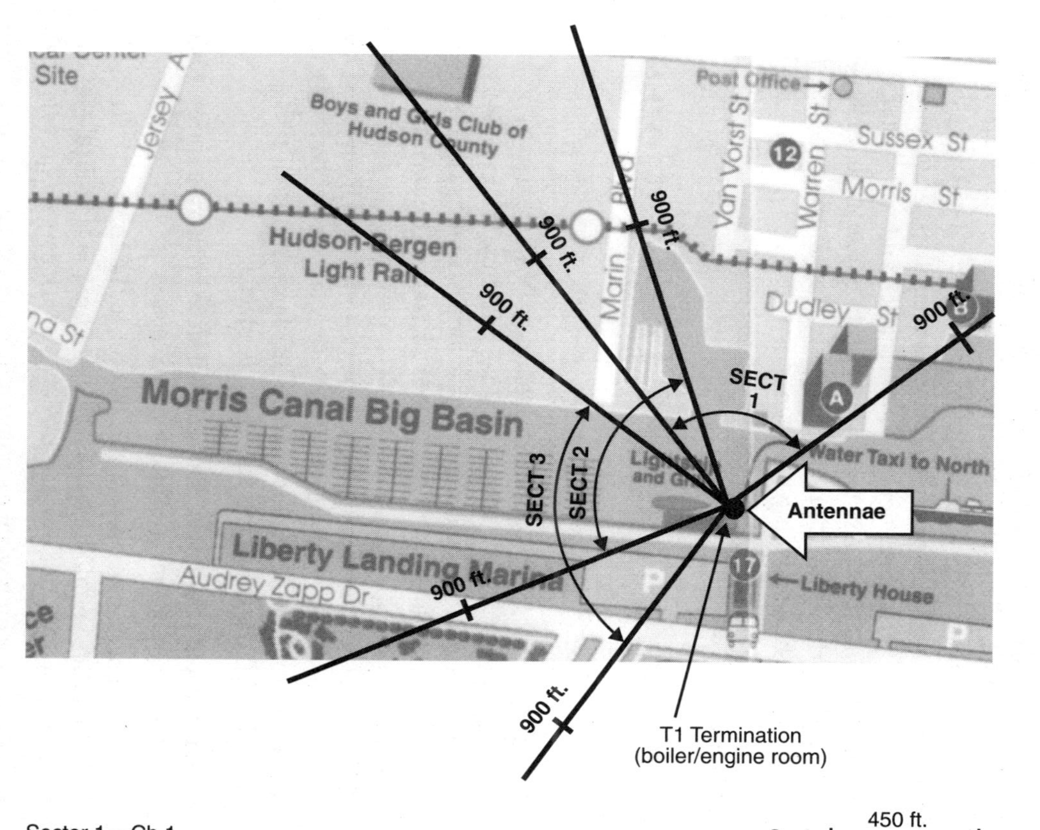

Figure 8-18 Specific topological example of hotspot installation

is quite astronomical. In fact, the two systems are not even seen as being competitive. A 100-foot yacht may use the satellite-based system while at sea (with hardware costs of around $10,000 compared to $100 for an IEEE 802.11b network interface card [NIC], and with an asymmetric link of much less uplink bandwidth and perhaps 256 Kbps on the downlink on a shared Very Small Aperture Terminal [VSAT] system), why would its owner spend that amount while sitting comfortably in the marina surfing the net for business or pleasure, when a subscription to a hotspot service can provide connectivity for $0.10 per hour? Work is afoot to provide roaming capabilities between picocells, microcells, macrocells, and worldcells, enabling the use of LAN-based WLAN technology while in a hotspot or business office and then, perhaps, shifting to a 3G system of much lower bandwidth and higher cost while traveling between hotspots.

Network Address Translation (NAT)

As noted, a typical hotspot network employs NAT to support nomadic users. Several types of NAT methods are described by the Internet Engineering Task Force (IETF) in various Request for Comments (RFCs); the key ones are listed in Table 8-3. RFC 3022 is the baseline document.

A technique called *overloading* (also known also as Port Address Translation [PAT], single-address NAT, or port-level multiplexed NAT), is the recommended form of dynamic NAT to map multiple unregistered IP addresses to a single registered IP address via different ports.

With PAT, the following actions take place:

- An internal network of APs and wireless bridges is set up with nonregistered/nonroutable IP addresses.
- The hotspot carrier sets up the router to the Internet with NAT capabilities enabled. The router has a unique IP address given to the carrier by IANA or by an Internet network service provider (NSP).
- A computer on a wireless domain attempts to connect to a computer on the Internet, such as a web server.
- The router receives the packet from the station on the wireless domain.
- The router saves the station's nonroutable IP address and port number to an address translation table. The router replaces the sending

station's nonroutable IP address with the router's IP address. The router replaces the sending station's source port with the port number that matches where the router saved the sending station's address information in the address translation table. The translation table now has a mapping of the wireless station's nonroutable IP address and port number along with the router's IP address.

- When a packet returns from the destination host on the Internet, the router looks at the destination port on the packet. It then scans the address translation table to see which station on the wireless domain the packet belongs to. It changes the destination address and destination port to the one saved in the address translation table and sends it to that station.
- The station receives the packet from the router and the process repeats as needed.
- Since the NAT router now has the wireless station's source address (likely provided via DHCP) and source port saved to the address translation table, it will continue to use that same port number for the duration of the connection. A timer is reset each time the router accesses an entry in the table. If the entry is not accessed again before the timer expires, the entry is removed from the table.

Note that it's perfectly possible to maintain some devices on the private domain that use dedicated IP addresses (infrastructure nodes or high-usage nodes). One can create an access list of IP addresses that instructs the router as to which devices on the network require NAT; the other IP addresses will pass through untranslated.

The number of simultaneous translations that a router can support is determined in large measure by the amount of Dynamic Random Access Memory (DRAM) it possesses. A typical entry in the address translation table requires about 160 bytes; hence, a router with 4 MB of DRAM can theoretically process 26,214 simultaneous translations.

IANA has actually set aside specific ranges of IP addresses for use as nonroutable internal network addresses. These addresses are considered unregistered, (see RFC 1918: Address Allocation for Private Internets).[4] A range is recommended for each of the three classes of IP addresses used for networking:

- Range 1/Class A: 10.0.0.0 through 10.255.255.255
- Range 2/Class B: 172.16.0.0 through 172.31.255.255
- Range 3/Class C: 192.168.0.0 through 192.168.255.255

Table 8-3

Key NAT-related RFCs

Number	Title	Authors or Editors	Date	More Info (obs/upd)	Status
RFC1631	The IP Network Address Translator (NAT)	K. Egevang and P. Francis	May 1994	Obsoleted by RFC3022	Informational
RFC2391	Load Sharing Using IP Network Address Translation (LSNAT)	P. Srisuresh and D. Gan	August 1998		Informational
RFC2428	FTP Extensions for IPv6 and NATs	M. Allman, S. Ostermann, and C. Metz	September 1998		Proposed standard
RFC2663	IP Network Address Translator (NAT) Terminology and Considerations	P. Srisuresh and M. Holdrege	August 1999		Informational
RFC2709	Security Model with Tunnel-mode IPSec for NAT Domains	P. Srisuresh	October 1999		Informational
RFC2766	Network Address Translation—Protocol Translation (NAT-PT)	G. Tsirtsis and P. Srisuresh	February 2000	Updated by RFC3152	Proposed standard
RFC2766	Network Address Translation—Protocol Translation (NAT-PT)	G. Tsirtsis and P. Srisuresh	February 2000	Updated by RFC3152	Proposed standard
RFC2962	An SNMP Application Level Gateway for Payload Address Translation	D. Raz, J. Schoenwaelder, and B. Sugla	October 2000		Informational
RFC2993	Architectural Implications of NAT	T. Hain	November 2000		Informational
RFC3022	Traditional IP Network Address Translator (Traditional NAT)	P. Srisuresh and K. Egevang	January 2001	Obsoletes RFC1631	Informational
RFC3102	Realm Specific IP: Framework	M. Borella, J. Lo, D. Grabelsky, and G. Montenegro	October 2001		Experimental
RFC3103	Realm Specific IP: Protocol Specification	M. Borella, D. Grabelsky, J. Lo, and K. Taniguchi	October 2001		Experimental
RFC3104	RSIP Support for End-to-end IPSec	G. Montenegro and M. Borella	October 2001		Experimental
RFC3105	Finding an RSIP Server with SLP	J. Kempf and G. Montenegro	October 2001		Experimental
RFC3188	Using National Bibliography Numbers as Uniform Resource Names	J. Hakala	October 2001		Informational
RFC3235	Network Address Translator (NAT)-Friendly Application Design Guidelines	D. Senie	January 2002		Informational

Working with NAT

The NAT operation described in RFC 3022[1] extends the address translation introduced in RFC 1631 and includes a new type of network address and Transmission Control Protocol/User Datagram Protocol (TCP/UDP) port translation. Basic NAT is a method by which IP addresses are mapped from one group to another, transparent to end users. Network Address Port Translation (NAPT) is a method by which many network addresses and their TCP/UDP ports are translated into a single network address and its TCP/UDP ports. Together, these two operations, referred to as traditional NAT, provide a mechanism to connect a realm that has private addresses to an external realm with globally unique registered addresses. The discussion that follows is with reference to the RFC.[2]

Introduction to NAT Engineering The need for IP address translation arises when a network's internal IP addresses cannot be used outside the network either for privacy reasons or because they are invalid for use outside the network.

Network topology outside a local domain can change in many ways. Customers may change providers, company backbones may be reorganized, or providers may merge or split. Whenever external topology changes with time, address assignments for nodes within the local domain must also change to reflect the external changes. Changes of this type can be hidden from users within the domain by centralizing changes to a single address translation router.

Basic address translation would (in many cases, except as noted in Srisuresh and Holdrege)[3] enable hosts in a private network to transparently access the external network and enable access to selective local hosts from the outside. Organizations with a network set up predominantly for internal use, with a need for occasional external access, are good candidates for this scheme.

Many Small Office, Home Office (SOHO) users and telecommuting employees have multiple network nodes in their office running TCP/UDP applications, but they have a single IP address assigned to their remote access router by their service provider to access remote networks. This ever-increasing community of remote access users would be benefited by NAPT, which would permit multiple nodes in a local network to simultaneously access remote networks using the single IP address assigned to their router.

Using the translation method has its limitations, however. It is mandatory that all requests and responses pertaining to a session be routed via the same NAT router. One way to ascertain this would be to have NAT based on a border router that is unique to a stub domain, where all IP packets are either originated from the domain or destined to the domain. There are other ways to ensure this with multiple NAT devices. For example, a private domain could have two distinct exit points to different providers, and the session flow from the hosts in a private network could traverse through whichever NAT device has the best metric for an external host. When one of the NAT routers fails, the other could route traffic for all the connections. However, the caveat with this approach is that rerouted flows could fail at the time of switchover to the new NAT router. A way to overcome this potential problem is that the routers share the same NAT configuration and exchange state information to ensure a failsafe backup for each other.

Address translation is application independent and often accompanied by application-specific gateways (application-level gateways [ALGs]) to perform payload monitoring and alterations. File Transfer Protocol (FTP) is the most popular ALG resident on NAT devices. Applications requiring ALG intervention must not have their payload encoded, as doing that would effectively disable the ALG, unless the ALG has the key to decrypt the payload.

This solution has the disadvantage of taking away the end-to-end significance of an IP address and making up for it with increased state in the network. As a result, end-to-end IP network-level security assured by IP Security (IPSec) cannot be assumed to end hosts, with a NAT device en route. The advantage of this approach, however, is that it can be installed without changes to hosts or routers.

Definitions of terms such as address realm, transparent routing, TCP/UDP (TU) ports, ALG, and others, used throughout the section, may be found in Srisuresh and Holdrege's "IP Network Address Translator (NAT) Terminology and Considerations."[3]

Traditional NAT

The address translation operation presented in RFC 3022 is referred to as traditional NAT. Other variations of NAT exist but will not be explored here. In most cases, traditional NAT enables hosts within a private network to access hosts in the external network transparently. Sessions are unidirectional, outbound from the private network. Sessions in the opposite direction may be allowed on an exceptional basis using static address maps for preselected hosts. Basic NAT and NAPT are two variations of traditional NAT, in that translation in Basic NAT is limited to IP addresses alone, whereas translation in NAPT is extended to include the IP address and transport identifier (such as the TCP/UDP port or Internet Control Message Protocol [ICMP] query ID).

Unless specifically mentioned, usage of the terms address translation or NAT throughout this chapter should be read as traditional NAT, namely Basic NAT and NAPT. Only the stub border routers, as described in Figure 8-19, may be configured to perform address translation.

Summary of Basic NAT Basic NAT operation is as follows. A stub domain with a set of private network addresses could be enabled to communicate with the external network by dynamically mapping the set of private addresses to a set of globally valid network addresses. If the number of local nodes is less than or equal to addresses in the global set, each local address is guaranteed a global address to map to. Otherwise, nodes permitted simultaneous access to the external network are limited by the number of addresses in the global set. Individual local addresses may be statically mapped to specific global addresses to ensure guaranteed access to the outside or to enable access to the local host from external hosts via a fixed public address. Multiple simultaneous sessions may be initiated from a local node, using the same address mapping.

Addresses inside a stub domain are local to that domain and not valid outside the domain. Thus, addresses inside one stub domain can be reused by other stub domains. Single Class A addresses are used by many stub domains. NAT is installed at each exit point between a stub domain and a backbone. If more than one exit point exists, it is essential to ensure that each NAT has the same translation table.

For instance, in the example of Figure 8-20, both stubs A and B internally use class A private address block 10.0.0.0/8.[4] Stub A's NAT is assigned the class C address block 198.76.29.0/24, and stub B's NAT is assigned the class C address block 198.76.28.0/24. The class C addresses are globally unique; no other NAT boxes can use them.

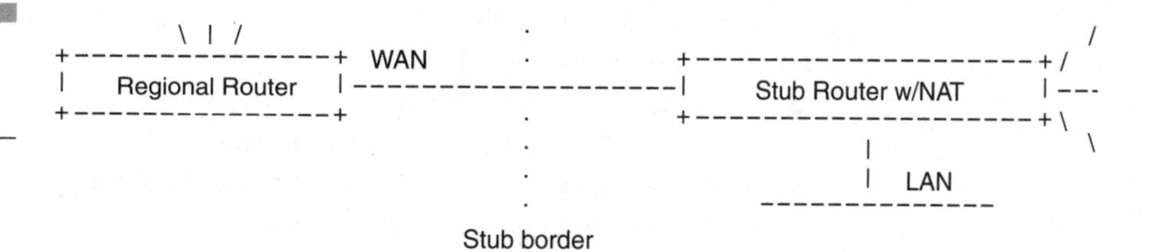

Figure 8-19 Traditional NAT configuration

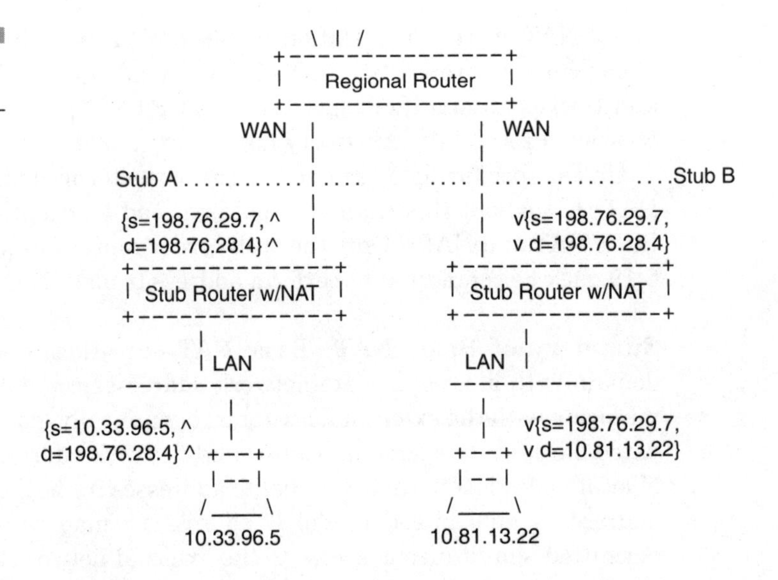

Figure 8-20
Basic NAT operation

When stub A host 10.33.96.5 wishes to send a packet to stub B host 10.81.13.22, it uses the globally unique address 198.76.28.4 as the destination and sends the packet to its primary router. The stub router has a static route for net 198.76.0.0, so the packet is forwarded to the WAN link. However, NAT translates the source address 10.33.96.5 of the IP header to the globally unique 198.76.29.7 before the packet is forwarded. Likewise, IP packets on the return path go through similar address translations.

Notice that this requires no changes to hosts or routers. For instance, as far as the stub A host is concerned, 198.76.28.4 is the address used by the host in stub B. The address translations are transparent to end hosts in most cases. Of course, this is just a simple example. There are numerous issues to be explored.

Summary of NAPT Let's say an organization has a private IP network and a WAN link to a service provider. The private network's stub router is assigned a globally valid address on the WAN link and the remaining nodes in the organization have IP addresses that have only local significance. In such a case, nodes on the private network could be allowed simultaneous access to the external network, using the single registered IP address with the aid of NAPT. NAPT would enable the mapping of tuples with the local IP addresses and the local TU port number to tuples with the registered IP address and assigned TU port number.

This model fits the requirements of most SOHO groups to access external networks using a single service-provider-assigned IP address. This model could be extended to enable inbound access by statically mapping a local node per each service TU port of the registered IP address.

In the example of Figure 8-21, stub A uses class A address block 10.0.0.0/8 internally. The stub router's WAN interface is assigned IP address 138.76.28.4 by the service provider.

When stub A host 10.0.0.10 sends a telnet packet to host 138.76.29.7, it uses the globally unique address 138.76.29.7 as the destination and sends the packet to its primary router. The stub router has a static route for the subnet 138.76.0.0/16 so the packet is forwarded to the WAN link. However, NAPT translates the tuple of source address 10.0.0.10 and source TCP port 3017 in the IP and TCP headers into the globally unique 138.76.28.4 and a uniquely assigned TCP port, such as 1024, before the packet is forwarded. Packets on the return path go through similar address and TCP port translations for the target IP address and target TCP port. Once again, notice that this requires no changes to hosts or routers. The translation is completely transparent.

In this setup, only TCP/UDP sessions are allowed and must originate from the local network. However, services such as DNS demand inbound access.

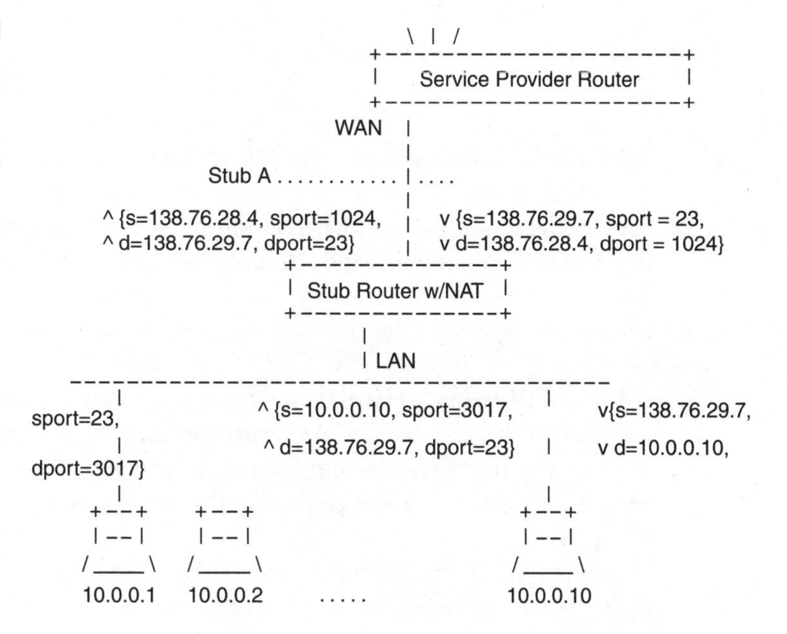

Figure 8-21 Network address port translation (NAPT) operation

There may be other services for which an organization enables inbound session access. It is possible to statically configure a well-known TU port service[5] on the stub router to be directed to a specific node in the private network.

In addition to TCP/UDP sessions, ICMP messages, with the exception of the REDIRECT message type, may also be monitored by a NAPT router. ICMP-query-type packets are translated similar to that of TCP/UDP packets, in that the identifier field in ICMP message header will be uniquely mapped to a query identifier of the registered IP address. The identifier field in ICMP query messages is set by the query sender and returned unchanged in the response message from the query responder. So, the tuple of the local IP address and the local ICMP query identifier is mapped to a tuple of the registered IP address and the assigned ICMP query identifier by the NAPT router to uniquely identify ICMP queries of all types from any of the local hosts. Modifications to ICMP error messages are discussed in a later section, as that involves modifications to the ICMP payload as well as the IP and ICMP headers.

In NAPT setup, where the registered IP address is the same as the IP address of the stub router WAN interface, the router has to be sure to make a distinction between TCP, UDP, or ICMP query sessions originated from itself versus those originated from the nodes on the local network. The end node of the inbound sessions (including TCP, UDP, and ICMP query sessions) is assumed to be the NAT router, unless the target service port is statically mapped to a different node on the local network. Sessions other than the TCP, UDP, and ICMP query types are simply not permitted from local nodes serviced by a NAPT router.

Translation Phases of a Session

The translation phases with traditional NAT are the same as those described in Srisuresh and Holdrege's "RFC 2663: IP Network Address Translator (NAT) Terminology and Considerations."[3] The following subsections identify items that are specific to traditional NAT.

Address binding With Basic NAT, a private address is bound to an external address when the first outgoing session is initiated from the private host. Subsequent to that, all other outgoing sessions originating from the same private address will use the same address binding for packet translation.

In the case of NAPT, where many private addresses are mapped to a single globally unique address, the binding would be from the tuple of the private address and the private TU port to the tuple of the assigned address and the assigned TU port. As with Basic NAT, this binding is determined when the first outgoing session is initiated by the tuple of the private address and the private TU port on the private host. Although not a common practice, it is possible to have an application on a private host establish multiple simultaneous sessions originating from the same tuple of the private address and the private TU port. In such a case, a single binding for the tuple of the private address and the private TU port may be used for the translation of packets pertaining to all sessions originating from the same tuple on a host.

Address Lookup and Translation After an address binding or an address and TU port tuple binding in case NAPT is established, a soft state may be maintained for each of the connections using the binding. Packets belonging to the same session will be subject to session lookup for translation purposes.

Address Unbinding When the last session based on an address, or an address and TU port tuple binding, is terminated, the binding itself may be terminated.

Packet Translations

Packets pertaining to NAT-managed sessions undergo translation in either direction. Individual packet translation issues are covered in detail in the following subsections.

IP, TCP, UDP, and ICMP Header Manipulations In the Basic NAT model, the IP header of every packet must be modified. This modification includes the IP address (the source IP address for outbound packets and the destination IP address for inbound packets) and the IP checksum.

For TCP[6] and UDP[7] sessions, modifications must include updating the checksum in the TCP/UDP headers. This is because the TCP/UDP checksum also covers a pseudoheader containing the source and destination IP addresses. As an exception, UDP headers with a 0 checksum should not be modified. As for ICMP query packets,[8] no further changes in the ICMP header are required since the checksum in the ICMP header does not cover IP addresses.

In the NAPT model, modifications to the IP header are similar to that of Basic NAT. For TCP/UDP sessions, modifications must be extended to include the translation of the TU port (the source TU port for outbound packets and the destination TU port for inbound packets) in the TCP/UDP header. The ICMP header in ICMP query packets must also be modified to replace the query ID and ICMP header checksum. The private host query ID must be translated into an assigned ID on the outbound packet and the exact reverse on the inbound packet. The ICMP header checksum must be corrected to account for query ID translation.

Checksum Adjustment NAT modifications are per-packet-based and can be very computation intensive, as they involve one or more checksum modifications in addition to simple field translations. Luckily, we have an algorithm that makes checksum adjustments to IP, TCP, UDP, and ICMP headers very simple and efficient. Since all these headers use a one's complement sum, it is sufficient to calculate the arithmetic difference between the before-translation and after-translation addresses and add this to the checksum. The following algorithm is applicable only for even offsets (optr must be at an even offset from the start of the header) and even lengths (olen and nlen below must be even).

ICMP Error Packet Modifications Changes to the ICMP error message[8] will include changes to IP and ICMP headers on the outer layer as well as changes to headers of the packet embedded within the ICMP error message payload.

In order for NAT to be transparent to the end host, the IP address of the IP header embedded within the payload of the ICMP error message must be modified, the checksum field of the embedded IP header must be modified, and lastly the ICMP header checksum must also be modified to reflect changes to the payload.

In a NAPT setup, if the IP message embedded within ICMP happens to be a TCP, UDP, or ICMP query packet, you will also need to modify the appropriate TU port number within the TCP/UDP header or the Query Identifier field in the ICMP query header. Lastly, the IP header of the ICMP packet must also be modified.

FTP Support One of the most popular applications, FTP, requires an ALG to monitor the control session payload to determine the ensuing data session parameters. FTP ALG is an integral part of most NAT implementations.

The FTP ALG would require a special table to correct the TCP sequence and acknowledge numbers with source port FTP or destination port FTP. The table entries should have a source address, destination address, source port, destination port, a delta for sequence numbers, and a timestamp. New entries are created only when FTP PORT commands or PASV responses are seen. The sequence number delta may be increased or decreased for every FTP PORT command or passive (PASV) response. Sequence numbers are incremented on the outbound, and acknowledge numbers are decremented on the inbound by this delta.

FTP payload translations are limited to private addresses and their assigned external addresses (encoded as individual octets in ASCII) for Basic NAT. For NAPT setup, however, the translations must be extended to include the TCP port octets (in ASCII) following the address octets.

DNS Support Considering that sessions in a traditional NAT are predominantly outbound from a private domain, DNS ALG may be obviated from use in conjunction with traditional NAT as follows. A DNS server(s) internal to the private domain maintains the mapping of names to IP addresses for internal hosts and possibly some external hosts. External DNS servers maintain name mapping for external hosts alone and not for any of the internal hosts. If the private network does not have an internal DNS server, all DNS requests may be directed to an external DNS server to find address mapping for the external hosts.

IP Option Handling An IP datagram with any of the IP options Record Route, Strict Source Route, or Loose Source Route would involve recording or using IP addresses of intermediate routers. A NAT intermediate router may choose not to support these options or leave the addresses untranslated while processing the options. The result of leaving the addresses untranslated would be that private addresses along the source route are exposed end to end. This should not jeopardize the traversal path of the packet per se, as each router is supposed to look at the next hop router only.

Miscellaneous Issues

This section looks at a number of issues related to NAT.

Partitioning of Local and Global Addresses For NAT to operate as described in this chapter, it is necessary to partition the IP address space into two parts: the private addresses used internal to the stub domain, and the globally unique addresses. Any given address must either be a private address or a global address. There is no overlap.

The problem with overlap is the following. Let's say a host in stub A wants to send packets to a host in stub B, but the global addresses of stub

B overlap with the private addresses of stub A. In this case, the routers in stub A would not be able to distinguish the global addresses of stub B from its own private addresses.

Private Address Space Recommendation RFC 1918, "Address Allocation for Private Internets," contains recommendations on address space allocation for private networks.[4] IANA has three blocks of IP address space, namely 10.0.0.0/8, 172.16.0.0/12, and 192.168.0.0/16 for private internets. In pre-Classless Interdomain Routing (CIDR) notation, the first block is nothing but a single class A network number, while the second block is a set of 16 contiguous class B networks, and the third block is a set of 256 contiguous class C networks.

An organization that decides to use IP addresses in the address space defined previously can do so without any coordination with IANA or an Internet registry. The address space can thus be used privately by many independent organizations at the same time, with NAT operation enabled on their border routers.

Routing across NAT The router running NAT should not advertise the private networks to the backbone. Only the networks with global addresses may be known outside the stub. However, global information that NAT receives from the stub border router can be advertised in the stub the usual way.

Typically, the NAT stub router will have a static route configured to forward all external traffic to the service provider router over a WAN link, and the service provider router will have a static route configured to forward NAT packets (those whose destination IP addresses fall within the range of the NAT-managed global address list) to the NAT router over a WAN link.

Switch-over from Basic NAT to NAPT In Basic NAT setup, when private network nodes outnumber global addresses available for mapping (say, a class B private network mapped to a class C global address block), external network access to some of the local nodes is abruptly cut off after the last global address from the address list is used up. This is very inconvenient and constraining. Such an incident can be safely avoided by optionally allowing the Basic NAT router to switch over to NAPT setup for the last global address in the address list. Doing this will ensure that hosts on the private network will have continued, uninterrupted access to the external nodes and services for most applications. Note, however, that it could be confusing if some of the applications that used to work with Basic NAT suddenly break due to the switchover to NAPT.[9]

Dynamic Host Configuration Protocol (DHCP)

As we've previously noted, a hotspot network is likely to use DHCP to support nomadic users.

Overview of DHCP

DHCP is an Internet protocol for automating the configuration of computers that use TCP/IP. The DHCP protocol is described in RFC 2131 (RFC 2131 obsoletes RFC 1541). DHCP can be used to automatically assign IP addresses to wireless stations in hotspots, to deliver TCP/IP stack configuration parameters such as the subnet mask and default router, and to provide other configuration information. DHCP uses UDP as its transport protocol. The client sends messages to the server on port (67), while the server sends messages to the client on port (68). DHCP is an extension of the Bootstrap Protocol (BOOTP) mechanism.

DHCP works well when one has to manage a large number of mobile users: users with laptops working in and out of the office, visiting branch offices, entering hotspot environments, and so on. Once the mobile stations configure a laptop to use DHCP, it can be automatically configured on any network with a DHCP server.

DHCP stores a list of addresses in a table for each of the subnets it is serving. When a DHCP client starts, it requests an address from the server. The server looks up an available address and assigns the address to the client. DHCP can also assign static addresses to clients if needed. In DHCP terms, clients "lease" IP addresses. DHCP leases only last a certain amount of time; the default period is one day, but one can modify this parameter. Clients can request leases of a specific duration, but to prevent any machine from holding onto the lease forever, you can configure a maximum allowable lease time on your server. DHCP has broad appeal, but it is particularly useful in hotspot services.

DHCP is also used to enable hosts (DHCP clients) on an IP network to obtain their configurations from a server (DHCP server). This reduces the work necessary to administer an IP network. The most significant configuration option the client receives from the server is its IP address, but other configuration parameters (timers, SSIDs, encryption keys, subnet masks, routers, domains, and Domain Name Servers [DNSs]) can also be downloaded.

Three mechanisms are used to assign an IP address to the client:

- **Automatic allocation** DHCP assigns a permanent IP address to a client.
- **Manual allocation** A client's IP address is assigned by the administrator, and DHCP conveys the address to the client.
- **Dynamic allocation** DHCP assigns an IP address to the client for a limited period of time (lease).

In general terms, the client (such as a wireless nomadic terminal entering a hotspot area) sends a request for an IP address. The server responds with an available IP address. Next, the client sends a request to the selected server for its configuration options. Thirdly, the server responds with the client's committed IP address along with other options (see Figure 8-22).

DHCP in Some Detail

DHCP provides configuration parameters to Internet hosts, and this is described in RFC 2131. DHCP consists of two components: a protocol for

Figure 8-22
DHCP mechanism

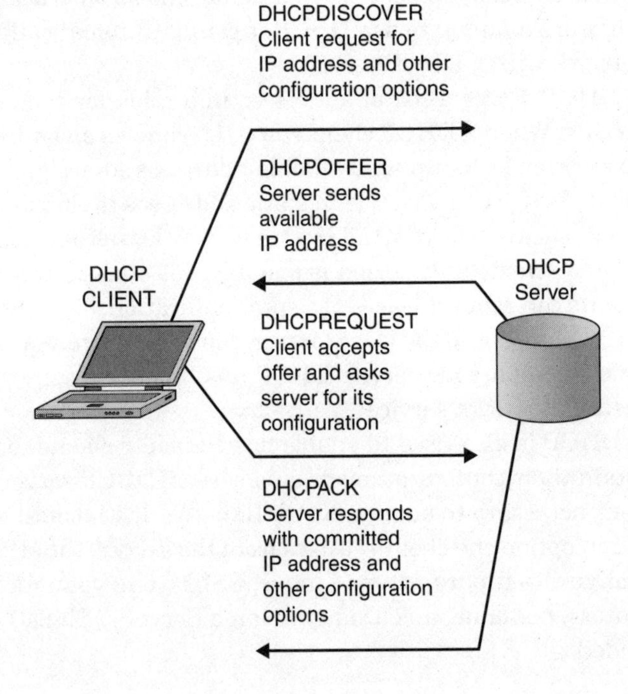

delivering host-specific configuration parameters from a DHCP server to a host, and a mechanism for the allocation of network addresses to hosts.[10]

DHCP is built on a client-server model, where designated DHCP server hosts allocate network addresses and deliver configuration parameters to dynamically configured hosts. In DHCP parlance, the term "server" refers to a host providing initialization parameters through DHCP, and the term "client" refers to a host requesting initialization parameters from a DHCP server.[11]

A host should not act as a DHCP server unless explicitly configured to do so by a system administrator. The diversity of hardware and protocol implementations on the Internet would preclude reliable operations if random hosts were allowed to respond to DHCP requests. For example, IP requires the setting of many parameters within the protocol implementation software. Because IP can be used on many dissimilar kinds of network hardware, values for those parameters cannot be guessed or be assumed to have correct defaults. Also, distributed address allocation schemes depend on a polling/defense mechanism for the discovery of addresses that are already in use. IP hosts may not always be able to defend their network addresses, so that such a distributed address allocation scheme cannot be guaranteed to avoid the allocation of duplicate network addresses.

DHCP supports three mechanisms for IP address allocation. In the first, known as *automatic allocation*, DHCP assigns a permanent IP address to a client. In *dynamic allocation*, DHCP assigns an IP address to a client for a limited period of time (or until the client explicitly relinquishes the address). In *manual allocation*, a client's IP address is assigned by the network administrator, and DHCP is used simply to convey the assigned address to the client. A particular network will use one or more of these mechanisms, depending on the policies of the network administrator.

Dynamic allocation is the only one of the three mechanisms that enables the automatic reuse of an address that is no longer needed by the client to which it is assigned. Thus, dynamic allocation is particularly useful for assigning an address to a client that will be connected to the network only temporarily or for sharing a limited pool of IP addresses among a group of clients that do not need permanent IP addresses. Dynamic allocation may also be a good choice for assigning an IP address to a new client being permanently connected to a network where IP addresses are sufficiently scarce so that it is important to reclaim them when old clients are retired. Manual allocation enables DHCP to be used to eliminate the error-prone process of manually configuring hosts with IP addresses in environments where (for whatever reason) it is desirable to manage IP address assignment outside of the DHCP mechanisms.

The format of DHCP messages is based on the format of BOOTP messages in order to capture the BOOTP relay agent behavior described as part of the BOOTP specification[12,13] and to enable the interoperability of existing BOOTP clients with DHCP servers. Using BOOTP relay agents eliminates the necessity of having a DHCP server on each physical network segment.

The RFC uses the following terms:

- **DHCP client** A DHCP client is an Internet host using DHCP to obtain configuration parameters such as a network address.
- **DHCP server** A DHCP server is an Internet host that returns configuration parameters to DHCP clients.
- **BOOTP relay agent** A BOOTP relay agent or relay agent is an Internet host or router that passes DHCP messages between DHCP clients and DHCP servers. DHCP is designed to use the same relay agent behavior as specified in the BOOTP protocol specification.
- **Binding** A binding is a collection of configuration parameters, including at least an IP address associated with or bound to a DHCP client. DHCP servers manage bindings.

The following list gives the general design goals for DHCP:

- DHCP should be a mechanism rather than a policy. It must enable local system administrators control over configuration parameters where desired; for example, local system administrators should be able to enforce local policies concerning allocation and access to local resources where desired.
- Clients should require no manual configuration. Each client should be able to discover the appropriate local configuration parameters without user intervention and incorporate those parameters into its own configuration.
- Networks should require no manual configuration for individual clients. Under normal circumstances, the network manager should not have to enter any per-client configuration parameters.
- DHCP should not require a server on each subnet. To allow for scale and economy, DHCP must work across routers or through the intervention of BOOTP relay agents.
- A DHCP client must be prepared to receive multiple responses to a request for configuration parameters. Some installations may include multiple, overlapping DHCP servers to enhance reliability and increase performance.
- DHCP must coexist with statically configured, nonparticipating hosts and with existing network protocol implementations.

- DHCP must interoperate with the BOOTP relay agent behavior as described by RFC 951 and RFC 1542.[13]
- DHCP must provide service to existing BOOTP clients.

The following list gives design goals specific to the transmission of the network-layer parameters. DHCP must

- Guarantee that any specific network address will not be in use by more than one DHCP client at a time.
- Retain DHCP client configuration across DHCP client reboots. A DHCP client should, whenever possible, be assigned the same configuration parameters (such as the network address) in response to each request.
- Retain DHCP client configuration across server reboots and, whenever possible, a DHCP client should be assigned the same configuration parameters despite restarts of the DHCP mechanism.
- Enable the automated assignment of configuration parameters to new clients to avoid hand configuration (manual configuration) for new clients.
- Support fixed or permanent allocation of configuration parameters to specific clients.

Protocol Summary

From the client's point of view, DHCP is an extension of the BOOTP mechanism. This behavior allows existing BOOTP clients to interoperate with DHCP servers without requiring any change to the clients' initialization software. RFC 1542 describes interactions between BOOTP and DHCP clients and servers.[15] New, optional transactions that optimize the interaction between DHCP clients and servers are described in upcoming sections on DHCP's client/server protocol and some of its specifications.

Figure 8-23 gives the format of a DHCP message and Table 8-4 describes each of the fields in the DHCP message. The numbers in parentheses indicate the size of each field in octets. The names for the fields in the figure will be used throughout this document to refer to the fields in DHCP messages.

Two primary differences exist between DHCP and BOOTP. First, DHCP defines mechanisms through which clients can be assigned a network address for a finite lease, allowing for the serial reassignment of network addresses to different clients. Second, DHCP provides the mechanism for a client to acquire all of the IP configuration parameters that it needs in order to operate.

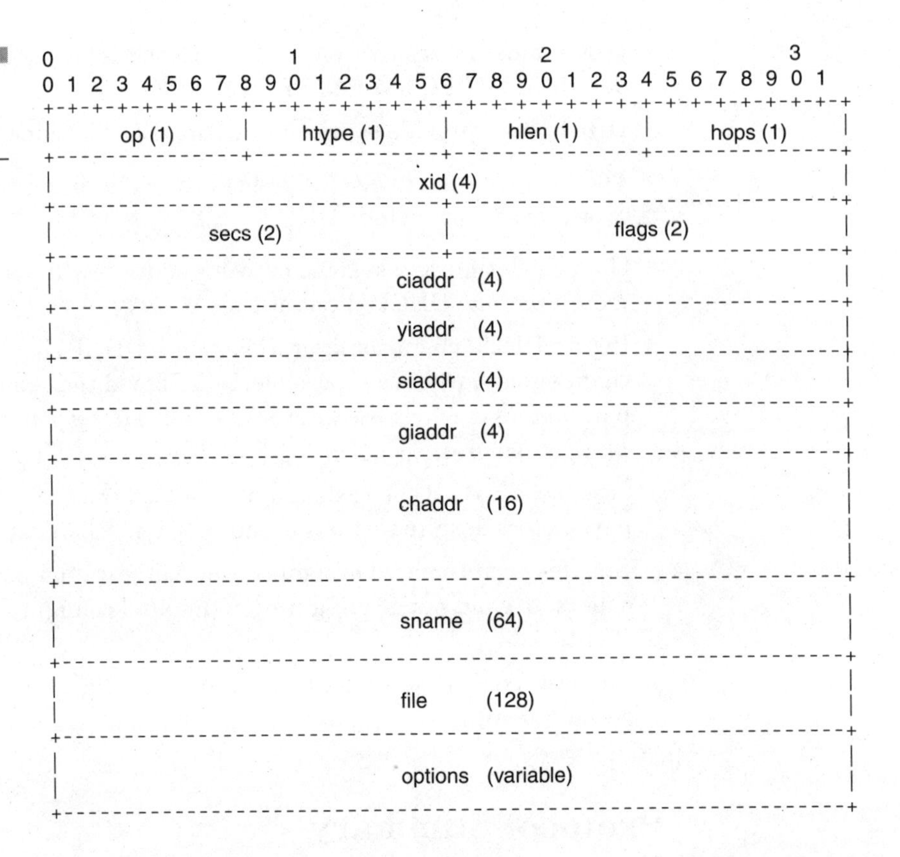

Figure 8-23 Format of a DHCP message

DHCP introduces a small change in terminology intended to clarify the meaning of one of the fields. What was the Vendor extensions field in BOOTP has been renamed the Options field in DHCP. Similarly, the tagged data items that were used inside the BOOTP Vendor extensions field, which were formerly referred to as vendor extensions, are now termed simply options.

DHCP defines a new Client identifier option that is used to pass an explicit identifier to a DHCP server. This change eliminates the overloading of the chaddr field in BOOTP messages, where chaddr is used both as a hardware address for the transmission of BOOTP reply messages and as a client identifier. The client identifier is an opaque key, not to be interpreted by the server. For example, the client identifier may contain a hardware address, identical to the contents of the chaddr field, or it may contain another type of identifier, such as a DNS name. The client identifier chosen by a DHCP client must be unique to that client within the subnet to which

Table 8-4

Description of fields in a DHCP message

Field	Octets	Description
op	1	Message op code / message type. 1 = BOOTREQUEST, 2 = BOOTREPLY.
htype	1	Hardware address type, see ARP section in "Assigned Numbers" RFC; for example, 1 = 10 MB Ethernet.
hlen	1	Hardware address length (6 for 10 MB Ethernet).
hops	1	Client sets to 0, optionally used by relay agents when booting via a relay agent.
xid	4	Transaction ID, a random number chosen by the client and used by the client and server to associate messages and responses between a client and a server.
secs	2	Filled in by client, seconds elapsed since client began address acquisition or renewal process.
flags	2	Flags (see Figure 8-24).
ciaddr	4	Client IP address; only filled in if client is in BOUND, RENEW, or REBINDING state and can respond to Address Resolution Protocol (ARP) requests.
yiaddr	4	Your (client) IP address.
siaddr	4	IP address of next server to use in bootstrap; returned in DHCPOFFER, DHCPACK by server.
giaddr	4	Relay agent IP address, used in booting via a relay agent.
chaddr	16	Client hardware address.
sname	64	Optional server host name, null-terminated string.
file	128	Boot file name, null-terminated string; generic name or null in DHCPDISCOVER, fully qualified directory path name in DHCPOFFER.
options	var	Optional parameters field. See the options documents for a list of defined options.

the client is attached. If the client uses a client identifier in one message, it must use that same identifier in all subsequent messages to ensure that all servers correctly identify the client.

DHCP clarifies the interpretation of the siaddr field as the address of the server to use in the next step of the client's bootstrap process. A DHCP server may return its own address in the siaddr field if the server is prepared to supply the next bootstrap service (such as the delivery of an operating system executable image). A DHCP server always returns its own address in the server identifier option.

The options field is now variable length. A DHCP client must be prepared to receive DHCP messages with an options field of at least 312 octets long. This requirement implies that a DHCP client must be prepared to receive a message of up to 576 octets, the minimum IP datagram size an IP host must be prepared to accept.[16] DHCP clients may negotiate the use of larger DHCP messages through the maximum DHCP message size option. The options field may be further extended into the file and sname fields.

In the case of a client using DHCP for initial configuration (before the client's TCP/IP software has been completely configured), DHCP requires the creative use of the client's TCP/IP software and liberal interpretation of RFC 1122. The TCP/IP software should accept and forward to the IP layer any IP packets delivered to the client's hardware address before the IP address is configured. DHCP servers and BOOTP relay agents may not be able to deliver DHCP messages to clients that cannot accept hardware unicast datagrams before the TCP/IP software is configured.

To work around some clients that cannot accept IP unicast datagrams before the TCP/IP software is configured, as discussed in the previous paragraph, DHCP uses the flags field.[13] The leftmost bit is defined as the BROADCAST (B) flag. The semantics of this flag are discussed in the "Constructing and Sending DHCP Messages" section. The remaining bits of the flags field are reserved for future use. They must be set to zero by clients and ignored by servers and relay agents. Figure 8-24 gives the format of the flags field.

Configuration Parameters Repository

The first service provided by DHCP is to provide persistent storage of network parameters for network clients. The model of DHCP persistent stor-

Figure 8-24
Format of the flags field

```
0                   1 1 1 1 1 1
0 1 2 3 4 5 6 7 8 9 0 1 2 3 4 5
+-+-+-+-+-+-+-+-+-+-+-+-+-+-+-+-+
|B|             MBZ             |
+-+-+-+-+-+-+-+-+-+-+-+-+-+-+-+-+
```

B: BROADCAST flag

MBZ: MUST BE ZERO (reserved for future use)

age is that the DHCP service stores a key-value entry for each client, where the key is some unique identifier (for example, an IP subnet number and a unique identifier within the subnet) and the value contains the configuration parameters for the client.

For example, the key might be the pair of IP-subnet-number and hardware-address, allowing for the serial or concurrent reuse of a hardware address on different subnets, and for hardware addresses that may not be globally unique. Note that the hardware-address should be typed by the type of hardware to accommodate the possible duplication of hardware addresses resulting from bit-ordering problems in a mixed-media, bridged network. Alternately, the key might be the pair of IP-subnet-number and hostname, allowing the server to assign parameters intelligently to a DHCP client that has been moved to a different subnet or has changed hardware addresses (perhaps because the network interface failed and was replaced). The protocol defines that the key will be IP-subnet-number and hardware-address unless the client explicitly supplies an identifier using the client identifier option. A client can query the DHCP service to retrieve its configuration parameters. The client interface to the configuration parameters repository consists of protocol messages to request configuration parameters and responses from the server carrying the configuration parameters.

Dynamic Allocation of Network Addresses

The second service provided by DHCP is the allocation of temporary or permanent network (IP) addresses to clients. The basic mechanism for the dynamic allocation of network addresses is simple: a client requests the use of an address for some period of time. The allocation mechanism (the collection of DHCP servers) guarantees not to reallocate that address within the requested time and attempts to return the same network address each time the client requests an address. In this document, the period over which a network address is allocated to a client is referred to as a lease.[17] The client may extend its lease with subsequent requests. The client may issue a message to release the address back to the server when the client no longer needs the address. The client may ask for a permanent assignment by asking for an infinite lease. Even when assigning "permanent" addresses, a server may choose to give out lengthy but noninfinite leases to enable the detection of the fact that the client has been retired.

In some environments, it will be necessary to reassign network addresses due to the exhaustion of available addresses. In such environments, the allocation mechanism will reuse addresses whose leases have expired. The server should use whatever information is available in the configuration information repository to choose an address to reuse. For example, the server may choose the least recently assigned address. As a consistency check, the allocating server should probe the reused address before allocating the address, such as with an ICMP echo request, and the client should probe the newly received address, such as with the Address Resolution Protocol (ARP).

The Client-Server Protocol

DHCP uses the BOOTP message format defined in RFC 951 and given in Table 8-5 and Figure 8-25. The op field of each DHCP message sent from a client to a server contains BOOTREQUEST. BOOTREPLY is used in the op field of each DHCP message sent from a server to a client.

The first four octets of the options field of the DHCP message contain the (decimal) values 99, 130, 83, and 99, respectively (this is the same magic cookie as defined in RFC 1497[18]). The remainder of the options field consists of a list of tagged parameters that are called options. All of the vendor extensions listed in RFC 1497 are also DHCP options. RFC 1533 gives the complete set of options defined for use with DHCP.

Several options have been defined so far. One particular option, the DHCP message type option, must be included in every DHCP message. This option defines the type of the DHCP message. Additional options may be allowed, required, or not allowed, depending on the DHCP message type.

Throughout this document, DHCP messages that include a DHCP message type option will be identified by the type of the message. For example, a DHCP message with DHCP message type option type 1 will be referred to as a DHCPDISCOVER message.

Client-server Interaction: Allocating a Network Address

The following summary of the protocol exchanges between clients and servers refers to the DHCP messages described in Table 8-5. The timeline in Figure 8-25 shows the timing relationships in a typical client-server interaction. If the client already knows its address, some steps may be omitted; this abbreviated interaction is described in the next subsection.

The sequence of protocol exchanges is as follows:

Table 8-5

DHCP messages

Message	Use
DHCPDISCOVER	Client broadcast to locate available servers.
DHCPOFFER	Server to client in response to DHCPDISCOVER with offer of configuration parameters.
DHCPREQUEST	Client message to servers either (a) requesting offered parameters from one server and implicitly declining offers from all others, (b) correctness of previously allocated address after system reboot, for example, or (c) extending the lease on a particular network address.
DHCPACK	Server to client with configuration parameters, including committed network address.
DHCPNAK	Server to client indicating client's notion of network address is incorrect (client has moved to new subnet) or client's lease has expired.
DHCPDECLINE	Client to server indicating network address is already in use.
DHCPRELEASE	Client to server relinquishing network address and cancelling remaining lease.
DHCPINFORM	Client to server, asking only for local configuration parameters; client already has externally configured network address.

1. The client broadcasts a DHCPDISCOVER message on its local physical subnet. The DHCPDISCOVER message may include options that suggest values for the network address and lease duration. BOOTP relay agents may pass the message on to DHCP servers not on the same physical subnet.
2. Each server may respond with a DHCPOFFER message that includes an available network address in the yiaddr field (and other configuration parameters in DHCP options). Servers need not reserve the offered network address, although the protocol will work more efficiently if the server avoids allocating the offered network address to another client. When allocating a new address, servers should check that the offered network address is not already in use; for example, the server may probe the offered address with an ICMP echo request. Servers should be implemented so that network administrators may choose to disable probes of newly allocated addresses. The server transmits the DHCPOFFER message to the client, using the BOOTP relay agent if necessary.

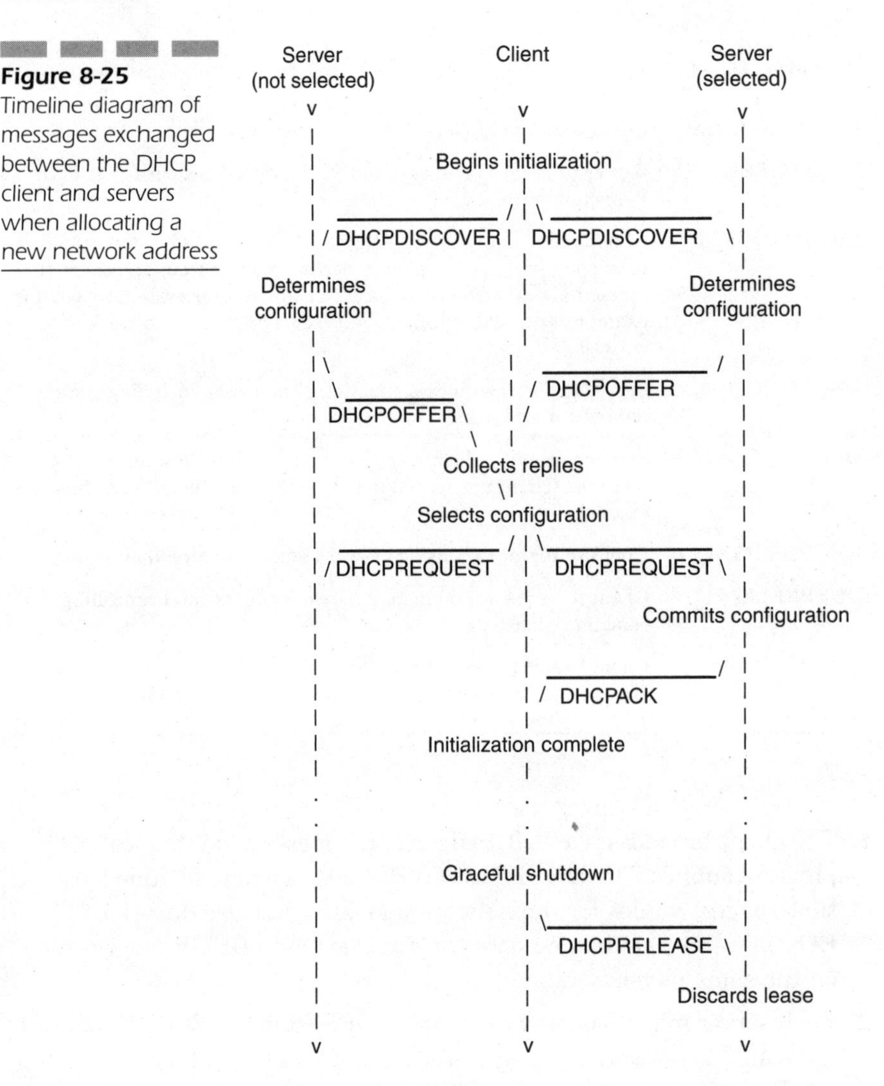

Figure 8-25 Timeline diagram of messages exchanged between the DHCP client and servers when allocating a new network address

3. The client receives one or more DHCPOFFER messages from one or more servers. The client may choose to wait for multiple responses. The client chooses one server from which to request configuration parameters, based on the configuration parameters offered in the DHCPOFFER messages. The client broadcasts a DHCPREQUEST message that must include the server identifier option to indicate which server it has selected, and that may include other options specifying desired configuration values. The requested IP address

option must be set to the value of yiaddr in the DHCPOFFER message from the server. This DHCPREQUEST message is broadcast and relayed through DHCP/BOOTP relay agents. To help ensure that any BOOTP relay agents forward the DHCPREQUEST message to the same set of DHCP servers that received the original DHCPDISCOVER message, the DHCPREQUEST message must use the same value in the DHCP message header's secs field and be sent to the same IP broadcast address as the original DHCPDISCOVER message. The client times out and retransmits the DHCPDISCOVER message if the client receives no DHCPOFFER messages.

4. The servers receive the DHCPREQUEST broadcast from the client. Those servers not selected by the DHCPREQUEST message use the message as notification that the client has declined that server's offer. The server selected in the DHCPREQUEST message commits the binding for the client to persistent storage and responds with a DHCPACK message containing the configuration parameters for the requesting client. The combination of client identifier or chaddr and assigned network address constitute a unique identifier for the client's lease and are used by both the client and server to identify a lease referred to in any DHCP message. Any configuration parameters in the DHCPACK message should not conflict with those in the earlier DHCPOFFER message to which the client is responding. The server should not check the offered network address at this point. The yiaddr field in the DHCPACK messages is filled in with the selected network address.

   If the selected server is unable to satisfy the DHCPREQUEST message (if the requested network address has been allocated), the server should respond with a DHCPNAK message.

   A server may choose to mark addresses offered to clients in DHCPOFFER messages as unavailable. The server should mark an address offered to a client in a DHCPOFFER message as available if the server receives no DHCPREQUEST message from that client.

5. The client receives the DHCPACK message with configuration parameters. The client should perform a final check on the parameters (ARP for allocated network address) and notes the duration of the lease specified in the DHCPACK message. At this point, the client is configured. If the client detects that the address is already in use (through the use of ARP), the client must send a DHCPDECLINE message to the server and restart the configuration process. The client should wait a minimum of 10 seconds before restarting the

configuration process to avoid excessive network traffic in case of looping.

If the client receives a DHCPNAK message, the client restarts the configuration process.

The client times out and retransmits the DHCPREQUEST message if the client receives neither a DHCPACK nor a DHCPNAK message. The client retransmits the DHCPREQUEST according to the retransmission algorithm in the section on constructing and sending DHCP messages. The client should choose to retransmit the DHCPREQUEST enough times to give adequate probability of contacting the server without causing the client (and the user of that client) to wait overly long before giving up. For example, a client retransmitting as described in the upcoming section on message construction might retransmit the DHCPREQUEST message 4 times, for a total delay of 60 seconds, before restarting the initialization procedure. If the client receives neither a DHCPACK nor a DHCPNAK message after employing the retransmission algorithm, the client reverts to INIT state and restarts the initialization process. The client should notify the user that the initialization process has failed and is restarting.

6. The client may choose to relinquish its lease on a network address by sending a DHCPRELEASE message to the server. The client identifies the lease to be released with its client identifier or chaddr and network address in the DHCPRELEASE message. If the client used a client identifier when it obtained the lease, it must use the same client identifier in the DHCPRELEASE message.

Client-server Interaction: Reusing a Previously Allocated Network Address

If a client remembers and wishes to reuse a previously allocated network address, a client may choose to omit some of the steps described in the previous section. The timeline diagram in Figure 8-26 shows the timing relationships in a typical client-server interaction for a client reusing a previously allocated network address.

The sequence of protocol exchanges is as follows:

1. The client broadcasts a DHCPREQUEST message on its local subnet. The message includes the client's network address in the requested IP

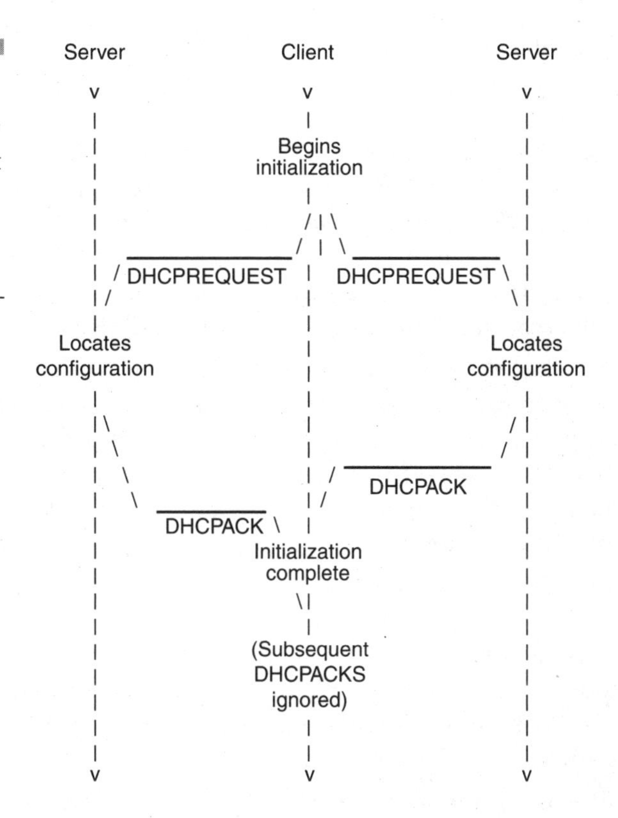

Figure 8-26 Timeline diagram of messages exchanged between DHCP client and servers when reusing a previously allocated network address

address option. As the client has not received its network address, it must not fill in the ciaddr field. BOOTP relay agents pass the message on to DHCP servers not on the same subnet. If the client used a client identifier to obtain its address, the client must use the same client identifier in the DHCPREQUEST message.

2. Servers with knowledge of the client's configuration parameters respond with a DHCPACK message to the client. Servers should not check that the client's network address is already in use; the client may respond to ICMP echo request messages at this point.

   If the client's request is invalid (the client has moved to a new subnet), servers should respond with a DHCPNAK message to the client. Servers should not respond if their information is not guaranteed to be accurate. For example, a server that identifies a request for an expired binding that is owned by another server should not respond with a DHCPNAK unless the servers are using an explicit mechanism to maintain coherency among them.

If giaddr is 0x0 in the DHCPREQUEST message, the client is on the same subnet as the server. The server must broadcast the DHCPNAK message to the 0xffffffff broadcast address because the client may not have a correct network address or subnet mask, and the client may not be answering ARP requests. Otherwise, the server must send the DHCPNAK message to the IP address of the BOOTP relay agent, as recorded in giaddr. The relay agent will, in turn, forward the message directly to the client's hardware address, so that the DHCPNAK can be delivered even if the client has moved to a new network.

3. The client receives the DHCPACK message with configuration parameters. The client performs a final check on the parameters, as in the previous section, and notes the duration of the lease specified in the DHCPACK message. The specific lease is implicitly identified by the client identifier or chaddr and the network address. At this point, the client is configured.

   If the client detects that the IP address in the DHCPACK message is already in use, the client must send a DHCPDECLINE message to the server and restart the configuration process by requesting a new network address. This action corresponds to the client moving to the INIT state in the DHCP state diagram, described in more detail when we discuss DHCP client behavior.

   If the client receives a DHCPNAK message, it cannot reuse its remembered network address. It must instead request a new address by restarting the configuration process, this time using the (nonabbreviated) procedure from the last section. This action also corresponds to the client moving to the INIT state in the DHCP state diagram.

   The client times out and retransmits the DHCPREQUEST message if it receives neither a DHCPACK nor a DHCPNAK message. A retransmission uses the retransmission algorithm. The client must retransmit the DHCPREQUEST enough times to give adequate probability of contacting the server without causing the client (and the user of that client) to wait overly long before giving up. For example, a client might retransmit the DHCPREQUEST message 4 times, for a total delay of 60 seconds, before restarting the initialization procedure. If the client receives neither a DHCPACK nor a DHCPNAK message after employing the retransmission algorithm, it may choose to use the previously allocated network address and configuration parameters for the remainder of the unexpired lease. This corresponds to moving to a BOUND state in the client state transition diagram shown in Figure 8-27.

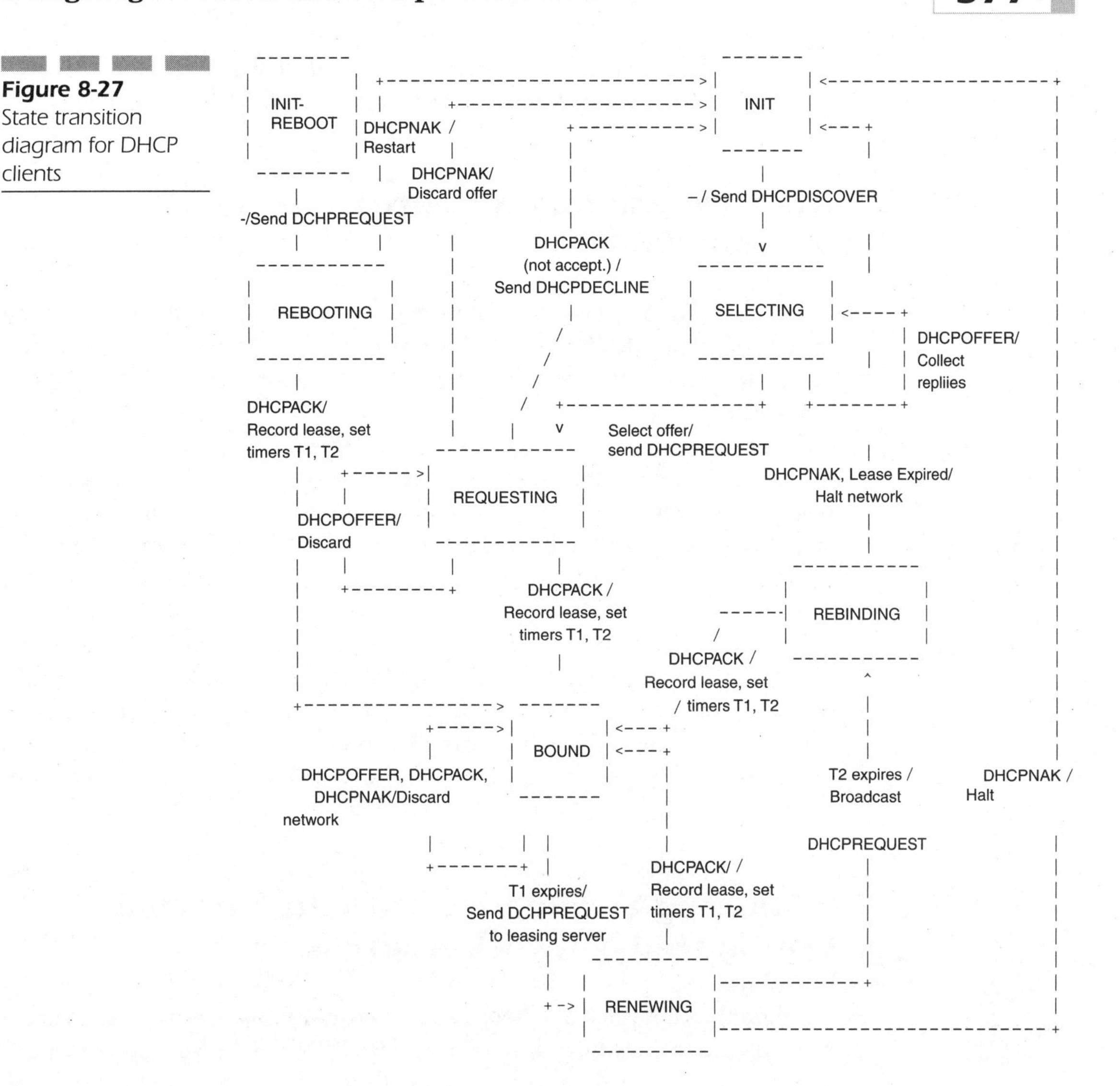

Figure 8-27 State transition diagram for DHCP clients

4. The client may choose to relinquish its lease on a network address by sending a DHCPRELEASE message to the server. The client identifies the lease to be released with its client identifier or chaddr and network address in the DHCPRELEASE message.

   Note that in this case, where the client retains its network address locally, the client will not normally relinquish its lease during a graceful shutdown. Only in the case where the client explicitly needs to relinquish its lease (when the client is about to be moved to a different subnet, for example) will the client send a DHCPRELEASE message.

Interpretation and Representation of Time Values

A client acquires a lease for a network address for a fixed period of time (which may be infinite). Throughout the protocol, times are to be represented in units of seconds. The time value of 0xffffffff is reserved to represent infinity.

As clients and servers may not have synchronized clocks, times are represented in DHCP messages as relative times, to be interpreted with respect to the client's local clock. Representing relative times in units of seconds in an unsigned, 32-bit word gives a range of relative times from 0 to approximately 100 years, which is sufficient for the relative times to be measured using DHCP.

The algorithm for lease duration interpretation given in the previous paragraph assumes that client and server clocks are stable relative to each other. If there is a drift between the two clocks, the server may consider the lease expired before the client does. To compensate, the server may return a shorter lease to the client than it commits to its local database of client information.

Obtaining Parameters with an Externally Configured Network Address

If a client has obtained a network address through some other means (such as manual configuration), it may use a DHCPINFORM request message to obtain other local configuration parameters. Servers receiving a DHCPINFORM message construct a DHCPACK message with any local configuration parameters appropriate for the client without allocating a new address, checking for an existing binding, filling in yiaddr, or including lease time parameters. The servers should unicast the DHCPACK reply to the address given in the ciaddr field of the DHCPINFORM message.

The server should check the network address in a DHCPINFORM message for consistency, but it must not check for an existing lease. The server forms a DHCPACK message containing the configuration parameters for the requesting client and sends the DHCPACK message directly to the client.

Client Parameters in DHCP

Not all clients require the initialization of all parameters listed in the RFC. Two techniques are used to reduce the number of parameters transmitted from the server to the client. First, most of the parameters have defaults defined in the Host Requirements RFCs; if the client receives no parameters from the server that override the defaults, a client uses those default values. Second, in its initial DHCPDISCOVER or DHCPREQUEST message, a client may provide the server with a list of specific parameters the client is interested in. If the client includes a list of parameters in a DHCPDISCOVER message, it must include that list in any subsequent DHCPREQUEST messages.

Clients should include the maximum DHCP message size option to let the server know how large the server may make its DHCP messages. The parameters returned to a client may still exceed the space allocated to options in a DHCP message. In this case, two additional options flags (which must appear in the options field of the message) indicate that the file and sname fields are to be used for options.

The client can inform the server as to which configuration parameters the client is interested in by including the parameter request list option. The data portion of this option explicitly lists the options requested by tag number.

In addition, the client may suggest values for the network address and lease time in the DHCPDISCOVER message. The client may include the requested IP address option to suggest that a particular IP address be assigned, and it may include the IP address lease time option to suggest the lease time it would like. Other options representing hints at configuration parameters are allowed in a DHCPDISCOVER or DHCPREQUEST message. However, additional options may be ignored by servers, and multiple servers may therefore not return identical values for some options. The requested IP address option is to be filled in only in a DHCPREQUEST message when the client is verifying network parameters obtained previously. The client fills in the ciaddr field only when correctly configured with an IP address in a BOUND, RENEWING, or REBINDING state.

If a server receives a DHCPREQUEST message with an invalid requested IP address option, the server should respond to the client with a DHCPNAK message and may choose to report the problem to the system administrator. The server may include an error message in the message option.

Use of DHCP in Clients with Multiple Interfaces

A client with multiple network interfaces must use DHCP through each interface independently to obtain configuration information parameters for them.

When Should Clients Use DHCP? A client should use DHCP to reacquire or verify its IP address and network parameters whenever the local network parameters may have changed. This may be at system boot time or after a disconnection from the local network, as the local network configuration may change without the client or user's knowledge.

If a client has knowledge of a previous network address and is unable to contact a local DHCP server, the client may continue to use the previous network address until the lease for that address expires. If the lease expires before the client can contact a DHCP server, the client must immediately discontinue use of the previous network address and may inform local users of the problem.

Specification of the DHCP Client-Server Protocol In this section, we assume that a DHCP server has a block of network addresses from which it can satisfy requests for new addresses. Each server also maintains a database of allocated addresses and leases in local permanent storage.

Constructing and Sending DHCP Messages DHCP clients and servers both construct DHCP messages by filling in fields in the fixed format section of the message and appending tagged data items in the variable length option area. The options area includes a four-octet "magic cookie," followed by the options. The last option must always be the end option.

DHCP uses UDP as its transport protocol. DHCP messages from a client to a server are sent to the DHCP server port (67), and DHCP messages from a server to a client are sent to the DHCP client port (68). A server with multiple network addresses (a multihomed host) may use any of its network addresses in outgoing DHCP messages.

The server identifier field is used both to identify a DHCP server in a DHCP message and as a destination address from clients to servers. A server with multiple network addresses must be prepared to accept any of its network addresses as identifying that server in a DHCP message. To accommodate potentially incomplete network connectivity, a server must choose an address as a server identifier that, to the best of the server's knowledge, is reachable from the client.

For example, if the DHCP server and the DHCP client are connected to the same subnet (the giaddr field in the message from the client is zero), the server should select the IP address the server is using for communication on that subnet as the server identifier. If the server is using multiple IP addresses on that subnet, any such address may be used. If the server has received a message through a DHCP relay agent, the server should choose an address from the interface on which the message was received as the server identifier (unless the server has other, better information on which to make its choice). DHCP clients must use the IP address provided in the server identifier option for any unicast requests to the DHCP server. DHCP messages broadcast by a client prior to that client obtaining its IP address must have the source address field in the IP header set to zero.

If the giaddr field in a DHCP message from a client is nonzero, the server sends any return messages to the DHCP server port on the BOOTP relay agent whose address appears in giaddr. If the giaddr field is zero and the ciaddr field is nonzero, then the server unicasts DHCPOFFER and DHCPACK messages to the address in ciaddr. If giaddr is zero and ciaddr is zero, and the broadcast bit is set, then the server broadcasts DHCPOFFER and DHCPACK messages to 0xffffffff. If the broadcast bit is not set and giaddr is zero and ciaddr is zero, then the server unicasts DHCPOFFER and DHCPACK messages to the client's hardware address and yiaddr address. In all cases, when giaddr is zero, the server broadcasts any DHCPNAK messages to 0xffffffff.

If the options in a DHCP message extend into the sname and file fields, the option overload option must appear in the options field with value 1, 2, or 3, as specified in RFC 1533. If the option overload option is present in the options field, the options in the options field must be terminated by an end option and may contain one or more pad options to fill the options field. The options in the sname and file fields (if in use as indicated by the options overload option) must begin with the first octet of the field, must be terminated by an end option, and must be followed by pad options to fill the remainder of the field. Any individual option in the options, sname, and file fields must be entirely contained in that field. The options in the options field must be interpreted first, so that any option overload options may be interpreted. The file field must be interpreted next (if the option overload option indicates that the file field contains DHCP options), followed by the sname field.

The values to be passed in an option tag may be too long to fit in the 255 octets available to a single option (a list of routers in a router option[13]). Options may appear only once, unless otherwise specified in the options document. The client concatenates the values of multiple instances of the same option into a single parameter list for configuration.

DHCP clients are responsible for all message retransmission. The client must adopt a retransmission strategy that incorporates a randomized exponential backoff algorithm to determine the delay between retransmissions. The delay between retransmissions should be chosen to allow sufficient time for replies from the server to be delivered based on the characteristics of the internetwork between the client and the server.

For example, in a 10 Mbps Ethernet internetwork, the delay before the first retransmission should be 4 seconds randomized by the value of a uniform random number chosen from the range −1 to +1. Clients with clocks that provide resolution granularity of less than one second may choose a noninteger randomization value. The delay before the next retransmission should be 8 seconds randomized by the value of a uniform number chosen from the range −1 to +1. The retransmission delay should be doubled with subsequent retransmissions up to a maximum of 64 seconds. The client may provide an indication of retransmission attempts to the user as an indication of the progress of the configuration process.

The client uses the xid field to match incoming DHCP messages with pending requests. A DHCP client must choose xids in such a way as to minimize the chance of using an xid that is identical to one used by another client. For example, a client may choose a different, random initial xid each time the client is rebooted and subsequently use sequential xids until the next reboot. Selecting a new xid for each retransmission is an implementation decision. A client may choose to reuse the same xid or select a new 'one for each retransmitted message.

Normally, DHCP servers and BOOTP relay agents attempt to deliver DHCPOFFER, DHCPACK, and DHCPNAK messages directly to the client using unicast delivery. The IP destination address (in the IP header) is set to the DHCP yiaddr address, and the link-layer destination address is set to the DHCP chaddr address. Unfortunately, some client implementations are unable to receive such unicast IP datagrams until the implementation has been configured with a valid IP address (leading to a deadlock in which the client's IP address cannot be delivered until the client has been configured with an IP address).

A client that cannot receive unicast IP datagrams until its protocol software has been configured with an IP address should set the BROADCAST bit in the Flags field to 1 in any DHCPDISCOVER or DHCPREQUEST messages that the client sends. The BROADCAST bit will provide a hint to the DHCP server and BOOTP relay agent to broadcast any messages to the client on the client's subnet. A client that can receive unicast IP datagrams before its protocol software has been configured should clear the BROADCAST bit to 0. The BOOTP clarifications document discusses the ramifications of the use of the BROADCAST bit.[13]

A server or relay agent sending or relaying a DHCP message directly to a DHCP client (not to a relay agent specified in the giaddr field) should examine the BROADCAST bit in the Flags field. If this bit is set to 1, the DHCP message should be sent as an IP broadcast using an IP broadcast address (preferably 0xffffffff) as the IP destination address and the link-layer broadcast address as the link-layer destination address. If the BROADCAST bit is cleared to 0, the message should be sent as an IP unicast to the IP address specified in the yiaddr field and the link-layer address specified in the chaddr field. If unicasting is not possible, the message may be sent as an IP broadcast using an IP broadcast address (preferably 0xffffffff) as the IP destination address and the link-layer broadcast address as the link-layer destination address.

DHCP Server Administrative Controls

DHCP servers are not required to respond to every DHCPDISCOVER and DHCPREQUEST message they receive. For example, a network administrator, to retain stringent control over the clients attached to the network, may choose to configure DHCP servers to respond only to clients that have been previously registered through some external mechanism. The DHCP specification describes only the interactions between clients and servers when the clients and servers choose to interact; it is beyond the scope of the DHCP specification to describe all the administrative controls that system administrators might want to use. Specific DHCP server implementations may incorporate any controls or policies desired by a network administrator. In some environments, a DHCP server will have to consider the values of the vendor class options included in DHCPDISCOVER or DHCPREQUEST messages when determining the correct parameters for a particular client.

A DHCP server needs to use some unique identifier to associate a client with its lease. The client may choose to explicitly provide the identifier through the client identifier option. If the client supplies a client identifier, the client must use the same client identifier in all subsequent messages, and the server must use that identifier to identify the client. If the client does not provide a client identifier option, the server must use the contents of the chaddr field to identify the client. It is crucial for a DHCP client to use an identifier that is unique within the subnet to which the client is attached in the client identifier option. Use of chaddr as the client's unique identifier may cause unexpected results, as that identifier may be associated with a hardware interface that could be moved to a new client. Some sites may choose to use a manufacturer's serial number as the client identifier to avoid unexpected changes in a client's network address due to the transfer of hardware interfaces among computers. Sites may also choose to use a DNS name as the client identifier, causing address leases to be associated with the DNS name rather than a specific hardware box.

DHCP clients are free to use any strategy in selecting a DHCP server among those from which the client receives a DHCPOFFER message. The client implementation of DHCP should provide a mechanism for the user to select directly the vendor class identifier values.

DHCP Server Behavior

A DHCP server processes incoming DHCP messages from a client based on the current state of the binding for that client. A DHCP server can receive the following messages from a client:

- DHCPDISCOVER
- DHCPREQUEST
- DHCPDECLINE
- DHCPRELEASE
- DHCPINFORM

Table 8-6 shows the use of fields and options in a DHCP message sent by a server. The remainder of this section describes the action of the DHCP server for each possible incoming message.

DHCPDISCOVER Message When a server receives a DHCPDISCOVER message from a client, the server chooses a network address for the requesting client. If no address is available, the server may choose to report the problem to the system administrator. If an address is available, the new address should be chosen as follows:

1. Use the client's current address as recorded in the client's current binding.
2. Otherwise, use the client's previous address as recorded in the client's (now expired or released) binding, if that address is in the server's pool of available addresses and not already allocated.

Table 8-6

Fields and options used by DHCP servers

Field	DHCPOFFER	DHCPACK	DHCPNAK
op	BOOTREPLY	BOOTREPLY	BOOTREPLY
htype	(From "Assigned Numbers" RFC)		
hlen	(Hardware address length in octets)		
hops	0	0	0
xid	xid from client DHCPDISCOVER message	xid from client DHCPREQUEST message	xid from client DHCPREQUEST message
secs	0	0	0
ciaddr	0	ciaddr from DHCPREQUEST or 0	0
yiaddr	IP address offered to client	IP address assigned to client	0
siaddr	IP address of next bootstrap server	IP address of next bootstrap server	0
flags	flags from client DHCPDISCOVER message	flags from client DHCPREQUEST message	flags from client DHCPREQUEST message
giaddr	giaddr from client DHCPDISCOVER message	giaddr from client DHCPREQUEST message	giaddr from client DHCPREQUEST message
chaddr	chaddr from client DHCPDISCOVER message	chaddr from client DHCPREQUEST message	chaddr from client DHCPREQUEST message
sname	Server host name or options	Server host name or options	(unused)
file	Client boot file name or options	Client boot file name or options	(unused)
options	options	options	

Option	DHCPOFFER	DHCPACK	DHCPNAK
Requested IP address	Must not	Must not	Must not
IP address lease time	Must	Must (DHCPREQUEST)	Must not Must not (DHCPINFORM)
Use file/sname fields	May	May	Must not
DHCP message type	DHCPOFFER	DHCPACK	DHCPNAK

Table 8-6 cont.

Fields and options used by DHCP servers

Option	DHCPOFFER	DHCPACK	DHCPNAK
Parameter request list	Must not	Must not	Must not
Message	Should	Should	Should
Client identifier	Must not	Must not	May
Vendor class identifier	May	May	May
Server identifier	Must	Must	Must
Maximum message size	Must not	Must not	Must not
All others	May	May	Must not

3. Otherwise, use the address requested in the Requested IP Address option, if that address is valid and not already allocated.
4. Otherwise, use a new address allocated from the server's pool of available addresses. The address is selected based on the subnet from which the message was received (if giaddr is 0) or on the address of the relay agent that forwarded the message (giaddr when not 0).

As described in the last section, a server may, for administrative reasons, assign an address other than the one requested or it may refuse to allocate an address to a particular client even though free addresses are available.

Note that in some network architectures (Internets with more than one IP subnet assigned to a physical network segment), it may be the case that the DHCP client should be assigned an address from a different subnet than the address recorded in giaddr. Thus, DHCP does not require that the client be assigned as an address from the subnet in giaddr. A server is free to choose some other subnet, and it is beyond the scope of the DHCP specification to describe ways in which the assigned IP address might be chosen.

Although not required for the correct operation of DHCP, the server should not reuse the selected network address before the client responds to the server's DHCPOFFER message. The server may choose to record the address as offered to the client.

The server must also choose an expiration time for the lease, as follows:

1. If the client has not requested a specific lease in the DHCPDISCOVER message and the client already has an assigned network address, the server returns the lease expiration time previously assigned to that

address (note that the client must explicitly request a specific lease to extend the expiration time on a previously assigned address).

2. Otherwise, if the client has not requested a specific lease in the DHCPDISCOVER message and the client does not have an assigned network address, the server assigns a locally configured default lease time.
3. Otherwise, if the client has requested a specific lease in the DHCPDISCOVER message (regardless of whether the client has an assigned network address), the server may choose either to return the requested lease (if the lease is acceptable to local policy) or select another lease.

Once the network address and lease have been determined, the server constructs a DHCPOFFER message with the offered configuration parameters. It is important for all DHCP servers to return the same parameters (with the possible exception of a newly allocated network address) to ensure predictable client behavior regardless of which server the client selects. The configuration parameters must be selected by applying the following rules in the specified order. The network administrator is responsible for configuring multiple DHCP servers to ensure uniform responses from those servers. The server must return to the client

- The client's network address, as determined by the rules given earlier in this section
- The expiration time for the client's lease, as determined by the rules given earlier in this section
- Parameters requested by the client, according to the following rules:
  - If the server has been explicitly configured with a default value for the parameter, the server must include that value in an appropriate option in the option field.
  - Otherwise, if the server recognizes the parameter as a parameter defined in the Host Requirements document, the server must include the default value for that parameter as given in the Host Requirements document in an appropriate option in the option field.
  - Otherwise, the server must not return a value for that parameter. The server must supply as many of the requested parameters as possible and must omit any parameters it cannot provide. The server must include each requested parameter only once unless explicitly allowed in the DHCP Options and BOOTP Vendor Extensions document.

- Return any parameters from the existing binding that differ from the Host Requirements document are by default.
- Return any parameters that are specific to this client (as identified by the contents of chaddr or client identifier in the DHCPDISCOVER or DHCPREQUEST message), such as those configured by the network administrator.
- Return any parameters that are specific to this client's class (as identified by the contents of the vendor class identifier option in the DHCPDISCOVER or DHCPREQUEST message), such as those configured by the network administrator. The parameters must be identified by an exact match between the client's vendor class identifiers and the client's classes identified in the server.
- Return any parameters that have nondefault values on the client's subnet.

The server may choose to return the vendor class identifier used to determine the parameters in the DHCPOFFER message to assist the client in selecting which DHCPOFFER to accept. The server inserts the xid field from the DHCPDISCOVER message into the xid field of the DHCPOFFER message and sends the DHCPOFFER message to the requesting client.

DHCPREQUEST Message A DHCPREQUEST message may come from a client responding to a DHCPOFFER message from a server, from a client verifying a previously allocated IP address, or from a client extending the lease on a network address. If the DHCPREQUEST message contains a server identifier option, the message is in response to a DHCPOFFER message. Otherwise, the message is a request to verify or extend an existing lease. If the client uses a client identifier in a DHCPREQUEST message, it must use that same client identifier in all subsequent messages. If the client included a list of requested parameters in a DHCPDISCOVER message, it must include that list in all subsequent messages.

Any configuration parameters in the DHCPACK message should not conflict with those in the earlier DHCPOFFER message to which the client is responding. The client should use the parameters in the DHCPACK message for configuration.

Clients send DHCPREQUEST messages as follows:

- **DHCPREQUEST generated during the SELECTING state** The client inserts the address of the selected server in server identifier, ciaddr must be zero, and requested IP address must be filled in with the yiaddr value from the chosen DHCPOFFER.

Note that the client may choose to collect several DHCPOFFER messages and select the best offer. The client indicates its selection by identifying the offering server in the DHCPREQUEST message. If the client receives no acceptable offers, the client may choose to try another DHCPDISCOVER message. Therefore, the servers may not receive a specific DHCPREQUEST from which they can decide whether or not the client has accepted the offer. Because the servers have not committed any network address assignments on the basis of a DHCPOFFER, servers are free to reuse offered network addresses in response to subsequent requests. As an implementation detail, servers should not reuse offered addresses and may use an implementation-specific timeout mechanism to decide when to reuse an offered address.

- **DHCPREQUEST generated during the INIT-REBOOT state** server identifier must not be filled in, but requested IP address option must be filled in with the client's notion of its previously assigned address. ciaddr must be zero. The client is seeking to verify a previously allocated, cached configuration. The server should send a DHCPNAK message to the client if the requested IP address is incorrect or is on the wrong network.

Determining whether a client in the INIT-REBOOT state is on the correct network is done by examining the contents of giaddr, the requested IP address option, and a database lookup. If the DHCP server detects that the client is on the wrong net (the result of applying the local subnet mask or remote subnet mask, if giaddr is not zero, to the requested IP address option value doesn't match reality), then the server should send a DHCPNAK message to the client.

If the network is correct, then the DHCP server should check if the client's notion of its IP address is correct. If not, then the server should send a DHCPNAK message to the client. If the DHCP server has no record of this client, then it must remain silent and may output a warning to the network administrator. This behavior is necessary for peaceful coexistence of noncommunicating DHCP servers on the same wire.

If giaddr is 0x0 in the DHCPREQUEST message, the client is on the same subnet as the server. The server must broadcast the DHCPNAK message to the 0xffffffff broadcast address because the client may not have a correct network address or subnet mask, and the client may not be answering ARP requests.

If giaddr is set in the DHCPREQUEST message, the client is on a different subnet. The server must set the broadcast bit in the DHCPNAK, so that the relay agent will broadcast the DHCPNAK to the client, because the client may not have a correct network address or subnet mask, and the client may not be answering ARP requests.

- **DHCPREQUEST generated during the RENEWING state** The server identifier and requested IP address options must not be filled in, but ciaddr must be filled in with the client's IP address. In this situation, the client is completely configured and is trying to extend its lease. This message will be unicast, so no relay agents will be involved in its transmission. Because giaddr is therefore not filled in, the DHCP server will trust the value in ciaddr and use it when replying to the client.

A client may choose to renew or extend its lease prior to T1. The server may choose not to extend the lease (as a policy decision by the network administrator) but should return a DHCPACK message regardless.

- **DHCPREQUEST generated during the REBINDING state** The server identifier and requested IP address options must not be filled in, but ciaddr must be filled in with the client's IP address. In this situation, the client is completely configured and is trying to extend its lease. This message must be broadcast to the 0xffffffff IP broadcast address. The DHCP server should check ciaddr for correctness before replying to the DHCPREQUEST.

The DHCPREQUEST from a REBINDING client is intended to accommodate sites that have multiple DHCP servers and a mechanism for maintaining consistency among leases managed by multiple servers. A DHCP server may extend a client's lease only if it has the local administrative authority to do so.

DHCPDECLINE Message If the server receives a DHCPDECLINE message, the client has discovered through some other means that the suggested network address is already in use. The server must mark the network address as not available and should notify the local system administrator of a possible configuration problem.

DHCPRELEASE Message Upon receipt of a DHCPRELEASE message, the server marks the network address as not allocated. The server should retain a record of the client's initialization parameters for possible reuse in response to subsequent requests from the client.

DHCPINFORM Message The server responds to a DHCPINFORM message by sending a DHCPACK message directly to the address given in the ciaddr field of the DHCPINFORM message. The server must not send a lease expiration time to the client and should not fill in yiaddr. The server includes other parameters in the DHCPACK message as defined in the section on DHCPDISCOVER.

Client Messages Table 8-7 details the differences between messages from clients in various states.

DHCP Client Behavior

Figure 8-27 is a state-transition diagram for a DHCP client. A client can receive the following messages from a server:

- DHCPOFFER
- DHCPACK
- DHCPNAK

Note that the figure does not show a DHCPINFORM message. A client simply sends the DHCPINFORM and waits for DHCPACK messages. Once the client has selected its parameters, it has completed the configuration process.

Table 8-8 shows the use of the fields and options in a DHCP message by a client. The remainder of this section describes the action of the DHCP client for each possible incoming message. The description in the following section corresponds to the full configuration procedure previously described in allocating a network address. The text in the subsequent section corresponds to the abbreviated configuration procedure described in reusing a previously allocated address.

Table 8-7

Client messages from different states

	INIT-REBOOT	SELECTING	RENEWING	REBINDING
broad/unicast	Broadcast	Broadcast	Unicast	Broadcast
server-ip	Must not	Must	Must not	Must not
requested-ip	Must	Must	Must not	Must not
ciaddr	Zero	Zero	IP address	IP address

Table 8-8

Fields and options used by DHCP clients

Field	DHCPDISCOVER/ DHCPINFORM	DHCPREQUEST	DHCPDECLINE/ DHCPRELEASE
op	BOOTREQUEST	BOOTREQUEST	BOOTREQUEST
htype	(From "Assigned Numbers" RFC)		
hlen	(Hardware address length in octets)		
hops	0	0	0
xid	selected by client	xid from server DHCPOFFER message	selected by client
secs	0 or seconds since DHCP process started	0 or seconds since DHCP process started	0
flags	Set BROADCAST flag if client requires broadcast reply	Set BROADCAST flag if client requires broadcast reply	0
ciaddr	0 (DHCPDISCOVER) client's network address (DHCPINFORM)	0 or client's network address (BOUND/ RENEW/REBIND)	0 (DHCPDECLINE) client's network address (DHCPRELEASE)
yiaddr	0	0	0
siaddr	0	0	0
giaddr	0	0	0
chaddr	client's hardware address	client's hardware address	client's hardware address
sname	options, if indicated in sname/file option; otherwise unused	options, if indicated in sname/file option; otherwise unused	client's hardware address (unused)
file	options, if indicated in sname/file option; otherwise unused	options, if indicated in sname/file option; otherwise unused	(unused)
options	options	options	(unused)

Option	DHCPDISCOVER/ DHCPINFORM	DHCPREQUEST	DHCPDECLINE/ DHCPRELEASE
Requested IP address	May (DISCOVER), Must not (INFORM)	Must (in SELECTING or INIT-REBOOT), Must not (in BOUND or RENEWING)	Must (DHCPDECLINE), Must not (DHCPRELEASE)
IP address lease time	May (DISCOVER), Must not (INFORM)	May	Must not

Table 8-8 cont.

Fields and options used by DHCP clients

Option	DHCPDISCOVER/ DHCPINFORM	DHCPREQUEST	DHCPDECLINE/ DHCPRELEASE
Use file/ sname fields	May	May	May
DHCP message type	DHCPDISCOVER/ DHCPINFORM	DHCPREQUEST	DHCPDECLINE/ DHCPRELEASE
Client identifier	May	May	May
Vendor class identifier	May	May	Must not
Server identifier	Must not	Must (after SELECTING), Must not (after INIT-REBOOT, BOUND, RENEWING or REBINDING)	Must
Parameter request list	May	May	Must not
Maximum message size	May	May	Must not
Message	Should not	Should not	Should
Site-specific	May	May	Must not
All others	May	May	Must not

Initialization and Allocation of Network Addresses

The client begins in the INIT state and forms a DHCPDISCOVER message. The client should wait a random time between 1 and 10 seconds to desynchronize the use of DHCP at startup. The client sets ciaddr to 0x00000000. The client may request specific parameters by including the parameter request list option. The client may suggest a network address and/or lease time by including the requested IP address and IP address lease time options. The client must include its hardware address in the chaddr field, if necessary for the delivery of DHCP reply messages. The client may include a different unique identifier in the client identifier option, as noted in the discussion of server administrative controls. If the client included a list of requested parameters in a DHCPDISCOVER message, it must include that list in all subsequent messages.

The client generates and records a random transaction identifier and inserts that identifier into the xid field. The client records its own local time for later use in computing the lease expiration. The client then broadcasts the DHCPDISCOVER on the local hardware broadcast address to the 0xffffffff IP broadcast address and DHCP server UDP port.

If the xid of an arriving DHCPOFFER message does not match the xid of the most recent DHCPDISCOVER message, the DHCPOFFER message must be silently discarded. Any arriving DHCPACK messages must be silently discarded.

The client collects DHCPOFFER messages over a period of time, selects one DHCPOFFER message from the (possibly many) incoming DHCPOFFER messages (such as the first DHCPOFFER message or the DHCPOFFER message from the previously used server), and extracts the server address from the server identifier option in the DHCPOFFER message. The time over which the client collects messages and the mechanism used to select one DHCPOFFER are implementation dependent.

If the parameters are acceptable, the client records the address of the server that supplied the parameters from the server identifier field and sends that address in the server identifier field of a DHCPREQUEST broadcast message. Once the DHCPACK message from the server arrives, the client is initialized and moves to BOUND state. The DHCPREQUEST message contains the same xid as the DHCPOFFER message. The client records the lease expiration time as the sum of the time at which the original request was sent and the duration of the lease from the DHCPACK message. The client should perform a check on the suggested address to ensure that the address is not already in use.

For example, if the client is on a network that supports ARP, the client may issue an ARP request for the suggested request. When broadcasting an ARP request for the suggested address, the client must fill in its own hardware address as the sender's hardware address, and fill in 0 as the sender's IP address, to avoid confusing ARP caches in other hosts on the same subnet. If the network address appears to be in use, the client must send a DHCPDECLINE message to the server. The client should broadcast an ARP reply to announce the client's new IP address and clear any outdated ARP cache entries in hosts on the client's subnet.

Initialization with a Known Network Address

The client begins in the INIT-REBOOT state and sends a DHCPREQUEST message. The client must insert its known network address as a requested IP address option in the DHCPREQUEST message. The client may request specific configuration parameters by including the parameter request list option. The client generates and records a random transaction identifier and inserts that identifier into the xid field. The client records its own local time for later use in computing the lease expiration. The client must not include a server identifier in the DHCPREQUEST message. The client then broadcasts the DHCPREQUEST on the local hardware broadcast address to the DHCP server UDP port.

Once a DHCPACK message with an xid field matching the field in the client's DHCPREQUEST message arrives from any server, the client is initialized and moves to BOUND state. The client records the lease expiration time as the sum of the time at which the DHCPREQUEST message was sent and the duration of the lease from the DHCPACK message.

Initialization with an Externally Assigned Network Address

The client sends a DHCPINFORM message. The client may request specific configuration parameters by including the parameter request list option. The client generates and records a random transaction identifier and inserts that identifier into the xid field. The client places its own network address in the ciaddr field. The client should not request lease time parameters.

The client then unicasts the DHCPINFORM to the DHCP server if it knows the server's address; otherwise, it broadcasts the message to the limited (all 1s) broadcast address. DHCPINFORM messages must be directed to the DHCP server UDP port.

Once a DHCPACK message with an xid field matching the field in the client's DHCPINFORM message arrives from any server, the client is initialized.

If the client does not receive a DHCPACK within a reasonable period of time (60 seconds or 4 tries if using the timeout suggested in the "Constructing and Sending DHCP Messages" section), then the client should display a message informing the user (the client's user) of the problem. It should then begin network processing using suitable defaults (as provided in the RFC).

Use of Broadcast and Unicast The DHCP client broadcasts DHCPDISCOVER, DHCPREQUEST, and DHCPINFORM messages, unless the client knows the address of a DHCP server. The client unicasts DHCPRELEASE messages to the server. Because the client is declining the use of the IP address supplied by the server, the client broadcasts DHCPDECLINE messages.

When the DHCP client knows the address of a DHCP server in either an INIT or REBOOTING state, the client may use that address in the DHCPDISCOVER or DHCPREQUEST rather than the IP broadcast address. The client may also use unicast to send DHCPINFORM messages to a known DHCP server. If the client receives no response to DHCP messages sent to the IP address of a known DHCP server, the DHCP client reverts to using the IP broadcast address.

Reacquisition and Expiration The client maintains two times, T1 and T2, which specify when the client requests a lease extension on its network address. T1 is the time at which the client enters the RENEWING state and attempts to contact the server that originally issued the client's network address. T2 is the time at which the client enters the REBINDING state and attempts to contact any server. T1 must be earlier than T2, which, in turn, must be earlier than the time at which the client's lease will expire. To avoid the need for synchronized clocks, T1 and T2 are expressed in options as relative times.[14]

At time T1, the client moves to a RENEWING state and sends (via unicast) a DHCPREQUEST message to the server to extend its lease. The client sets the ciaddr field in the DHCPREQUEST to its current network address. The client records the local time at which the DHCPREQUEST message is sent for computation of the lease expiration time. The client must not include a server identifier in the DHCPREQUEST message.

Any DHCPACK messages that arrive with an xid that does not match the xid of the client's DHCPREQUEST message are silently discarded. When the client receives a DHCPACK from the server, the client computes the lease expiration time as the sum of the time at which the client sent the DHCPREQUEST message and the duration of the lease in the DHCPACK message. The client has successfully reacquired its network address, returns to BOUND state, and may continue network processing.

If no DHCPACK arrives before time T2, the client moves to a REBINDING state and sends (via broadcast) a DHCPREQUEST message to extend its lease. The client sets the ciaddr field in the DHCPREQUEST to its current network address. The client must not include a server identifier in the DHCPREQUEST message.

Times T1 and T2 are configurable by the server through options. T1 defaults to (0.5 × duration of lease). T2 defaults to (0.875 × duration of lease). Times T1 and T2 should be chosen with some random fuzz around a fixed value to avoid synchronization of client reacquisition.

A client may choose to renew or extend its lease prior to T1. The server may choose to extend the client's lease according to a policy set by the network administrator. The server should return T1 and T2, and their values should be adjusted from their original values to take account of the time remaining on the lease.

In both RENEWING and REBINDING states, if the client receives no response to its DHCPREQUEST message, the client should wait half of the remaining time until T2 (in RENEWING state) and half of the remaining lease time (in REBINDING state), down to a minimum of 60 seconds, before retransmitting the DHCPREQUEST message.

If the lease expires before the client receives a DHCPACK, the client moves to the INIT state, must immediately stop any other network processing, and requests network initialization parameters as if the client were uninitialized. If the client then receives a DHCPACK allocating its previous network address, the client should continue network processing. If the client is given a new network address, it must not continue using the previous network address and should notify the local users of the problem.

DHCPRELEASE If the client no longer requires the use of its assigned network address (for example, the client is gracefully shut down), the client sends a DHCPRELEASE message to the server. Note that the correct operation of DHCP does not depend on the transmission of DHCPRELEASE messages.[19]

Appendix A: Case Study of a Plethora of Antenna Types from Cisco

This appendix offers a case study of antenna products by Cisco Systems.

Cisco Aironet Antennas and Accessories—Complete the Wireless Solution

Cisco offers a complete range of antennas for client adapter, access point (AP), and bridge equipment that provide a customized wireless solution for almost any installation.

Cisco Aironet Antennas and Accessories

Every wireless local area network (WLAN) deployment is different. When engineering an in-building solution, varying facility sizes, construction materials, and interior divisions raise a host of transmission and multipath considerations. When implementing a building-to-building solution, distance, physical obstructions between facilities, and number of transmission points involved must be accounted for. Cisco is committed to providing not only the best APs, client adapters, and bridges in the industry, but it is also committed to providing a complete solution for any WLAN deployment. This is why Cisco has the widest range of antennas, cables, and accessories available from any wireless manufacturer.

With the Cisco FCC-approved directional[20] and omnidirectional[21] antennas, low-loss cable, mounting hardware, and other accessories, installers can customize a wireless solution that meets the requirements of even the most challenging applications.

Client Adapter Antennas

Cisco Aironet wireless client adapters come complete with standard antennas that provide sufficient range[22] for most applications at 11 Mbps. To extend the transmission range for more specialized applications, a variety of optional, higher-gain[23] antennas are provided that are compatible with selected client adapters. (See Table A-1.)

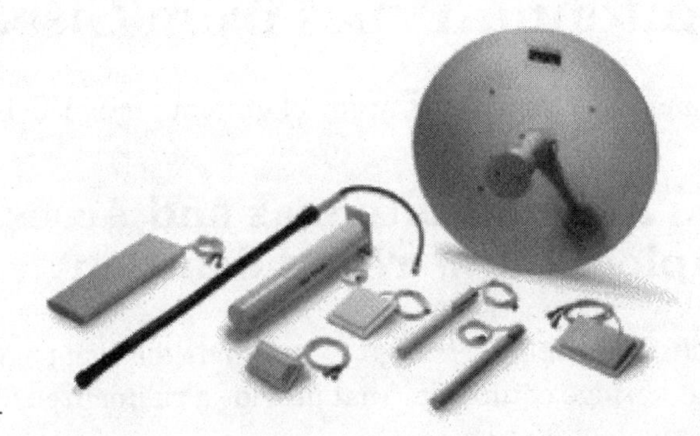

Figure A-1 Cisco offers a complete range of antennas for client adapter, AP, and bridge equipment that enable a customized wireless solution for almost any installation.

Table A-1

Cisco Aironet client adapter antenna features

Feature	AIR-ANT3351
Description	Packet over SONET (PoS) diversity dipole[a]
Application	Indoor diversity antenna[b] to extend the range of Aironet LMC client adapters
Gain	2.2 dBi[c]
Approximate indoor range at 1 Mbps[d]	350 ft. (107 m)
Approximate indoor range at 11 Mbps[d]	100 ft. (51 m)
Beam width	360° H 75° V
Cable length	5 ft. (1.5 m)
Dimensions	Base: 7×2 in. (18×5 cm) Height: 8 in. (20 cm)
Weight	9.2 oz. (261 g)

[a]A type of low-gain (2.2 dBi) antenna consisting of two (often internal) elements.

[b]An intelligent system of two antennas that continually senses incoming radio signals and automatically selects the antenna best positioned to receive it.

[c]A ratio of decibels to an isotropic antenna that is commonly used to measure antenna gain. The greater the dBi value, the higher the gain, and therefore, the more acute the angle of coverage.

[d]All range estimations are based on an integrated client adapter antenna associating with an AP under ideal indoor conditions. The distances referenced here are approximations and should be used for estimation purposes only.

AP Antennas

Cisco Aironet AP antennas are compatible with all Cisco Reverse Polarity Threaded Nut Connector (RP-TNC)-equipped APs. The antennas are available with different gain and range capabilities, beam widths,[24] and form factors. Coupling the right antenna with the right AP provides efficient coverage in any facility as well as better reliability at higher data rates. (See Figure A-2 and Table A-2.)

Table A-2

Cisco Aironet AP antenna features

Feature	AIR-ANT5959	AIR-ANT3195	AIR-ANT2012	AIR-ANT3213	AIR-ANT1728	AIR-ANT4941	AIR-ANT3549	AIR-ANT1729
Description	Diversity omnidirectional ceiling mount	3 dBi patch wall mount antenna	Diversity patch wall mount	Pillar mount diversity omnidirectional	High-gain omnidirectional ceiling mount	2.2 dBi dipole antenna	Patch wall mount	Patch wall mount
Application	Indoor unobtrusive antenna, which is best for ceiling mount; excellent throughput and coverage solution in high multipath cells and dense	Indoor/outdoor directional antenna	Indoor/outdoor unobtrusive medium-range antenna	Indoor unobtrusive medium-range antenna	Indoor medium-range antenna, which are typically hung from crossbars of drop ceilings	Indoor omnidirectional coverage	Indoor unobtrusive long-range antenna (may also be used as a medium-range bridge antenna)	Indoor unobtrusive medium-range antenna (may also be used as a medium-range bridge antenna)
Gain	Two separate 2 dBi omnidirectional elements: minimum gain 2.0 and maximum gain 2.35	3 dBi	6.5 dBi with two radiating elements	5.2 dBi	5.2 dBi	2.2 dBi	9 dBi	6 dBi
Approximate indoor range at 1 Mbps[e]	350 ft. (105 m)	AP: 271 ft. (82 m) Bridge: .5 miles (.9 km)	547 ft. (167 m)	497 ft. (151 m)	497 ft. (151 m)	350 ft.	AP: 700 ft. (213 m) Bridge: 2.0 miles (3.2 km)	AP: 542 ft. (165 m) Bridge: 1,900 ft. (580 m)
Approximate indoor range at 11 Mbps[e]	130 ft. (45 m)	AP: 80 ft. (24 m) Bridge: 950 ft. (290 m)	167 ft. (51 m)	142 ft. (44 m)	142 ft. (44 m)	130 ft.	AP: 200 ft. (61 m) Bridge: 3,390 ft. (1,032 m)	AP: 155 ft. (47 m) Bridge: 1,900 ft. (580 m)
Beam width	360° H 80° V	75° H 65° V	80° H 55° V	360° H 30° V	360° H 38° V	80 degrees	60° H 60° V	75° H 65° V
Cable length	3 ft. (0.91 m)	12 ft.	3 ft. (0.91 m)	3 ft. (0.91 m)	3 ft. (0.91 m)	N/A	3 ft. (0.91 m)	3 ft. (0.91 m)
Dimensions	5.3×2.8×0.9 in. (13.5×7.1×2.3 cm)	4×5 in. (9.7×13 cm)	4.78×6.66×.82 in. (12.14×16.92× 2.08 cm)	10×1 in. (25.4×2.5 cm)	Length: 9 in. (22.86 cm) Diameter: 1 in. (2.5 cm)	5.5 in.	5×5 in. (12.4×12.4 cm)	4×5 in. (9.7×13 cm)
Weight	0.3 lbs. (0.14 kg)	4.9 oz. (139 g)	9.6 oz. (272 g)	1 lb. (460 g)	4.6 oz. (131 g)	1.1 oz	5.3 oz. (150 g)	4.9 oz. (139 g)

[e]All range estimations are based on an external antenna associating with an integrated client adapter antenna under ideal indoor conditions. The distances referenced here are approximations and should be used for estimation purposes only.

Bridge Antennas

Cisco Aironet bridge antennas provide extraordinary transmission distances between two or more buildings. Cisco has a bridge antenna for every application. They are available in directional configurations for point-to-point transmission and omnidirectional configuration for point-to-multipoint implementations. (See Table A-3.)

Low-Loss/Ultralow-Loss Cables

Low-loss cable extends the length between any Cisco Aironet bridge and the antenna. With a loss of 6.7 dB per 100 feet (30 m) for the low-loss cable and 4.4 dB for the ultralow-loss cable, this provides installation flexibility without a significant sacrifice in range. (See Figure A-3 and Table A-4.)

Table A-3

Cisco Aironet bridge antenna features

Feature	AIR-ANT2506	AIR-ANT4121	AIR-ANT1949	AIR-ANT3338
Description	Omnidirectional mast mount	High-gain omnidirectional mast mount	Yagi mast mount	Solid dish
Application	Outdoor short-range point-to-multipoint applications	Outdoor medium-range point-to-multipoint applications	Outdoor medium-range directional connections	Outdoor long-range directional connections
Gain	5.2 dBi	12 dBi	13.5 dBi	21 dBi
Approximate range at 2 Mbps[f]	5,000 ft. (1,525 m)	4.6 miles (7.4 km)	6.5 miles (10.5 km)	25 miles (40 km)
Approximate range at 11 Mbps[f]	1,580 ft. (480 m)	1.4 miles (2.3 km)	2.0 miles (3.3 km)	11.5 miles (18.5 km)
Beam width	360° H 38° V	360° H 7° V	30° H 25° V	12.4° H 12.4° V
Cable length	3 ft. (0.91 m)	1 ft. (0.30 m)	3 ft. (0.91 m)	2 ft. (0.61 m)
Dimensions	Length: 13 in. (33 cm) Diameter: 1 in. (2.5 cm)	Length: 40 in. (101 cm) Diameter: 1.3 in. (3 cm)	Length: 18 in. (46 cm) Diameter: 3 in. (7.6 cm)	Diameter 24 in. (61 cm)
Weight	6 oz. (17 g)	1.5 lbs. (0.68 kg)	1.5 lbs. (0.68 kg)	11 lbs. (5 kg)

[f]All range estimations are based on use of 50 ft. (15 m) low-loss cable and the same type of antenna at each end of the connection under ideal outdoor conditions. The distances referenced here are approximations and should be used for estimation purposes only.

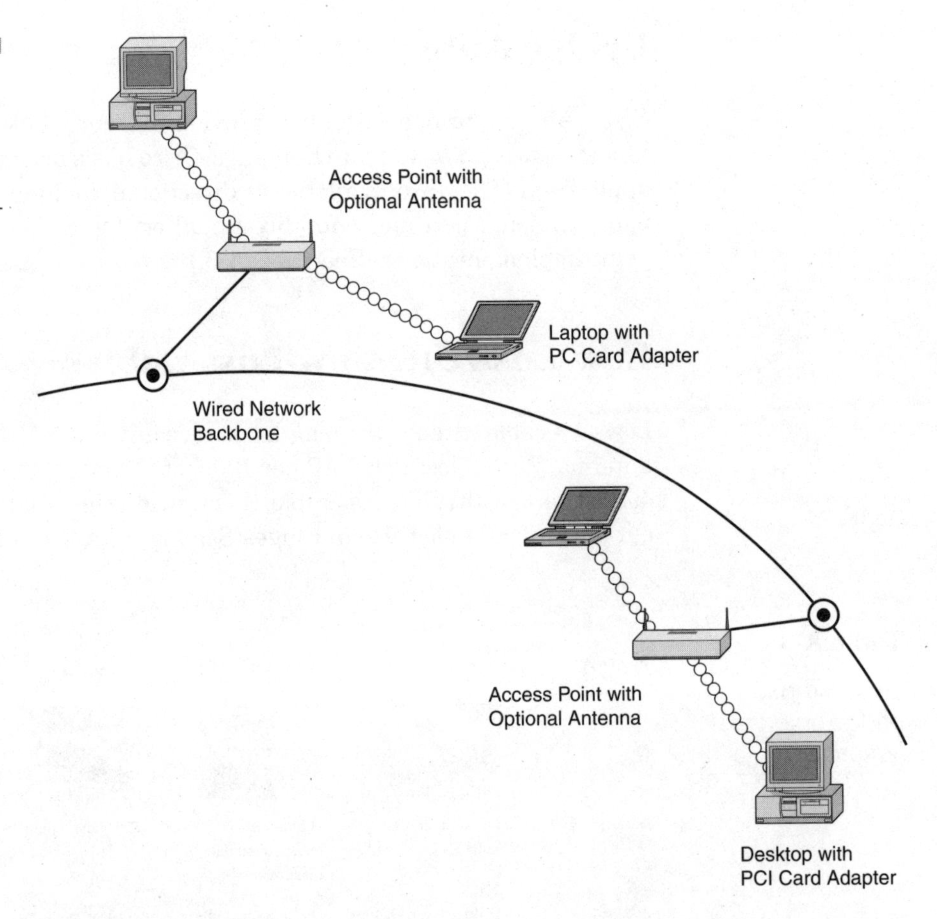

Figure A-2 Optional, higher-gain antennas can be used to extend the range of APs.

Table A-4 Cisco Aironet low-loss antenna cable features

Feature	AIR-CAB020LL-R	AIR-CAB050LL-R	AIR-CAB100ULL-R	AIR-CAB150ULL-R
Cable length	20 ft. (6 m)	50 ft. (15 m)	100 ft. (30 m)	150 ft. (46 m)
Transmission loss	1.3 dB	3.4 dB	4.4 dB	6.6 dB

Accessories

To complete an installation, Cisco provides a variety of accessories that offer increased functionality, safety, and convenience. (See Figure A-4 and Table A-5.)

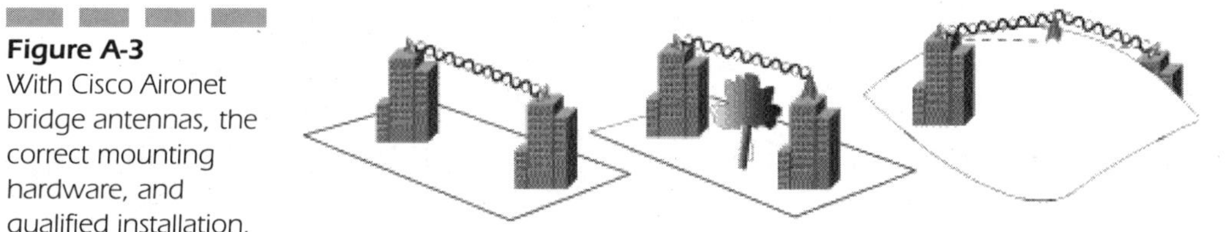

Figure A-3 With Cisco Aironet bridge antennas, the correct mounting hardware, and qualified installation, wireless links over great distances and obstacles are possible.

Figure A-4 Cisco Aironet antenna accessories

Table A-5 Cisco Aironet accessory features

Feature	AIR-ACC2537-060	AIR-ACC3354	AIR-ACC2662
Description	60 in. (152 cm) bulkhead extender	Lightning arrestor	Yagi articulating mount
Application	Flexible antenna cable that extends AP cabling typically within an enclosure	Helps prevent damage due to lightning-induced surges or static electricity	Adds swiveling capability to mast-mounted yagi antennas

End Notes

1. P. Srisuresh. "RFC 3022: Traditional IP Network Address Translator (Traditional NAT)." January 2001.
2. The remainder of this section is taken from P. Srisuresh's "RFC 3022: Traditional IP Network Address Translator (Traditional NAT)," January 2001.
3. Srisuresh, P. and M. Holdrege. "RFC 2663: IP Network Address Translator (NAT) Terminology and Considerations," August 1999.

4. Rekhter, Y., B. Moskowitz, D. Karrenberg, G. de Groot, and E. Lear. "BCP 5, RFC 1918: Address Allocation for Private Internets," February 1996.
5. Reynolds, J. and J. Postel. "STD 2, RFC 1700: Assigned Numbers," October 1994.
6. Defense Advanced Research Projects Agency Information Processing Techniques Office. "STD 7, RFC 793: Transmission Control Protocol (TCP) Specification," September 1981.
7. Postel, J. "STD 6, RFC 768, User Datagram Protocol," USC/Information Sciences Institute, August 1980.
8. ——. "STD 5, RFC 792: Internet Control Message (ICMP) Specification," September 1981.
9. RFC 3022 is Copyrighted (©) by The Internet Society (2001). This document and translations of it may be copied and furnished to others, and derivative works that comment on or otherwise explain it or assist in its implementation may be prepared, copied, published and distributed, in whole or in part, without restriction of any kind, provided that the above copyright notice and this paragraph are included on all such copies and derivative works.
10. R. Droms. "RFC 2131: Dynamic Host Configuration Protocol," March 1997.
11. The rest of this section is from R. Droms' "RFC 2131: Dynamic Host Configuration Protocol," March 1997.
12. Croft, B., and J. Gilmore. "RFC 951: Bootstrap Protocol (BOOTP)," Stanford and SUN Microsystems, September 1985.
13. Wimer, W. "RFC 1542: Clarifications and Extensions for the Bootstrap Protocol," Carnegie Mellon University, October 1993.
14. Alexander, S., and R. Droms. "RFC 1533: DHCP Options and BOOTP Vendor Extensions," Lachman Technology, Inc., Bucknell University, October 1993.
15. Droms, D. "RFC 1534: Interoperation between DHCP and BOOTP," Bucknell University, October 1993.
16. Braden, R. (ed.), "STD 3, RFC 1122: Requirements for Internet Hosts—Communication Layers," USC/Information Sciences Institute, October 1989.
17. Gray, C., and D. Cheriton. "Leases: An Efficient Fault-Tolerant Mechanism for Distributed File Cache Consistency." In Proceedings of the Twelfth ACM Symposium on Operating Systems Design, 1989.

18. Reynolds, J. "RFC 1497: BOOTP Vendor Information Extensions," USC/Information Sciences Institute, August 1993.

19. RFC 2131 is Copyrighted (©) by The Internet Society (1997). This document and translations of it may be copied and furnished to others, and derivative works that comment on or otherwise explain it or assist in its implementation may be prepared, copied, published and distributed, in whole or in part, without restriction of any kind, provided that the above copyright notice and this paragraph are included on all such copies and derivative works.

20. An antenna that concentrates transmission power into a direction that increases coverage distance at the expense of coverage angle. Directional antenna types include yagi, patch, and parabolic dish antennas. A yagi is a type of cylindrical directional antenna. A patch antenna is a type of flat antenna designed for flush wall mounting that radiates a hemispherical coverage area. A parabolic dish antenna is a concave or dish-shaped object, often refers to dish antennas. Parabolic dish antennas tend to provide the greatest gain and the narrowest beam width making them ideal for point-to-point transmission over the longest distances.

21. An antenna that provides a 360-degree transmission pattern. These types of antennas are used when coverage in all directions is required.

22. A linear measure of the distance that a transmitter can send a signal.

23. A method of increasing the transmission distance of a radio by the concentration of its signal in a single direction, typically through the use of a directional antenna. Gain does not increase the signal strength of a radio, but simply redirects it. Therefore, as gain increases, the decrease in angle of coverage is inversely proportional.

24. The angle of signal coverage provided by a radio; it may be decreased by a directional antenna to increase gain.

CHAPTER 9

Migrating to 3G WWANs

This short chapter looks at migration approaches to third generation (3G). The chapter will be short for three reasons: (1) a major body of literature exists discussing this topic; (2) wireless wide area networks (WWANs) are not as critical to hotspot/nomadic networks as it might appear at first; and (3) with the expense and bureaucracy involved in deploying these national-level networks (including spectrum allocation and acquisition), it will be several years before services become ubiquitous and the speeds are higher than 384 Kbps. In any event, the (W)WAN-based data speed will always be less than the speed achieved by local area network (LAN) or wireless LAN (WLAN) techniques, and WWAN services will always be more expensive than WLAN-based services. WLAN-based technology is as likely to replace WLAN technology, as cable-based WAN technology is likely to replace cable-based LAN technology (remember LAN Emulation [LANE]?) The two kinds of technologies are aimed at different applications and have complementary roles to play. Also, hotspot services are specific to a small geographic area, WISP services are specific to a city, and WWAN services are specific to a region (multiple areas, cities, or regions can be supported by providers, respectively.) The general outline of the possible migration paths was already provided in Chapter 1, "Introduction to Wireless Personal Area Networks (WPANs), Public Access Locations (PALs), and Hotspot Services," and Chapter 3, "Technologies for Hotspots."

The migration path is a function of geography and the embedded base. Figure 9-1 depicts the anticipated evolution path for WWANs in time, location, and technology. This evolution is expected to require billions of dollars ($20 to $50 billion), so it will not come overnight, nor will it be cheap. At a more macro level, the following transitions are anticipated:

- From one generic to many specialized radio access protocols
- From industry-specific protocols to Internet Engineering Task Force (IETF)/Internet Protocol (IP)-oriented protocols
- From hierarchical to dynamically routed backbone networks
- From narrowband links to broadband links
- From monolithic functions to distributed functions
- From closed architectures to open Internet-based architectures
- From 2G to 2.5G to 3G

To illustrate some of these points, Figure 9-2 depicts the transition from hierarchical networks to dynamically routed backbone networks. Figure 9–3 shows an all-IP WWAN of the future from a protocol and function point of view. Figure 9-4 explains how the control plane of such an all-IP network can be constructed using IETF/Institute of Electrical and Elec-

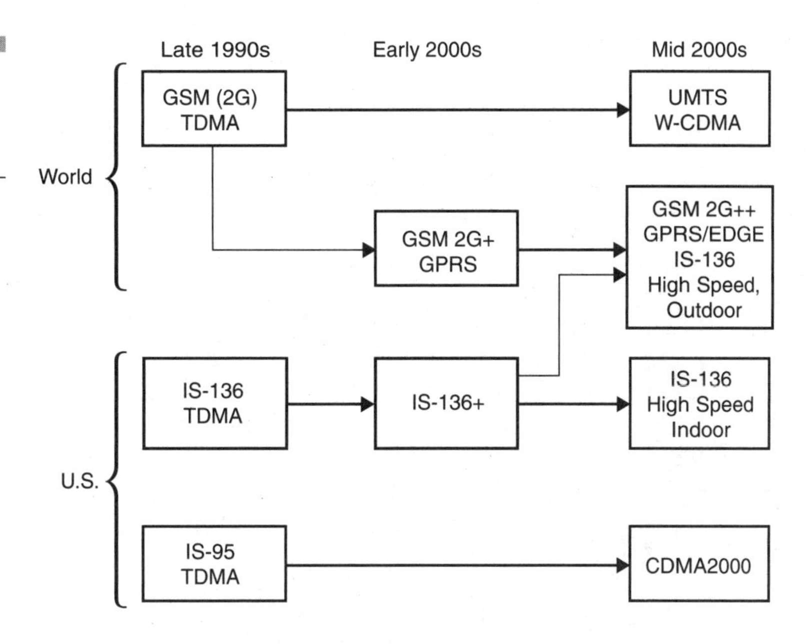

Figure 9-1 Evolution paths for WWAN voice and data

tronics Engineers (IEEE) standards. A core network with an all-IP architecture (see Figure 9-5) can be advantageous for a number of reasons:

- With the growth in the use of the Internet, new IP-oriented applications keep emerging. With an all-IP architecture, it will be simpler to bring the benefits of these applications to the 3G customers.
- Many IP-based protocols that are useful in wireless networks have already been developed or are in the process of development (such as the *intserv / diffserv* quality of service [QoS] protocols). Therefore, if the network is all IP, one can take advantage of these protocols.
- Good multimedia/voice over IP (VoIP) support exists in IP. This could enable multimedia terminals in a 3G system to use H.323 and/or H.324.

ITU IMT-2000

Recently, the International Telecommunication Union (ITU) has advanced the concept of Mobile Telecommunication in the year 2000 (IMT-2000) as its generalized view of future wireless/nomadic networks. To support high-speed data rates, and to support multimedia services, the ITU—Radio Com-

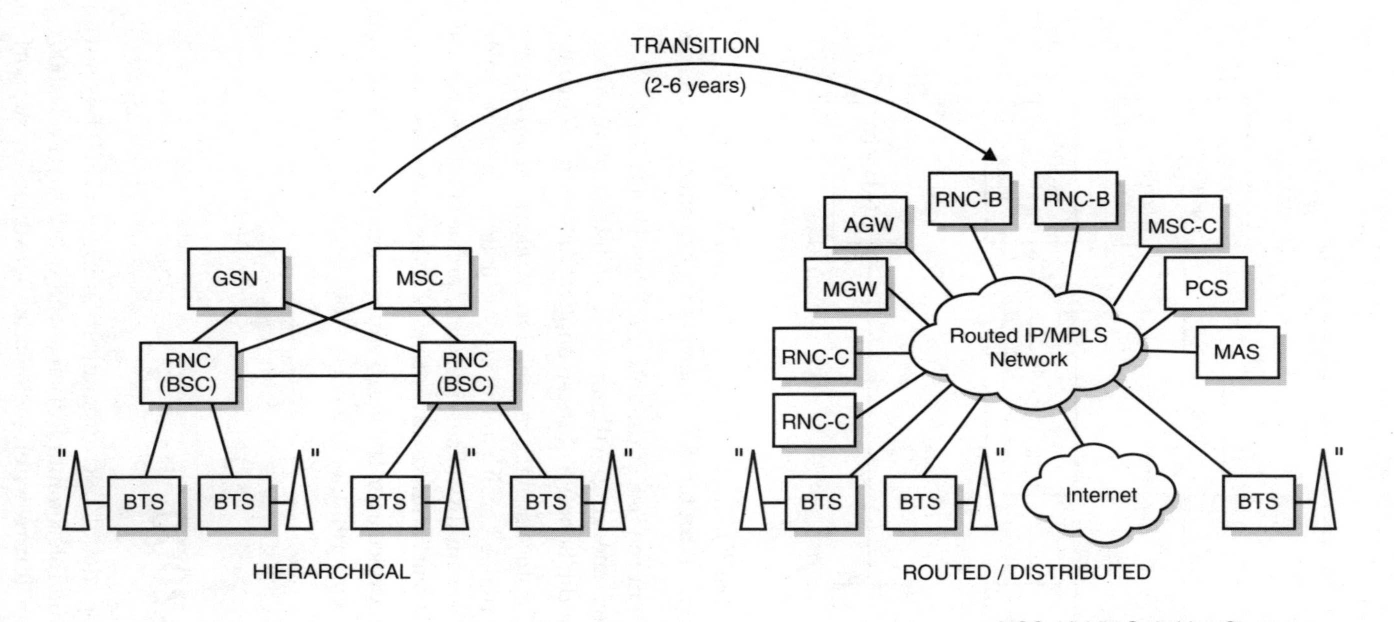

Figure 9-2 The transition of architecture in future networks

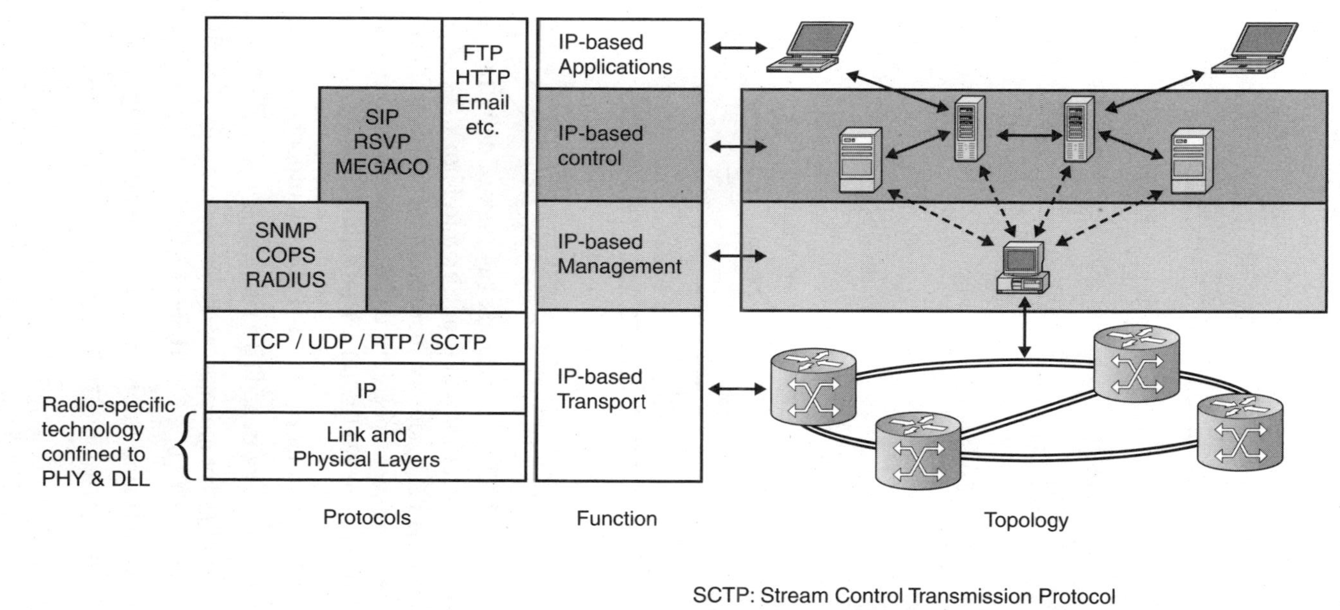

Figure 9-3 *An all-IP network*

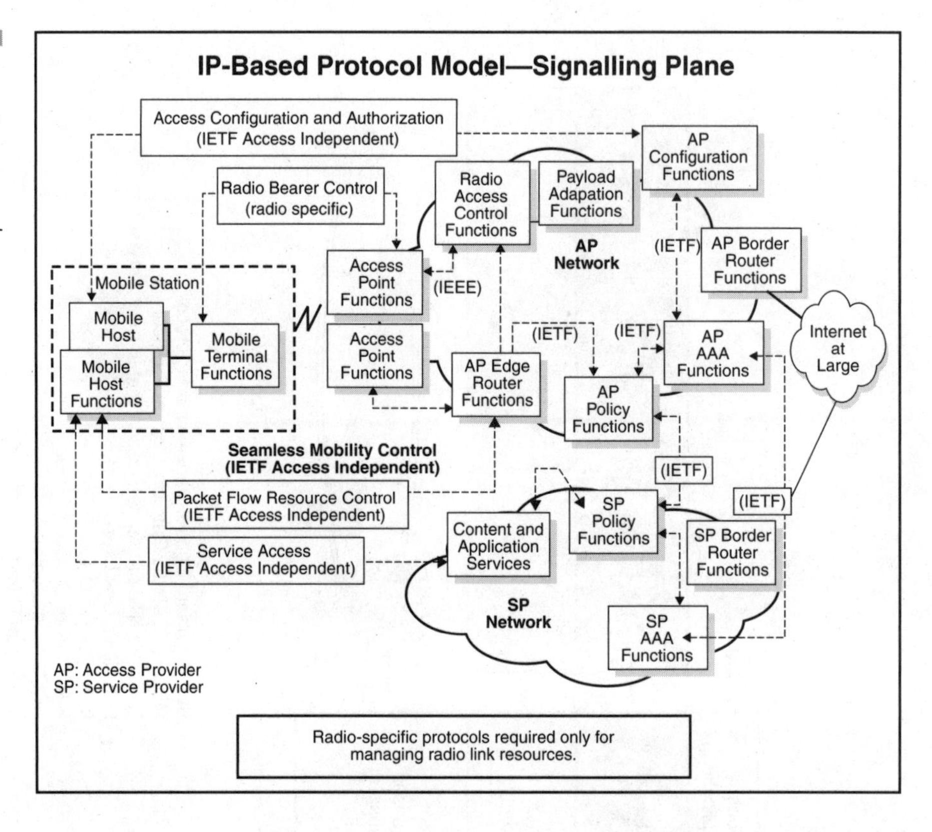

Figure 9-4 Control plane (signaling) of an all-IP wireless network (Source: Nortel Networks)[1]

munication Sector (ITU-R) undertook the task of defining a set of recommendations for IMT-2000, which is outlined in Figure 9-6.

IMT-2000 is "the ITU globally coordinated definition of 3G covering key issues such as frequency spectrum use and technical standards." Some of the basic desiderata for IMT-2000 are as follows:

- Speeds of 2 Mbps (indoors) and 144 Kbps (outdoors) or better
- Circuit- and packet-switched (IP) services
- Good voice quality comparable with wire-line quality (with Mean Opinion Score [MOS] of 4.0 or thereabouts)
- Increased capacity and improved spectrum efficiency
- Global roaming between different operational environments

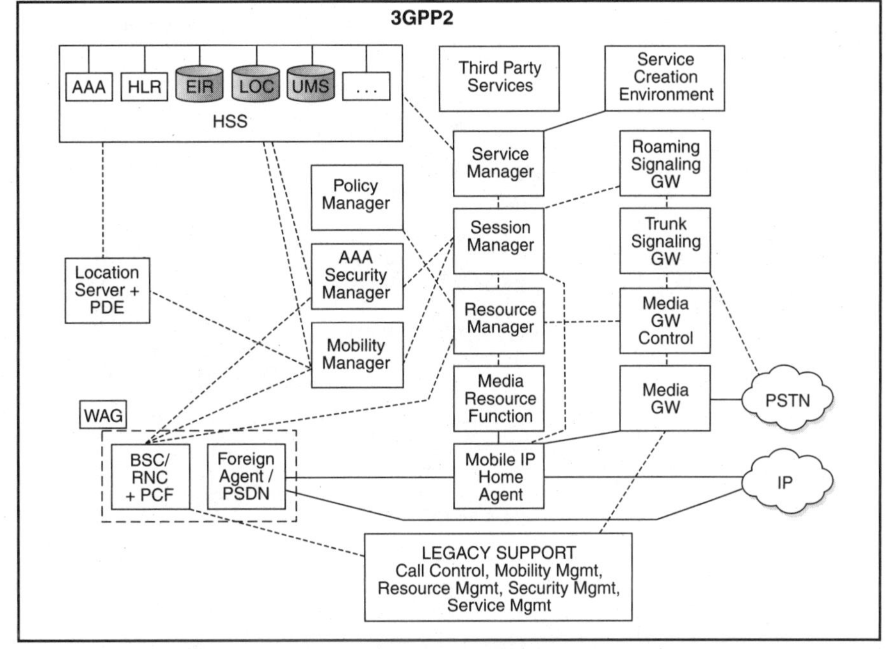

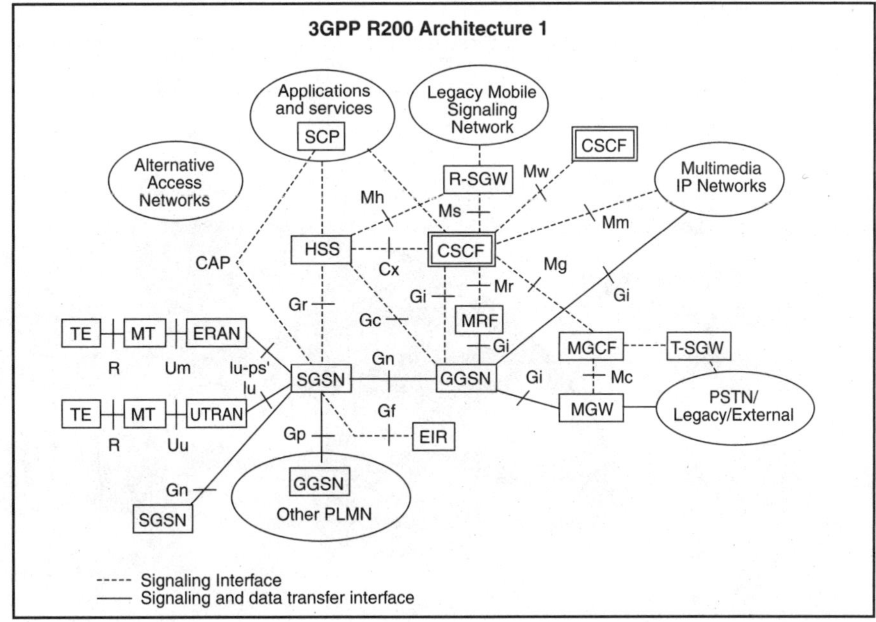

Figure 9-5
Movement to an all-IP networks (Source: Mark Tubinis Water Cove Network)

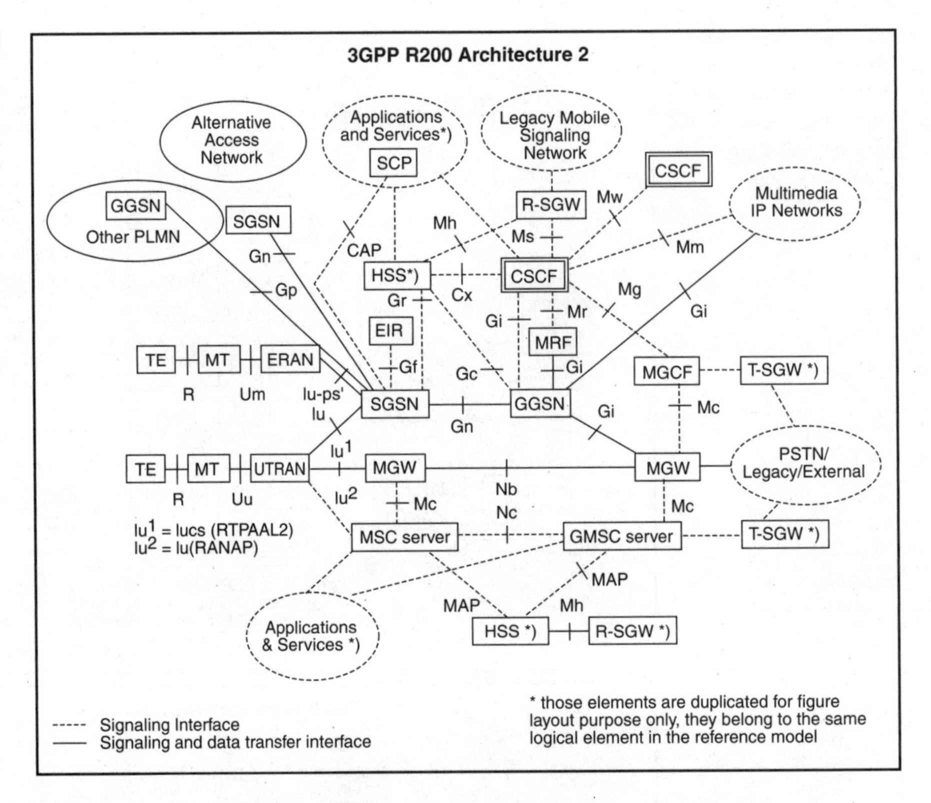

Figure 9-5 cont. Movement to an all-IP network

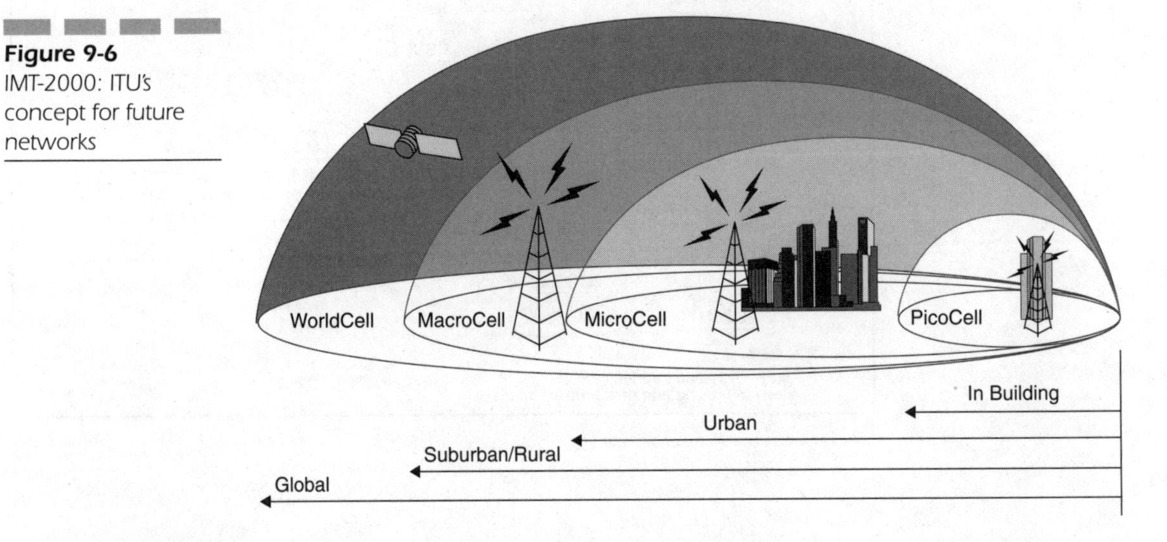

Figure 9-6 IMT-2000: ITU's concept for future networks

1G mobile telecommunication systems that were introduced in the 1980s were analog. These systems, which are still in use, do not intrinsically have data transport capabilities. To provide data services in these analog systems, a capability such as Cellular Digital Packet Data (CDPD) has to be added to the analog. Naturally, this arrangement supports only slow-speed data. 2G systems (IS-136, cdmaOne, and Global System for Mobile Communications [GSM]) are digital and have intrinsic data transport capabilities. However, the data support is still limited: GSM supports short messaging services and data at rates only up to 9.6 Kbps, and IS-95B provides data rates in the neighborhood of 64 to 115 Kbps in increments of 8 Kbps over a 1.25 MHz channel. 2.5G General Packet Radio Service (GPRS) was developed in the late 1990s as a further enhancement: in GPRS, each slot can handle up to 20 Kbps and since each user may be allocated up to 8 slots, data rates up to about 160 Kbps per user are therefore achievable. Yet none of these break the 256 Kbps barrier. The goal of 3G is to do just that.

IMT-2000 includes a variety standards, the most significant being: (1) IMT-DS (Direct Spread) (Wideband Code Division Multiple Access [W-CMDA]), (2) IMT-MC (Multicarrier) (CDMA2000), (3) IMT-TC (Time Code) (Time Division Synchronous Code Division Multiple Access [TD-SCDMA]), (4) IMT-SC (Single Carrier): UWC-136/Enhanced Data GSM Environment (EDGE), and (5) IMT-FT (Frequency Time) (Digital European Cordless Telephone [DECT]).

3G is not a technology *per se* but a term encompassing all aspects of future wireless networks. 3G adjoins high-speed radio access and IP-based services to enable subscribers to be "always on," that is, "always be online." This new technology is created to support a large numbers of users within one network. The 3G standards support different types of user traffic:

- Constant-bit-rate traffic, such as high-quality audio speech, video telephony, and video, that need QoS since they are sensitive to delays and delay variation.
- Real-time, variable-bit-rate traffic, such as variable-bit-rate audio, Moving Pictures Expert Group (MPEG)/International Organization for Standardization (ISO) video, and so on. This type of traffic also requires QoS, being sensitive to delays and delay variation.
- Non-real-time, variable-bit-rate traffic that can tolerate delays or delay variations.

Increased speeds may not be achieved in one shot in 3G: the data rate supported may initially be only 144 Kbps, it may be 384 Kbps in the second phase, and it may reach 2.048 Mbps in the final phase; phases are designed to be backwards compatible.

A number of national standards bodies have developed next-generation wireless standards. Eventually, four systems for 3G mobile communications materialized: CDMA2000, W-CDMA Universal Mobile Telecommunications System (UMTS) Frequency Division Duplex (FDD), W-CDMA UMTS Time Division Duplexing (TDD), and UWC-136. Recommendations on these systems were published by ITU-R as a harmonized standard with four modes in 1999. CDMA2000 must comply with Electronics Industry Association (EIA)/Telecommunications Industry Association (TIA) IS-41, and W-CDMA UMTS must comply with GSM Manufacturing Automation Protocol (MAP) intersystem networking standards. UWC-136 is based on Time Division Multiple Access (TDMA), while the other three use direct-sequence code division multiple access (DS-CDMA).

CDMA2000 is a multicarrier, direct-sequence CDMA FDD system with a single carrier that has a bandwidth of 1.25 MHz (later on it may have as many as three carriers.) CDMA2000 is an evolution of the existing North American CDMA system cdmaOne (IS-95 standards). CDMA2000 supports packet mode data services. UMTS W-CDMA FDD is a direct-sequence CDMA system with a nominal bandwidth of 5 MHz. UMTS W-CDMA TDD also uses CDMA with a bandwidth of 5 MHz; however, the frequency band is time-shared in both directions (half the time it is used for transmission in the forward direction, and the other half it's used in the reverse direction.)

The TDMA version of the 3G system for use in North America is known as UWC-136. UWC-136 is a TDMA scheme where each physical channel is partitioned into a number of fixed time slots; each user is assigned one or more slots. The UWC-136 system is planned to be introduced in three stages:

- IS-136+ with a bandwidth of 30 kHz (provides voice and up to 64 Kbps of data)
- IS-136 HS (vehicular/outdoor) with a bandwidth of 200 kHz (provides data rates up to 384 Kbps for outdoor/vehicular operations)
- IS-136 HS (indoor) with a bandwidth of 1.6 MHz (users can get a data rate of up to 2 Mbps)

IS-136+ is an enhancement of the existing IS-136 and uses improved modulation techniques. IS-136 HS (vehicular/outdoor) supports data rates up to 384 Kbps and has parameters similar to those of EDGE. IS-136 HS (indoor) provides data rates up to about 2 Mbps.[2]

All the various technologies build on a so-called *core network*. The technologies in the GSM evolution are based on a so-called GSM MAP core network and the others on the IS-41 core network. With the new IMT-2000 standards, all radio options (such as CDMA) should work on all core networks (such as GSM MAP—currently CDMA does not work over GSM MAP). This might make later changes in the network radio interfaces easier.[3,4]

As we have seen, several evolutionary paths to 3G exist. GSM operators can enhance their networks with GPRS and later deploy EDGE, which is already defined as a 3G technology by the IMT-2000. These networks will then evolve to future 3G networks based on W-CDMA, the standard technology for the UTMS band. TDMA operators can either switch to GSM and continue with that approach or go on to CDMA2000. The Japanese Personal Digital Cellular (PDC) standard will evolve to W-CDMA.[3] Specifications for 3G technologies are maintained by the Third Generation Partnership Project (3GPP) and the IMT-2000.

Several prototypes showing possible 3G terminal designs have appeared, but the first commercial introduction of 3G terminals has yet to take place. One important aspect is that terminals need to be dual-mode (3G and 2G) to enable efficient network usage.

Transitions to 3G for Wireless WANs (WWANs)

Without losing sight of the fact that hotspot services rely mostly on WLANs and WPANs, we close the technical part of this book by summarizing the anticipated evolution from the current-base WWAN system to the target system over the next three to five years. This evolution is expected to be as follows:

- Hotspot: IEEE 802.11b to IEEE 802.11a or IEEE 802.11b to IEEE 802.11g
- CDMAOne IS 95A (to cdmaOne IS-95B) to CDMA2000 1x to CDMA2000 1x EV-DO or EV-DV
- TDMA to CDMA2000 1x . . . or
- TDMA to GSM . . .
- GSM to GPRS to EDGE to W-CDMA

As just remarked, W-CDMA is the 3G technology called for in UMTS. W-CDMA is part of the ITU IMT-2000 standard. Thanks to the efficient use of the radio spectrum, a wealth of different services can be accessed simultaneously, including circuit and packet services. Data speeds can be as high as 2 Mbps. W-CDMA data speeds are expected to go up to 10.8 Mbps in 3GPP Release 5. W-CDMA is also the recommended upgrade path for GSM operators. Forecasters make the claim that with the upgrade to W-CDMA, GSM will likely hold an 80 percent market share in 2005.

Work is also under way to interwork 2.5G/3G WWANs with WLANs. Figure 9-7 suggests three scenarios. In the first, the WWAN network handles mobility, specifically GPRS. Here the WLAN is considered as one of the GPRS cells. This approach may require dual-mode network interface cards (NICs) to access the two different Layer 2 services. All traffic reaches the Serving GPRS Support Node (SGSN) or Gateway GPRS Support Node (GGSN) before reaching its intended final destination (this happens even if the final destination is the WLAN or LAN itself). In the second scenario, mobility is handled by the WLAN based on IEEE 802.11 rules. In the third scenario, a mobility gateway handles mobility. It also handles routing issues using Mobile IP (see Chapter 3) to handle mobility management. The mobility gateway is located between the GPRS and the WLAN networks. The gateway is a proxy that can be implemented in either the GPRS or the WLAN (both of these networks being considered peers).

As we covered in the text, IEEE 802.11a (also known as Wi-Fi 5) supports 54 Mbps in the 5 GHz band. There are already products on the market (both access points and PC/desktop NICs) from companies such as Proxim, Intel, D-Link, and Actiontec. However, it is a well-know fact that it is easier to reach longer distances at a lower transmission speed (also within the context of the IEEE 802.11b speed choices). At this time the consensus is that IEEE 802.11b will continue to be the hotspot technology of choice at least until 2004. Depending on field-strength validations with the IEEE 802.11a technology, this technology may or may not usurp to role for hotspots. In any event, the system bottleneck may well be on the terrestrial uplink into the Internet, so that the speed on the air link is, by itself, of limited value. Only when the uplink is increased to 10 Mbps or DS-3, the air link speed starts to make a difference. It should also be noted that, although the IEEE 802.11a standards was first approved in 1999, the technology has been slow taking off not only because engineering the 802.11a chipsets is proven challenging, but also because the success of 802.11b has made it a de facto standard.

IEEE 802.11g aims at supporting 54 Mbps in the 2.4 GHz band. It is designed to be backward compatible with existing Wi-Fi devices. Products

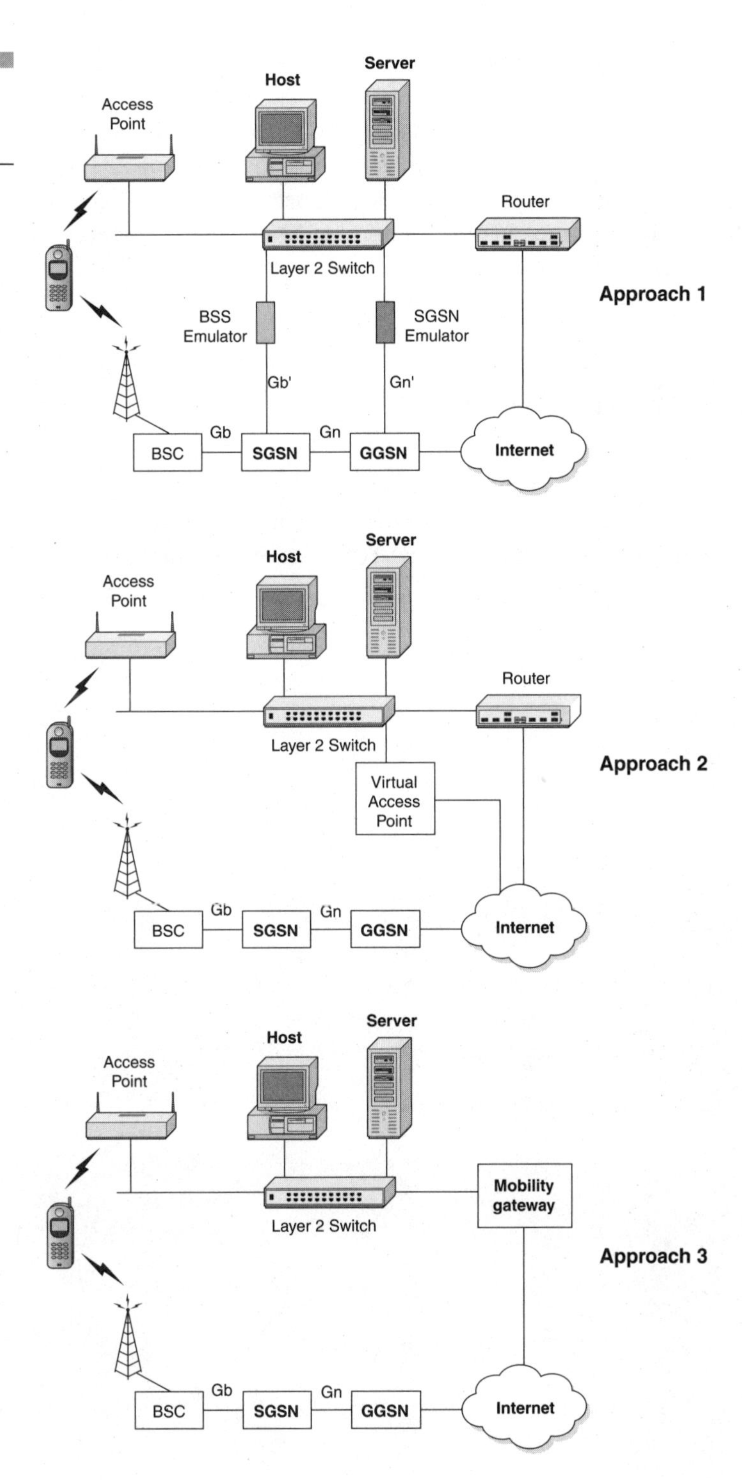

Figure 9-7 2.5G/3G-to-WLAN interworking

were expected for early 2003. The same observations just made with regards to IEEE 802.11a apply here. Furthermore, while both IEEE 802.11a and 802.11g may be beneficial for indoors WLAN technologies, they likely will have a harder time in open-space hotspot services. This is because higher speeds are more susceptible to noise and interference. Hence, an unprotected open ISM environment may prove problematic for these technologies.

End Notes

1. Dr. Al Javed, CTO, Wireless Internet, Nortel Networks.
2. M.R. Karim, and Mohsen Sarraf, *W-CDMA and cdma2000 for 3G Mobile Network*, New York: McGraw-Hill, 2002.
3. Oliver Thylmann, www.infosync.no/show.php?id=1246, February 2002.
4. www.imt-2000.org

REFERENCES

1. D. Minoli and K. Schneider, "A Technique for Establishing the Minimum Number of Frequencies Required for Urban Mobile Radio Communication," *IEEE Transaction on Communications* (September 1977): 1,054–1,056.
2. D. Minoli and I. Gitman, "On Connectivity in Mobile Packet Radio Networks," 28th IEEE Vehicular Technology Conference Record, March 1978, 105–109.
3. ———, "Analytical Models in Monitoring Mobile Packet Radio Devices," 28th IEEE Vehicular Technology Conference Record, March 1978, 110–118..
4. D. Minoli and K. Schneider, "An Optimal Receiver for Code Division Multiplexed Signals," *Alta Frequenza*, XLVII, no. 7 (July 1978): 587–591.
5. D. Minoli, "An Approximate Analytical Model for Initialization of Single Hop Packet Radio Networks," *IEEE Canadian Conference on Communications and Power* (1978): 107–110.
6. D. Minoli and I. Gitman, "Combinatorial Issues in Mobile Packet Radio," *IEEE Transactions on Communications* COMM-26 (December 1978): 1,821–1,826.
7. D. Minoli and I. Gitman, "Monitoring Mobile Packet Radio Devices," *IEEE Transactions on Communications* (concise paper) COMM-27 (February 1979): Part 2, 509–517.
8. D. Minoli, "A Closed Form Expression for Initialization Time of Packet Radio Networks," *Frequenz* 33, no. 5 (May 1979): 126–133.
9. D. Minoli, "Initialization Time for Packet Radio Networks with a Small Number of Buffers," *Alta Frequenza* XLVIII, no. 10 (October 1979): 653–628.
10. D. Minoli, "Packet Radio Monitoring via Repeater-On-Packets," *IEEE Transactions on Aerospace and Electronic Systems* 15, issue 4 (July 1979): 466–473.
11. D. Minoli, "Satellite On-Board Processing of Packetized Voice," ICC Conference Record, 1979, 58.4.1–58.4.5.
12. D. Minoli, "Exact Solution for the Initialization Time of Packet Radio Networks with Two Station Buffers, NCC Conference Record, 48, 1979, 875–885.
13. D. Minoli, I. Gitman, and D. Walters, "Analytical Models for Initialization of Single Hop Packet Radio Networks," *IEEE Transactions on*

Communications (special issue on digital radio) COMM-27 (December 1979): 1,959–1,967.

14. D. Minoli, E. Paterno, and W. Nakamine, "Packet Length Considerations in Carrier Sense Multiple Access Packet Radio Systems," INTELCOM 80 Conference Record, June 1980.
15. D. Minoli and E. Lipper, "Cost Implications for Survivability of Terrestrial Networks Under Malicious Failure," *IEEE Transactions on Communications* (special issue on military communications) COMM-28, no. 9 (September 1980): 1,668–1,674.
16. D. Minoli, "Aloha Channels Thruput Degradation," Computer Networking Symposium Conference Record, 1986, 151–159.
17. D. Minoli, "FM Subcarrier Communications," DataPro Report CA70-010-101, May 1986.
18. D. Minoli, "Making Use of Spare FM Band," *Computer World* (June 6, 1986): 25, 29.
19. D. Minoli, "An Overview of Radio Technologies for Data Communications," DataPro Report CA70-010-502, August 1987.
20. D. Minoli, *Telecommunication Technologies Handbook,* (Artech House, 1991). This contains three major chapters in radio applications to data transmission (40 pages), satellite communications (70 pages), and wireless over the air (15 pages).
21. D. Minoli and E. Minoli, *Delivering Voice over IP and the Internet*, 2nd edition, (New York: Wiley, 2002). This draws from the section on VoIP over wireless.
22. D. Minoli, P. Johnson, and E. Minoli, *Next-Generation SONET-Based Metro Area Networks—Planning and Designing the Provider Network* (New York: McGraw-Hill, 2002). This draws from the section on free-space optics.
23. William Stallings, *Wireless Communications and Networks*, (Upper Saddle River, NJ: Prentice Hall, 2001).
24. Rob Flickenger, *Building Wireless Community Networks*, (Boston: O'Reilly & Associates, 2001).
25. Brent A. Miller and Chatschik Bisdikian, *Bluetooth Revealed*, (Upper Saddle River, NJ: Prentice Hall, 2001).
26. Carl J. Weisman, *Essential Guide to RF and Wireless*, 2nd edition, (Upper Saddle River, NJ: Prentice Hall, 2002).
27. Jim Geier, *Wireless LANs: Implementing Interoperable Networks*, (San Francisco: Riders Publishing, 1998).
28. Ray Rischpater, *Wireless Web Development*, (Berkeley, CA: Apress Books, 2000).

INDEX

Symbols

A

B

C

D

E

F–G

H

I

J–L

M

N

O–P

Q–R

S

T

U

V–W

X–Z

ABOUT THE AUTHOR

Mr. Minoli is one of the most popular authors on the telecom and networking technology shelf, with several well-received guides to his credit. He is the newly appointed CTO and co-founder of Global Nautical Networks, a Wi-Fi ISP/ASP; the former CEO and CTO of InfoPort Communications, an optical & Gigabit Ethernet metro carrier; former Director of Engineering and Development, Data and Internet Services at Teleport Communications Group, now part of AT&T; and a former Senior Member of the technical staff at Bellcore, researching advanced broadband data services, such as Frame Relay and ATM. In the early 1980s he supported data networking for Wall Street firms; in the late 1970s he worked on data protocols at Bell Labs; and earlier he did research work on packet radio and packet voice.

Mr. Minoli's work in packet radio communications, packet radio repeaters, voice over packet radio, spread spectrum, satellite and VSAT communications spanned the 1970s and 80s. His pioneering packet radio and data-over-radio work is documented in two-dozen papers in IEEE and other journals. His work on LANs, broadband, and the Internet spanned the 1990s and 00s.

Mr. Minoli is able to take a cold-start or existing operation, department, function, network, or platform and develop it to planned targets. He has P&L-managed organizations of over 120 people with cumulative opex of $75M and capex of $250M. He has worked successfully at large and small companies, end-user organizations, and carriers.

This book is born out of real-life practice. Global Nautical Networks is a Wi-Fi carrier providing advanced hotspot services at hundreds of high-end marinas. Mr. Minoli's efforts entail: establishing technical direction of the company; designing, deploying, and turning up access points, bridges, routers, and AAA servers; assessing new technologies such as IEEE 802.11a, 802.11g, and 3G; supporting VoIP and Web-portal services; undertaking site surveys and producing engineering packages; producing prefabricated self-contained nodes that are installable in one hour; deploying 128-bit encryption, dynamic WEP, and LEAP-based security; configuring end-users and supporting them; establishing NOC functions; and reducing capex cost from $5,000 to $900 a node as well as reducing the site transmission MRC from $1,000 to $100.

Among his recent books are the highly rated *Ethernet-Based Metro Area Networks* and the companion *SONET-Based Metro Area Networks*, both written with Peter Johnson and Emma Minoli; as well as, the brand-new seminal book on *Voice Over MPLS*.